FUNGICIDE RESISTANCE

BCPC Monograph No 60

M.J. HOCART

FUNGICIDE RESISTANCE

BCPC Monograph No 60

Proceedings of a symposium organised by the British Society for Plant Pathology in association with the British Crop Protection Council and held at the University of Reading on 28-30 March 1994

Edited by: S Heaney, D Slawson, D W Hollomon, M Smith, P E Russell and D W Parry

BCPC BRITISH CROP PROTECTION COUNCIL

BSPP

BCPC Registered Office:
49 Downing Street
Farnham
Surrey GU9 7PH, UK.

49 Downing Street
Farnham, Surrey GU9 7PH

British Library Cataloguing in Publication Data

British Crop Protection Council
Fungicide Resistance
(Monograph Series, ISSN 0306-3941; No 60)
I Heaney, S et al
II Series 631

ISBN 0-948404-77-9

Cover design by Major Design & Production Ltd, Nottingham
Printed in Great Britain by Major Print Ltd, Nottingham

Contents

Page

Posters

SESSION 3:
BENZIMIDAZOLES AND DICARBOXIMIDES

Posters

Symposium Programme Committee

Chairman:	Dr D W Parry	Harper Adams Agricultural College, Shropshire, (BSPP Programme Secretary)
Members:	Dr K Brent	Formerly Long Ashton Research Station
	Dr J T Fletcher	Formerly ADAS, Wye
	Dr S Heaney	Zeneca Agrochemicals, Jealott's Hill Research Station
	Dr D W Hollomon	Long Ashton Reasearch Station
	Dr P E Russell	AgrEvo UK Ltd, Chesterford Park Research Station
	M Smith	Ciba Agriculture, Cambridge
	Dr D Slawson	MAFF, Pesticides Safety, Directorate

Chairmen and Session Organisers

Session 1		
Chairman and Organiser:	Dr S Heaney	Zenneca Agrochemicals
Session 2		
Chairman and Organiser:	M Smith	Ciba Agriculture
Session 3		
Chairman:	Dr D Slawson	MAFF, PSD
Organisers:	Dr D Slawson	MAFF, PSD
	Dr J T Fletcher	Formerly ADAS, Wye
Session 4		
Chairman:	Dr J Gilmour	SAC Advisory Service, Edinburgh
Organiser:	Dr D W Hollomon	Long Ashton Research Station
Session 5		
Chairman and Organiser:	Dr P E Russell	AgrEvo UK Ltd
Session 6		
Chairman:	Dr D W Hollomon	Long Ashton Research Station
Organisers:	Dr D W Hollomon	Long Ashton Research Station
	Dr K Brent	Formerly Long Ashton Research Station

Syndicate Workshop Session

Chairman	Syndicate 1:	Dr J A Lucas	University of Nottingham
	Syndicate 2:	Dr J Gilmour	SAC Advisory Service, Edinburgh
	Syndicate 3:	Dr M W Shaw	University of Reading
	Syndicate 4:	Dr J K M Brown	Cambridge Laboratory, Norwich
	Syndicate 5:	Dr C E Caten	University of Birmingham

Acknowledgements

The British Society for Plant Pathology gratefully acknowledges the financial assistance provided by:

AgrEvo UK Ltd
BASF Plc
Bayer Plc
Ciba Agriculture
Sandoz Agro Ltd

Abbreviations

acid equivalent	a.e.
active ingredient	AI
boiling point	b.p.
British Standards Institution	BSI
centimetre(s)	cm
concentration x time product	ct
concentration required to kill 50% of test organisms	LC50
correlation coefficient	*r*
cultivar	cv.
cultivars	cvs.
day(s)	d
days after treatment	DAT
degrees Celsius (centigrade)	°C
dose required to kill 50% of test organisums	LD50
dry matter	d.m.
Edition	Edn
Editor	Ed
Editors	Eds
emulsifiable concentrate	EC
freezing point	f.p.
gas chromatography-mass spectrometry	gcms
gas-liquid chromatography	glc
gram(s)	g
growth stage	GS
hectare(s)	ha
high performance (or pressure) liquid chromatography	hplc
hour	h
infrared	i.r.
International Standardisation Organisation	ISO
Kelvin	K
kilogram(s)	kg
least significant difference	LSD
litre(s)	Litre
litres per hectare	l/ha
mass	*m*
mass per mass	*m/m*
mass per volume	*m/V*
mass spectrometry	m.s.
maximum	max.
melting point	m.p.
metre(s)	m
milligram(s)	mg
millilitre(s)	ml
millimetre(s)	mm
minimum	min.
minute (time unit)	min
molar concentration	M

nuclear magnetic resonance	nmr
number average diameter	n.a.d.
number median diameter	n.m.d.
organic matter	o.m.
page	p.
pages	pp.
parts per million by volume	mg/l
parts per million by weight	mg/kg
pascal	Pa
percentage	%
post-emergence	post-em.
power take off	p.t.o.
pre-emergence	pre-em.
probability (statistical)	*P*
relative humidity	r.h.
revolutions per minute	rev./min
second (time unit)	s
standard error	SE
standard error of means	SEM
soluble powder	SP
species (singular)	sp.
species (plural)	spp.
square metre	m^2
subspecies	ssp.
surface mean diameter	s.m.d.
suspension concentrate	SC
temperature	temp.
thin-layer chromatography	tlc
tonne(s)	t
ultraviolet	u.v.
vapour pressure	v.p.
variety (wild plant use)	var.
volume	*V*
weight	*W*
weight by volume (mass by volume is more correct)	*W/V* (m/*V*)
weight by weight (mass by mass is more correct)	*W/W* *(m/m)*
wettable powder	WP
approximately	c.
less than	<
more than	>
not less than	<
not more than	>

Multiplying symbols-		Prefixes
mega	($x\ 10^6$)	M
kilo	($x\ 10^3$)	k
milli	($x\ 10^{-3}$)	m
micro	($x\ 10^{-6}$)	µ
nano	($x\ 10^{-9}$)	n
pico	($x\ 10^{-12}$)	p

Session 1
DMI Fungicides

Chairman &
Session Organiser S HEANEY

RESISTANCE TO FUNGICIDES WHICH INHIBIT STEROL 14α-DEMETHYLATION, AN HISTORICAL PERSPECTIVE

M.A. DE WAARD

Department of Phytopathology, Wageningen Agricultural University, P.O.Box 8025, Binnenhaven 9, 6700 EE Wageningen, The Netherlands

ABSTRACT

This paper gives an historical perspective on fungal resistance to fungicides which inhibit sterol 14α-demethylation (DMIs). Topics covered are resistance development in the field, the genetics of resistance and the biochemical mechanisms involved. At present, reduced sensitivity or resistance to DMIs has been reported for at least 13 plant pathogens. In most of these fungi, and in laboratory-generated mutants, resistance is polygenic. Cases of monogenic resistance have also been published. A variety of biochemical resistance mechanisms may operate. Resistance caused by decreased affinity of sterol 14α-demethylase to DMIs has not yet been established. Other mechanisms, such as increased efflux from fungal mycelium and changes in the sterol biosynthetic pathway have been reported, although the relevance of these mechanisms for resistance in plant pathogens remains to be elucidated.

INTRODUCTION

From 1969 onwards at least 36 sterol 14α-demethylation inhibitors (DMIs) were introduced as agricultural fungicides. Remarkable features of these type of compounds are their high protective and curative properties, low use rates, systemic and vapour phase activity, high chemical variability leading to diverse spectra of antifungal activity, and selective action between target and non-target organisms (De Waard, 1993). Nowadays, the use of fungicides in various crops is dominated by DMIs. Hence, the successful introduction of DMIs clearly represents a landmark in chemical disease control.

Development of resistance to fungicides with a specific mode of action is a major threat to effective chemical disease control. By 1969, serious resistance problems in many plant pathogens were already encountered with different site-specific fungicides. DMIs also have a site-specific mode of action. Therefore, it is understandable that these fungicides received timely attention with respect to the resistance risk. Fuchs and Drandarevski (1976) stated that development of resistance to DMIs under practical conditions would be rather unlikely. History has demonstrated that this statement was premature, since DMI-resistance has now been reported in various plant pathogens. However, resistance developed relatively slowly as compared with other classes of site-specific fungicides.

This paper gives an historical overview of resistance development to DMI fungicides in practice, its genetic basis, and the mechanisms of resistance involved.

RESISTANCE DEVELOPMENT

Assessment of resistance to DMIs is difficult since the level of resistance is often so low that its development could only be detected when proper background sensitivity studies were previously made. The presence of strains with decreased sensitivity does not necessarily imply

TABLE 1. Chronological list of reports on reduced sensitivity and/or field resistance to DMI fungicides in plant pathogens.

Pathogen	Crop	Authors
Erysiphe graminis f.sp. *hordei*	Barley	Fletcher and Wolfe, 1981
Sphaerotheca fuliginea	Cucumber	Schepers, 1983
Pyrenophora teres	Barley	Sheridan *et al.*, 1985
Venturia inaequalis	Apple	Stanis and Jones, 1985
Erysiphe graminis f.sp. *tritici*	Wheat	De Waard *et al.*, 1986
Rhynchosporium secalis	Barley	Hunter *et al.*, 1986
Penicillium digitatum	Citrus	Eckert, 1987
Uncinula necator	Grape	Steva *et al.*, 1990
Pseudocercosporella herpotrichoides	Wheat	Leroux and Marchegay, 1991
Botrytis cinerea	Vegetables	Elad, 1992
Mycosphaerella fijiensis	Banana	Anonymous, 1992
Puccinia horiana	Chrysanthemum	Cevat, 1992
Septoria tritici	Wheat	Hollomon (pers. comm., 1993)

loss of field control by a particular DMI (*e.g. Septoria tritici*). This depends on the level of resistance and the frequency of resistant strains. In this paper, the term "field resistance" is used for situations where these parameters coincide. The term "reduced sensitivity" is used for practical situations in which the presence of a pathogen population with decreased sensitivity to a DMI does not obviously result in decreased field performance.

Table 1 gives an overview of reports in which reduced sensitivity and/or resistance in field isolates of plant pathogens were described for the first time. Later reports on similar cases are not cited in this paper but often concur with the reported data or indicate that levels of resistance have increased over several years. The latter phenomenon has, for instance, been described for resistance to various DMIs in *Sphaerotheca fuliginea* in the Netherlands (Schepers, 1985), *Venturia inaequalis* in Canada (Hildebrand *et al.*, 1988), and *Erysiphe graminis* f.sp. *hordei* in the UK (Heaney, 1988). Despite this, the data mentioned in Table 1 have to be handled with care, since resistance may stabilize in different regions and under different conditions at different levels. For instance, despite the early report on resistance development, various DMIs still effectively control *S. fuliginea* in glasshouse cucumbers in the Netherlands, while failures of disease control in field-grown cucurbits are common in some Mediterranean countries (Huggenberger *et al.*, 1984). Barley powdery mildew in the UK became field resistant to some triazoles, while similar compounds are still effective against wheat powdery mildew in other European countries. Poor disease control or relatively short persistence of action by particular DMIs may also only become evident if disease conducive conditions coincide with the presence of pathogen populations with reduced sensitivity (De Waard *et al.*, 1986). Although cross-resistance usually extends to all DMIs, some DMIs, especially the recently introduced ones, may remain highly effective against cereal powdery mildews. This may be due to their relatively high activity, and to differences in levels of cross resistance to different DMIs (De Waard, 1992[a]).

Inclusion of some pathogens in Table 1 may be subject to dispute. This is the case for *Pseudocercosporella herpotrichoides* since Birchmore *et al.* reported in 1992 no major changes in sensitivity to prochloraz in populations of the pathogen in various European countries. Resistance of *Pyrenophora teres* to triadimenol could be a case of natural insensitivity, since insensitive isolates in New Zealand are present in the natural population of the pathogen (Sheridan *et al.*, 1985). Reduced sensitivity of *Botrytis cinerea* to DMIs was reported in Israel in field tests to study the efficacy of some DMIs against grey mould in cucumber. Field resistance at one site was associated with the occurrence of isolates with reduced sensitivity, and was attributed to the fact that DMIs had been used previously for control of cucumber powdery mildew at the same site (Elad, 1992). Background sensitivity data are lacking. Hence, one may wonder whether a fast selection of a small subpopulation of the pathogen with low sensitivity to DMIs from the natural population may play a role. Such a process would explain the low number of DMIs registered for grey mould control (De Waard, 1992[b]).

GENETICS OF RESISTANCE TO DMIs

Extensive studies on the genetic basis of resistance to DMIs have been carried out with laboratory-generated mutants of *Aspergillus nidulans* (Van Tuyl, 1977) and *Nectria haematococca* var. *cucurbitae* (Kalamarakis *et al.*, 1991). Genetic analysis of 202 mutants of *A. nidulans* with low levels of resistance to imazalil identified eight different loci allocated to six different linkage groups. Two additional loci conferring imazalil-resistance were identified in cycloheximide-resistant mutants. The mutants showed cross-resistance to fenarimol. These results demonstrate that resistance to DMIs in *A. nidulans* has a polygenic basis. Mutations for resistance to imazalil gave rise to pleiotropic effects such as resistance or hypersensitivity to unrelated toxicants, cold-sensitivity or reduced growth rates (Fuchs and De Waard, 1982). Allelic mutations in the same locus could result in different levels of resistance to imazalil. The effects of the two most frequently identified loci, *imaA* and *imaB*, together with a modifier gene, were additive. Genetic analysis of 51 mutants of *N. haematococca* var. *cucurbitae* with low levels of resistance to fenarimol also identified a polygenic system for resistance with at least nine chromosomal loci involved. The strains showed low levels of cross-resistance to some other DMI fungicides. All mutations appeared to be pleiotropic, having more or less adverse effects on saprophytic fitness and pathogenicity. Analysis of 30 mutants with high resistance to triadimenol only recognized one locus (Kalamarakis *et al.*, 1989). This mutation appeared to be highly mutable and did not lead to the pleiotropic mutations observed with the fenarimol-resistant mutants. The mutants were cross-resistant to other triazole DMIs, but showed no change in sensitivity to imidazole DMIs or even possessed an increased sensitivity. The authors suggest that resistance of *N. haematococca* var. *cucurbitae* is caused by major gene resistance. Genetic evidence, therefore, points to more than one mechanism of resistance operating against DMIs.

Only a few of the pathogens listed in Table 1 have been used to unravel the genetic basis for control of DMI resistance. This is understandable, since many of these pathogens lack a perfect stage or are more difficult to cross than *A. nidulans* and *N. haematococca* var. *cucurbitae*. Analysis of a low level of field resistance of *Venturia inaequalis* to DMIs showed that only a single gene was involved (Stanis and Jones, 1985). A more recent report suggests that resistance to DMIs in *V. inaequalis* is polygenic (Sholberg and Haag, 1993). For *E. graminis* f.sp. *hordei*, conflicting results have been described. Hollomon *et al.* (1984) suggest that a high level of resistance in this pathogen has a polygenic basis, while Brown *et al.* (1992) conclude that alleles at single loci control sensitivity and resistance in crosses of a sensitive isolate with a moderately resistant and a highly resistant isolate. Isolates of *P. teres* resistant to triadimenol may lack clear cross resistance to propiconazole (Peever and Milgroom, 1992). Resistance to triadimenol is conferred by alleles at the same locus (qualitative resistance), while resistance to propiconazole appeared to segregate in a quantitative manner. The variability in correlation coefficients between other DMI-DMI and DMI-fenpropimorph combinations have later been described in more detail

(Peever and Milgroom, 1993). Results indicate that inheritance of DMI resistance may not be the same for different pathogens and DMIs.

It is difficult to conclude whether the polygenic mechanism of resistance shown to be present in laboratory-generated isolates also plays a major role in DMI-resistant field isolates. On one hand, the stepwise development of resistance in *E. graminis* f.sp. *hordei*, *S. fuliginea*, and *V. inaequalis* points to the involvement of more than one gene. On the other hand, genetic analysis of DMI-resistant field isolates identified both monogenic and polygenic resistance. Laboratory-generated mutants often show reduced comparative fitness, and this phenomenon may counteract the development of polygenic resistance development under field conditions. However, loss in comparative fitness may be overcome by continued selection in natural populations for both DMI resistance and normal virulence.

MECHANISM OF NATURAL INSENSITIVITY OR RESISTANCE TO DMIs

A wide variety of mechanisms may operate in natural insensitivity or resistance to DMIs (Köller, 1992; Table 2). Reported mechanisms of natural insensitivity are detoxification (triforine), protonation of imidazole DMIs (imazalil), lack of activation (triadimefon), deposition in cell compartments (triadimenol) and tolerance of toxic sterols. In theory, these mechanisms may also be operative as mechanisms of resistance. However, this has not been observed for any of the DMI-resistant laboratory mutants or field isolates investigated. The mechanisms reported to operate in DMI-resistant laboratory mutants are: increased efflux from mycelium resulting in reduced accumulation, a defect in sterol 14α-demethylation, circumvention of toxic sterol formation, and overproduction of cytochrome-P450-dependent sterol 14α-demethylase ($P450_{14DM}$).

TABLE 2. Mechanisms of natural insensitivity and resistance to DMIs.

Mechanism	Authors
Detoxification	Gastonyi and Josepovits, 1975
Protonation	Siegel *et al.*, 1977
Increased efflux from mycelium	De Waard and Van Nistelrooy, 1979
Lack of activation	Gastonyi and Josepovits, 1979
Defect in sterol 14α-demethylation	Walsh and Sisler, 1982
Circumvention of toxic sterol formation	Taylor *et al.*, 1983
Overproduction of $P450_{14DM}$	Kalb *et al.*, 1986
Deposition in cell compartments	Hippe, 1987
Tolerance of toxic sterols	Weete and Wise, 1987
Induced resistance response	Smith and Köller, 1990
Decreased affinity of $P450_{14DM}$	Vanden Bossche *et al.*, 1990

Increased energy-dependent efflux in resistant mutants counteracts passive influx of DMIs in mycelium and results in a relatively low and constant level of accumulation. This will reduce complex formation between DMIs and their target site, $P450_{14DM}$, and hence may explain the relatively low levels of resistance observed. While efflux in the resistant mutants has a constitutive character, activity in wild-type isolates appears to be inducible, resulting in a transient accumulation pattern in time (De Waard and Van Nistelrooy, 1979). Fenarimol-efflux activity in wild-type isolates was inducible by pyrimidine, triazole and imidazole DMIs, but not by carbendazim, carboxin, and chloroneb; indicating a specific effect of DMIs (De Waard and Van Nistelrooy, 1981). In both wild-type isolates and resistant mutants, efflux activity could be inhibited by many metabolic inhibitors (De Waard and Van Nistelrooy, 1987). These may be regarded as potential synergists of DMI toxicity (De Waard and Van Nistelrooy, 1984[a]; 1984[b]). Increased energy-dependent efflux has now been described as a mechanism of resistance to various DMIs (De Waard and Van Nistelrooy, 1988) in *Aspergillus nidulans* (De Waard and Van Nistelrooy, 1979), *Penicillium italicum* (De Waard and Van Nistelrooy, 1984[b]), *Candida albicans* (Ryley *et al.*, 1984), *Monilia fructicola* (Ney, 1988), and *N. haematococca* var. *cucurbitae* (Kalamarakis *et al.*, 1991).

A defect in sterol 14α-demethylation and circumvention of toxic sterol formation are common mechanisms of resistance in *Ustilago maydis*, *Candida albicans* and *Saccharomyces cerevisiae*, but do not seem to operate in filamentous fungi. Demethylation defective mutants of *U. maydis* lack ergosterol but contain various C14 sterols (*e.g.* 14α-methylfecosterol) which are probably functional in membranes of the fungus (Walsh and Sisler, 1982). In demethylation-defective mutants of *S. cerevisiae*, a second mutation, a defect in C5-6 desaturation, circumvents the formation of toxic sterols (*e.g.* 14α-methyl-3,6 diol) by which these double mutants become viable (Taylor *et al.*, 1983). Mutants of *U. maydis* and *S. cerevisiae* have a considerably slower growth rate than the corresponding wild-types. Therefore, it is not likely that these mechanisms of resistance will be relevant in field-resistant isolates of plant pathogens. Resistance caused by overproduction of $P450_{14DM}$ has been observed in transformants of *S. cerevisiae* with multiple copies of the $P450_{14DM}$ coding gene (Kalb *et al*, 1986) and by $P450_{14DM}$ gene amplification (Vanden Bossche, pers. comm.). The result is compensation of the target site. It seems possible that these mechanisms may also play a role in decreased DMI sensitivity in plant pathogens.

The molecular mechanism leading to the induced resistance response to triadimenol in *Ustilago avenae* remains to be elucidated (Smith and Köller, 1990). Mechanisms which could be involved are: compensation by increased production of target sites, an altered lipid metabolism, and partitioning of the fungicide in cell compartments such as vacuoles (Hippe, 1987).

The most frequently observed mechanism of resistance to fungicides is decreased affinity of the target site for the fungicide. However, reports on resistance to DMIs caused by decreased affinity of $P450_{14DM}$ to these fungicides, without affecting functioning of the enzyme in sterol biosynthesis, are rare. It has been reported for resistance to some antimycotics in a *C. albicans* isolate, obtained from patients with chronic mucocutaneous candidosis who relapsed after prolonged treatment with ketoconazole (Vanden Bossche *et al.*, 1990). The reason for the reduced virulence of the isolate is not known. The potency of imazalil to inhibit ergosterol biosynthesis in cell-free preparations of laboratory-generated strains of *P. italicum* with different levels of resistance to imazalil is the same (Guan and De Waard, 1993). These results suggest that DMI resistance in this plant pathogen is not based on changes in affinity of $P450_{14DM}$. This mechanism of resistance has not yet been reported for other plant pathogens. The reason for this may be a technical one, since $P450_{14DM}$ is very unstable during preparation of cell-free extracts of filamentous fungi. The availability of proper methods now paves the way to do such experiments (Guan *et al.*, 1992; Stehmann *et al.*, 1994). It would, for instance, be feasible to study the affinity of $P450_{14DM}$ in imazalil-resistant packing-house isolates of *Penicillium digitatum* and in triadimenol-resistant mutants of *N. haematococca* var. *cucurbitae*. The method is not suitable for slow growing fungi or obligate parasites. In these instances a completely different approach, based on cloning and characterization of the sterol 14α-demethylase gene, should be adopted (De Waard, unpublished results; Holloman *et al.*, 1990).

CONCLUDING REMARKS

Extensive efforts have been made to elucidate the mechanism(s) of insensitivity and resistance to DMI fungicides. The fragmentary evidence available so far points to the operation of different mechanisms, and suggests that these mechanisms are not necessarily related to a decrease in affinity of the target site of DMIs, $P450_{14DM}$, in sterol biosynthesis. The variety in mechanisms of insensitivity and resistance also corroborates with the preliminary observations that the genetic basis of resistance may vary for different organisms. Differences in resistance mechanisms may be especially relevant for laboratory-generated mutants of model fungi and field-resistant isolates of plant pathogens, since reductions in comparative fitness or pathogenicity are often restricted to the first category. Much additional knowledge is needed to fully understand the phenomenon.

REFERENCES

Anonymous (1992) Sterol biosynthesis inhibitors - risk of resistance and recommended antiresistance strategies. Recommendations of the FRAC-DMI-working group for 1992. *Gesunde Pflanzen* **44**, 361-365.

Birchmore, R.J.; Ashman, P.I.; Stanley, S.; Russell, P.E. (1992) Effects of crop history on sensitivity to procloraz of *Pseudocercosporella herpotrichoides* isolates from cereals in Western Europe. *Brighton Crop Protection Conference; Pests and Diseases 1992* **1**, 183-188.

Brown, J.K.M.; Jessop, A.C.; Thomas, S.; Rezanoor, H.N. (1992) Genetic control of the response of *Erysiphe graminis* f.sp. *hordei* to ethirimol and triadimenol. *Plant Pathology* **41**, 126-135.

Cevat, H. (1992) Japanse roest terug van nooit weggeweest. Nu maatregelen nemen om aantasting in najaar te voorkomen. *Vakblad voor de Bloemisterij* **29**, 32-33.

De Waard, M.A. (1992[a]) Fungicide resistance management in winter wheat in the Netherlands. *German Phytomedical Society Series* **4**, 183-191.

De Waard, M.A. (1992[b]) Action of sterol biosynthesis inhibitors against *Botrytis cinerea*. In: K. Verhoeff, N.E. Malathrakis and B. Williamson (Eds), Wageningen, the Netherlands: Pudoc Scientific Publishers, pp. 223-227.

De Waard, M.A. (1993) Recent developments in fungicides. In: *Modern Crop Protection and Perspectives*, J.C. Zadoks (Ed.), Wageningen, the Netherlands: Wageningen Pers, pp. 11-19.

De Waard, M.A.; Kipp, E.M.C.; Horn, N.M.; Van Nistelrooy, J.G.M. (1986) Variation in sensitivity to fungicides which inhibit ergosterol biosynthesis in wheat powdery mildew. *Netherlands Journal of Plant Pathology* **92**, 21-32.

De Waard, M.A.; Van Nistelrooy, J.G.M. (1979) Mechanism of resistance to fenarimol in *Aspergillus nidulans*. *Pesticide Biochemistry and Physiology*, **10**, 219-229.

De Waard, M.A.; Van Nistelrooy, J.G.M. (1981) Induction of fenarimol-efflux in *Aspergillus nidulans* by fungicides inhibiting sterol biosynthesis. *Journal of General Microbiology* **126**, 483-489.

De Waard, M.A.; Van Nistelrooy, J.G.M. (1984[a]) Effects of phthalimide fungicides on the accumulatiom of fenarimol by *Aspergillus nidulans*. *Pesticide Science* **15**, 56-62.

De Waard, M.A.; Van Nistelrooy, J.G.M. (1984[b]) Differential accumulation of fenarimol by a wild-type isolate and fenarimol-resistant isolates of *Penicillium italicum*. *Netherlands Journal of Plant Pathology* **90**, 143-153.

De Waard, M.A.; Van Nistelrooy, J.G.M. (1987) Inhibitors of energy-dependent efflux of the fungicide fenarimol by *Aspergillus nidulans*. *Experimental Mycology* **11**, 1-10.

De Waard, M.A.; Van Nistelrooy, J.G.M. (1988) Accumulation of SBI fungicides in wild-type and fenarimol-resistant isolates of *Penicillium italicum*, *Pesticide Science* **22**, 371-382.

Eckert, J.W. (1987) *Penicillium digitatum* biotypes with reduced sensitivity to imazalil. *Phytopathology* **77**, 1728 (abstr.).

Elad, Y. (1992) Reduced sensitivity of *Botrytis cinerea* to two sterol biosynthesis-inhibiting fungicides: fenetrazole and fenethanil. *Plant Pathology* **41**, 47-54.

Fletcher, J.S.; Wolfe, M.S. (1981) Insensitivity of *Erysiphe graminis* f.sp. *hordei* to triadimefon, triadimenol and other fungicides. *Brighton Crop Protection Conference; Pests and Diseases - 1981* **2**, 633-640.

Fuchs, A.; De Waard, M.A. (1982) Resistance to ergosterol biosynthesis inhibitors. In: *Fungicide Resistance in Crop Protection*, J. Dekker and S.G. Georgopoulos (Eds), Wageningen, the Netherlands: Pudoc, pp. 71-100.

Fuchs, A.; Drandarevski, C.A. (1976) The likelihood of development of resistance to systemic fungicides which inhibit ergosterol biosynthesis. *Netherlands Journal of Plant Pathology* **82**, 85-87.

Gastonyi, M.; Josepovits, G. (1975) Biochemical and chemical factors of the selective antifungal action of triforine. *Acta Phytopathologica Academiae Scientiarum Hungaricae* **10**, 437-446.

Gastonyi, M.; Josepovits, G. (1979) The activation of triadimefon and its role in the selectivity of fungicide action. *Pesticide Science* **10**, 57-65.

Guan,J.; Stehmann, C.; Ellis, S.W.; Kerkenaar, A.; De Waard, M.A. (1992) Ergosterol biosynthesis in a cell free preparation of *Penicillium italicum* and its sensitivity to DMI fungicides. *Pesticide Biochemistry and Physiology* **42**, 262-270.

Guan, J.; De Waard, M.A. (1993) Inhibition of sterol 14α-demethylase activity in *Penicillium italicum* does not correlate with resistance to the DMI fungicide imazalil. *Netherlands Journal of Plant Pathology* **46**, 1-6.

Heaney, S.P. (1988) Population dynamics of DMI fungicide sensitivity changes in barley powdery mildew. In: *Fungicide Resistance in North America*, C.J. Delp (Ed.), St. Paul, Minnesota, USA: APS Press, pp. 89-92.

Hildebrand, P.D.; Lockart, C.L.; Newbery, R.J.; Ross, R.G. (1988) Resistance of *Venturia inaequalis* to bitertanol and other demethylation-inhibiting fungicides. *Canadian Journal of Plant Pathology* **10**, 311-316.

Hippe, S. (1987) Combined application of low temperature preparation and electron microscope autoradiography for the localization of systemic fungicides. *Histochemistry* **87**, 309-315.

Hollomon, D.W.; Butters, J.; Clark, J. (1984) Genetic control of triadimenol resistance in barley powdery mildew. *British Crop Protection Conference; Pests and Diseases - 1984* **2**, 477-482.

Hollomon, D.W.; Butters, J.; Hargreaves, J.A. (1990) Resistance to sterol biosynthesis-inhibiting fungicides. In: *Managing Resistance to Agrochemicals*, M.B. Green, H.M. LeBaron, and W.K. Moberg (Eds), Washington, USA: American Chemical Society, pp. 199 214.

Huggenberger, F.; Collins, M.A.; Skylakakis, G., (1984) Decreased sensitivity of *Sphaerotheca fuliginea* to fenarimol and other ergosterol-biosynthesis inhibitors. *Crop Protection* **3**, 137-149.

Hunter, T.; Jordan, V.W.; Kendall, S.J. (1986) Fungicide sensitivity changes in *Rhynchosporium secalis* in glasshouse experiments. *British Crop Protection Conference; Pests and Diseases - 1986* **2,** 523-536.

Kalamarakis, A.E.; De Waard, M.A.; Ziogas, B.N.; Georgopoulos, S.G. (1991) Resistance to fenarimol in *Nectria haematococca* var. *cucurbitae*. *Pesticide Biochemistry and Physiology* **40**, 212-220.

Kalamarakis, A.E.; Demopoulos, V.P.; Ziogas, B.N.; Georgopoulos, S.G. (1989) A highly mutable major gene for triadimenol resistance in *Nectria haematococca* var. *cucurbitae*. *Netherlands Journal of Plant Pathology* **95 suppl.1**, 109-120.

Kalb, V.F.; Loper, J.C.; Dey, C.R.; Woods, C.W.; Sutter, T.R. (1986) Isolation of a cytochrome P-450 structural gene from *Saccharomyces cerevisiae*. *Gene* **45**, 237-245.

Köller, W. (1992) Antifungal agents with target sites in sterol functions and biosynthesis. In: Target Sites of Fungicide Action, W. Köller (Ed.), Boca Raton, Florida, USA: CRC Press, pp. 119-206.

Leroux. P.; Marchegay, P. (1991) Caractérisation des souches de *Pseudocercosporella herpotrichoides*, agent du piétin-verse des céréales au prochloraze, isolées en France sur blé tendre d'hiver. *Agronomie* **11**, 767-776.

Ney, C. (1988) Untersuchungen zur Resistenz von *Monilia fructicola* (Wint.) Honey gegenüber Ergosterol-Biosynthese Hemmern. *Thesis University Basle*, 92 pp.

Peever, T.L.; Milgroom, M.G. (1992) Inheritance of triadimenol resistance in *Pyrenophora teres*. *Phytopathology* **82**, 821-828.

Peever, T.L.; Milgroom, M.G. (1993) Genetic correlations in resistance to sterol biosynthesis-inhibiting fungicides in *Pyrenophora teres*. *Phytopathology* **83**, 1076-1082.

Ryley, J.F.; Wilson, R.G.; Barrett-Bee, K.J. (1984) Azole resistance in *Candida albicans*. *Sabouraudia* **22**, 53-63.

Schepers, H.T.A.M. (1983) Decreased sensitivity of *Sphaerotheca fuliginea* to fungicides which inhibit ergosterol biosynthesis. *Netherlands Journal of Plant Pathology* **89**, 185-187.

Schepers, H.T.A.M. (1985) Changes during a three-year period in the sensitivity to ergosterol biosynthesis inhibitors of *Sphaerotheca fuliginea* in the Netherlands. *Netherlands Journal of Plant Pathology* **91**, 105-118.

Sheridan, J.E.; Grbavac, N.; Sheridan, M.H. (1985) Triadimenol insensitivity in *Pyrenophora teres*. *Transactions of the British Mycological Society* **85**, 338-341.

Sholberg, P.L.; Haag, P.D. (1993) Sensitivity of *Venturia inaequalis* isolates from British Columbia to flusilazole and myclobutanil. *Canadian Journal of Plant Pathology* **15**, 102-106.

Siegel, M.R.; Kerkenaar, A.; Kaars Sijpesteijn, A. (1977) Antifungal activity of the systemic fungicide imazalil. *Netherlands Journal of Plant Pathology* **83 suppl.1**, 121-133.

Smith, F.D.; Köller, W. (1990) The expression of resistance of *Ustilago avenae* to the sterol demethylation inhibitor triadimenol is an induced response. *Phytopathology* **80**, 584-590.

Stanis, V.F.; Jones, A.L. (1985) Reduced sensitivity to sterol-inhibiting fungicides in field isolates of *Venturia inaequalis*. *Phytopathology* **75**, 1098-1101.

Stehmann, C.; Kapteyn, J.C.; De Waard, M.A. (1994) Development of a cell-free assay from *Botrytis cinerea* as a biochemical screen for sterol biosynthesis inhibitors. *Pesticide Science*. *Pesticide Science* **40**, 1-8.

Steva, H.; Cartolaro, P.; Gomes da Silva, M.T. (1990) Tolerance of powdery mildew of SBI fungicides: situation for 1989. *Phytoma* **419**, 41-44.

Taylor, F.R.; Rodriguez, R.J.; Parks, L.W. (1983) Requirement for a second sterol biosynthetic mutation for viability of a sterol C-14 demethylation defect in *Saccharomyces cerevisiae*. *Journal of Bacteriology* **155**, 64-68.

Van Tuyl, J.M. (1977) Genetics of fungal resistance to systemic fungicides. *Mededelingen Landbouwhogeschool Wageningen* **77-2**, 1-136.

Vanden Bossche, H.; Marichal, P.; Gorrens, J.; Bellens, D.; Moereels, H.; Janssen, P.A. (1990) Mutation in cytochrome P-450-dependent 14α-demethylase results in decreased affinity for azole antifungals. *Biochemical Society Transactions* **18**, 56-59.

Walsh, R.C.; Sisler, H.D. (1982) A mutant of *Ustilago maydis* deficient in sterol C-14 demethylation: Characteristics and sensitivity to inhibitors of ergosterol biosynthesis. *Pesticide biochemistry and Physiology* **18**, 122-131.

Weete, J.D.; Wise, L. (1987) Effects of triazoles on fungi. V. Response by a naturally tolerant species, *Mucor rouxii*. *Experimental Mycology* **11**, 214-222.

SENSITIVITY BEHAVIOUR OF *SEPTORIA TRITICI* POPULATIONS ON WHEAT TO CYPROCONAZOLE

U. GISI and D. HERMANN

Agrobiological Research Station, SANDOZ AGRO Ltd., CH-4108 Witterswil, Switzerland

ABSTRACT

In a sensitivity monitoring program, bulk samples of *Septoria tritici* collected from wheat fields in the UK were tested against cyproconazole and flutriafol. The width of the sensitivity distribution (between lowest and highest EC 50 value) was a factor of about 30 and 60 for cyproconazole and flutriafol, respectively. No shifts in sensitivity were found when the 1992 and the 1993 populations were compared. In the two years mean EC 50 values of the populations were 0.2 mg/l and 0.1 mg/l for cyproconazole and 1.0 mg/l and 0.8 mg/l for flutriafol, respectively. No difference was observed in the sensitivity distributions between samples from DMI-treated fields and fields not treated with DMI fungicides. Sensitivity distributions of samples from different regions were similar and there was no correlation between sensitivity and sampling date. Positive cross resistance between cyproconazole and flutriafol was found for entire populations but also for most individual strains. So far, there is no indication of a decreased sensitivity of this pathogen against cyproconazole.

INTRODUCTION

In northern Europe, *Septoria tritici* (teleomorph *Mycosphaerella graminicola*), is one of the most important pathogens of wheat, causing leaf blotch on many cultivars. Since the control of this pathogen cannot easily be achieved by breeding programs, the use of pesticides has been common practice for many years. Amongst them, broad spectrum fungicides of the DMI class (Demethylation Inhibitors), mainly triazoles, are used successfully on a large acreage. The development of resistance to triazoles is known to have occurred in *Erysiphe gra/minis* and is also a potential risk with *S. tritici*. Therefore, a sensitivity monitoring was initiated two years ago for *S. tritici*. The objectives of the program were to evaluate the variability in triazole-sensitivity among field isolates (bulk samples) of *S. tritici* from different regions of the UK, and to compare isolates collected in fields treated or not treated with DMI's. In addition, a comparison of the sensitivity distribution was carried out between the years 1992 and 1993 for both cyproconazole and flutriafol.

MATERIALS AND METHODS

Leaf samples, infected with *S. tritici*, were collected between the end of May and beginning of August in different regions of England and Scotland. They were surface-sterilized with 2 % sodium hypochlorite solution, placed on water agar and exposed to UV light to stimulate pycnidia production. A spore suspension (bulk samples) was transferred to potato dextrose agar (PDA) agar containing 100 mg/l streptomycin and 100 mg/l rifampicin. After 2 days the fungus was transferred to malt yeast agar and incubated at 20° C until sporulation occurred. A spore suspension (10^4/ml) was spread on the surface of PDA-plates containing cyproconazole or flutriafol at concentrations of 0.005, 0.05, 0.5, 5, and 50 mg AI/l. After incubation for 6 days at 20° C, growth and sporulation of the isolates were assessed visually and compared to that on unamended agar. Dose-response correlations were used to calculate EC 50 values (effective concentrations resulting in 50% inhibition of fungal growth).

RESULTS AND DISCUSSION

Sensitivity distributions in 1992 and 1993

The EC 50 values for cyproconazole of the strains isolated in 1992 ranged from 0.03 to 0.9 mg/l, the average EC 50 was 0.2 mg/l. In 1993, the values ranged between 0.01 and 0.4 mg/l with an average EC 50 of 0.15 mg/l (Figs. 1 and 2). With flutriafol, the EC 50 values varied from 0.1 to 6.3 mg/l in 1992 (average 1.0 mg/l). In 1993, the EC 50 values ranged between 0.04 and 3.2 mg/l; the average EC 50 was 0.8 mg/l (Figs. 3 and 4). The width of the sensitivity distribution (between the lowest and highest EC 50 values) in both years was about the same (cyproconazole: factor of 30 in 1992, factor of 40 in 1993; flutriafol: factor of 60 in 1992, factor of 80 in 1993). The 1993 populations were somewhat more sensitive than those from 1992, but also the reference strains were more sensitive. The difference in sensitivity could partly be explained by slight changes in the testing method. No strain was found with a sensitivity level as low as that of the less sensitive reference strain S27. However, some strains proved to be even more sensitive than the sensitive reference strain RL2.

When the sensitivity distributions to flutriafol and cyproconazole of all samples were compared, flutriafol was about 6 to 10 times (1993 and 1992 figures) less active than cyproconazole. With individual strains, the difference in sensitivity to the two tested triazoles, expressed by dividing the EC 50 values of flutriafol and cyproconazole, varied by a factor of 1 to 34 (in 1993) and 1 to 105 (in 1992). Strains of the 1992 population with the most extreme EC 50 values were analysed, i.e. 5 % of the total population both at the upper and the lower end: The flutriafol/cyproconazole EC50-factor was between 1 and 15 (Table 1). Only 6 strains out of 252 resulted in factors higher than 50 (extreme: 105) and 16 strains had factors lower than 2 (extreme: 1.0). Therefore, a good positive cross resistance pattern between the two tested triazole fungicides was found at the population level and also for most of the individual strains. These results show clearly, how dangerous it is to do cross resistance studies with only a few strains.

Samples collected at different sites

No difference was observed in the sensitivity distribution of samples isolated in both years from fields that had not been treated with DMI-fungicides (Fig. 5) and from DMI-treated fields (Fig. 6). Comparing the sensitivity distribution of 1992 with 1993, and of DMI-treated with -untreated populations, no differences were detected by the non-parametric Kolmogoroff-Smirnoff test. The EC 50 values (1993) found with samples from untreated fields ranged from

FIGS. 1 and 2: Sensitivity distribution of bulk samples of *S. tritici* to cyproconazole, isolated from wheat fields in the UK in 1992 (Fig. 1) and in 1993 (Fig. 2). RL2 and S27 are sensitive and less sensitive reference strains, respectively; $\bar{x}$ and m are mean and median of EC50 value distribution

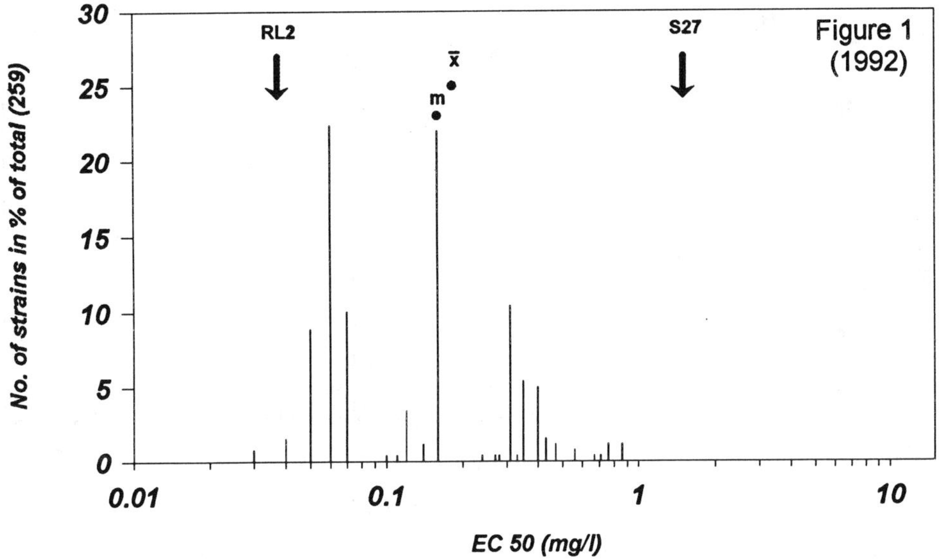

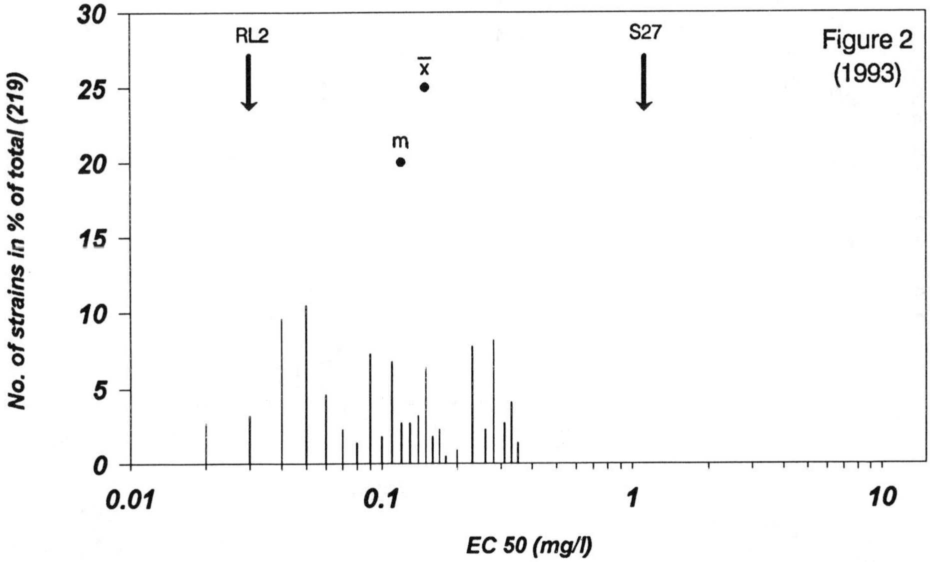

FIGS. 3 and 4: Sensitivity distribution of bulk samples of *S. tritici* to flutriafol isolated from wheat fields in the UK in 1992 (Fig. 3) and in 1993 (fig 4). (Symbol explanation see Fig. 1)

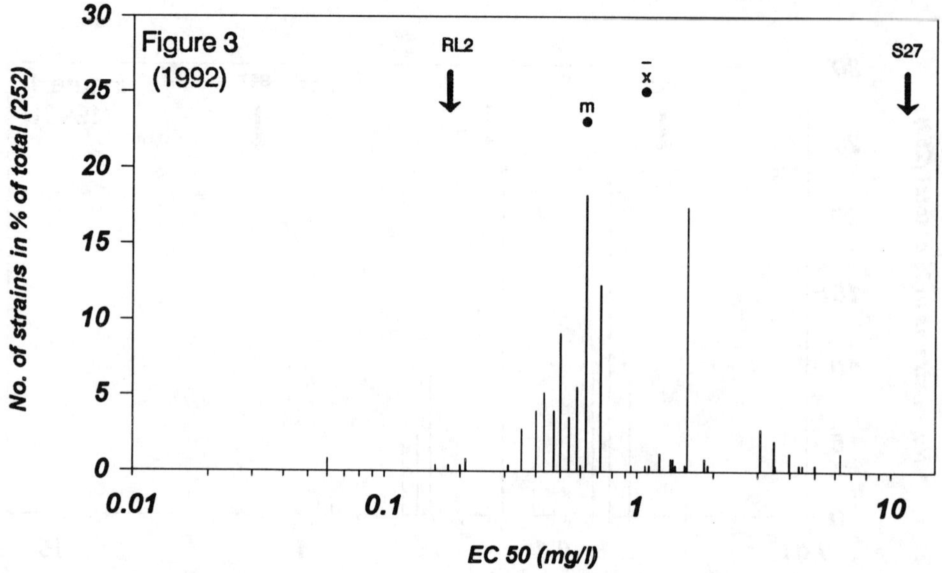

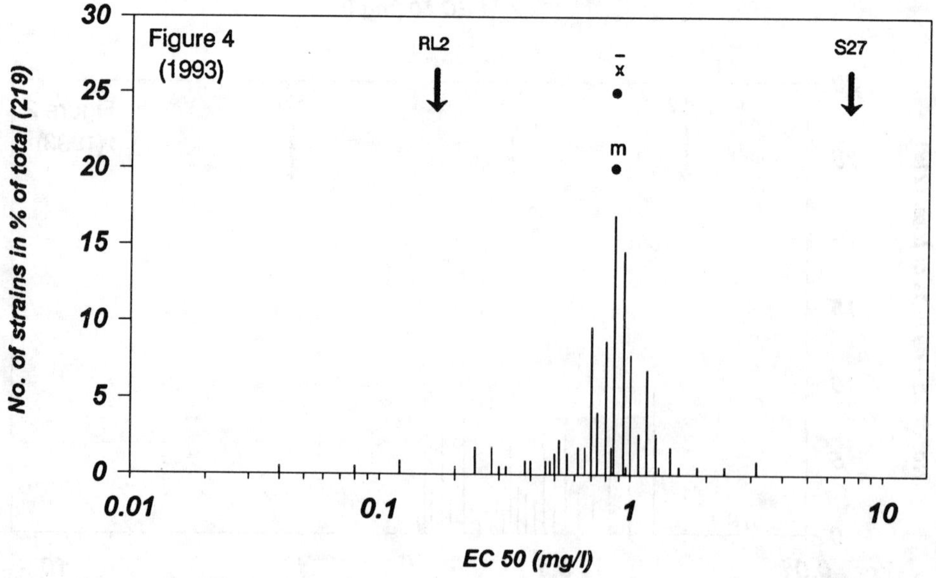

TABLE 1: Sensitivity of individual strains of *S. tritici* to cyproconazole and flutriafol *in vitro*

strain no.	sensitivity (EC 50 in mg/l) to cyproconazole	flutriafol	factor [a)] (EC 50 flu/EC 50 cyp)
highly sensitive to cyproconazole:			
183	0.03	0.4	13
28	0.04	0.4	10
176	0.04	0.4	10
55	0.04	0.4	10
203	0.04	0.6	15
244	0.05	0.2	4
134	0.05	0.4	8
222	0.05	0.4	8
151	0.05	0.4	8
139	0.05	0.4	8
81	0.05	0.4	8
177	0.05	0.4	8
95	0.05	0.5	10
less sensitive to cyproconazole:			
36	0.43	3.5	8
16	0.47	0.6	1
30	0.47	1.4	3
245	0.47	1.6	3
121	0.56	1.2	2
140	0.56	1.4	2
192	0.67	3.1	5
116	0.71	1.8	3
172	0.76	1.6	2
133	0.76	1.8	2
117	0.86	1.4	2
126	0.86	1.6	2
125	0.86	1.6	2

[a)] A factor of 5 should be considered as variation among tests (a total of 252 strains have been tested in 12 tests). This factor is based on the EC 50 values found for the reference strains in individual tests. In addition, a factor of 6 - 10 represents the difference in intrinsic activity between cyproconazole and flutriafol. Therefore, strains with factors (EC 50 flu/EC 50 cyp) between 2 and 50 are considered to represent positive cross resistance.

0.03 to 0.4 mg/l (average 0.18 mg/l). The values of samples from DMI-treated fields varied between 0.01 and 0.4 mg/l (average 0.14 mg/l). Three regions have been analyzed in more detail: East Anglia/Midlands (141 samples, Fig. 7), South/South West (37 samples, Fig. 8) and North/Scotland (41 samples, Fig. 9). There was no obvious difference in the sensitivity distributions of samples from the three regions. The average EC 50 values of the samples from the regions were: East Anglia/Midlands: 0.14 mg/l, South/SouthWest: 0.18 mg/l, North/ Scotland: 0.14 mg/l. The strains from East Anglia/Midlands were sampled over a period of approx. 11 weeks (from the end of May to the beginnig of August). A correlation was

FIGS 5 and 6: Sensitivity distribution of bulk samples of *S. tritici* to cyproconazole, isolated in 1993 in the UK from wheat fields not treated with DMI fungicides (Fig. 5) and treated with DMI fungicides (Fig. 6). (Symbol explanation see Fig. 1)

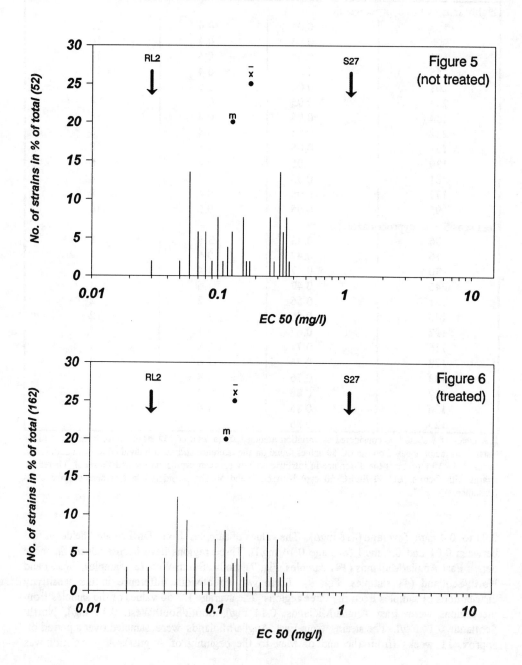

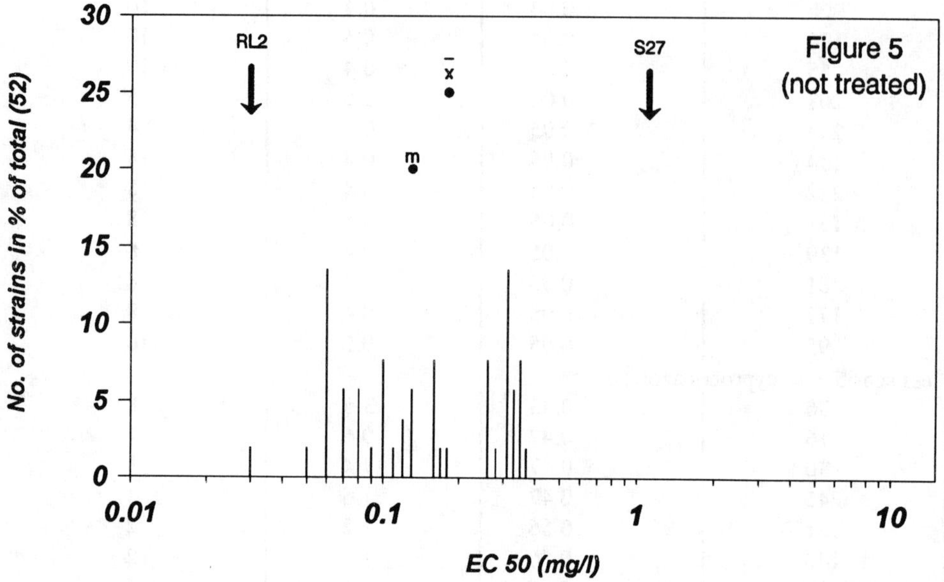

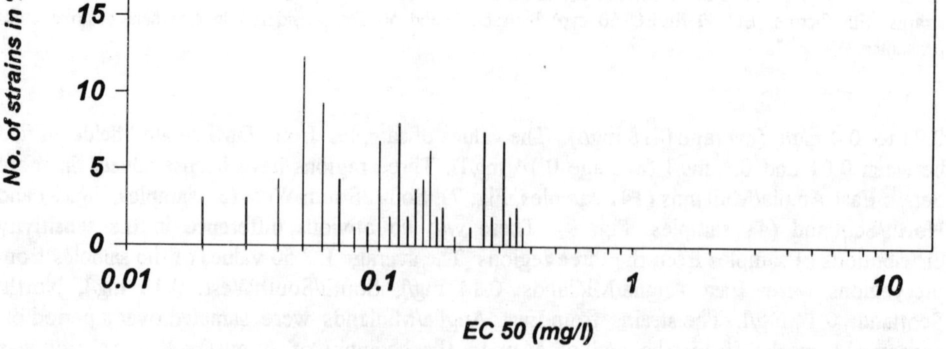

FIGS. 7, 8 and 9: Sensitivity distribution of bulk samples of *S. tritici* to cyproconazole, isolated in 1993 from wheat fields in East Anglia/Midlands (Fig. 7), in the South/South West (Fig. 8) and in North England/Scotland (Fig. 9) (Symbol explanation see Fig. 1)

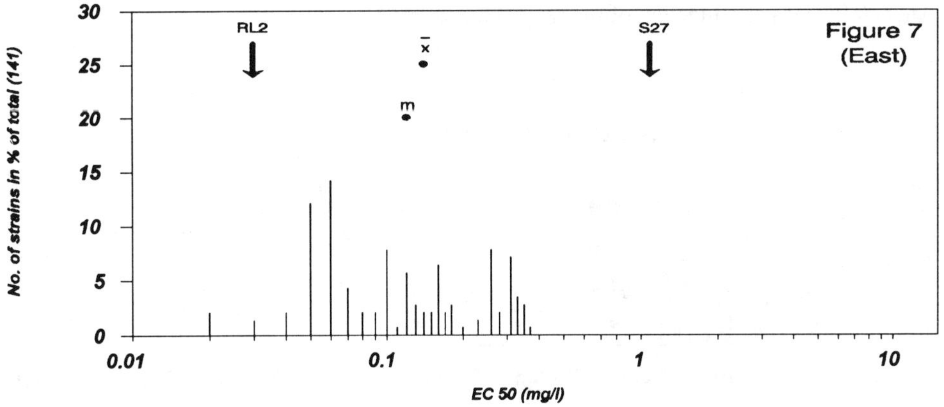

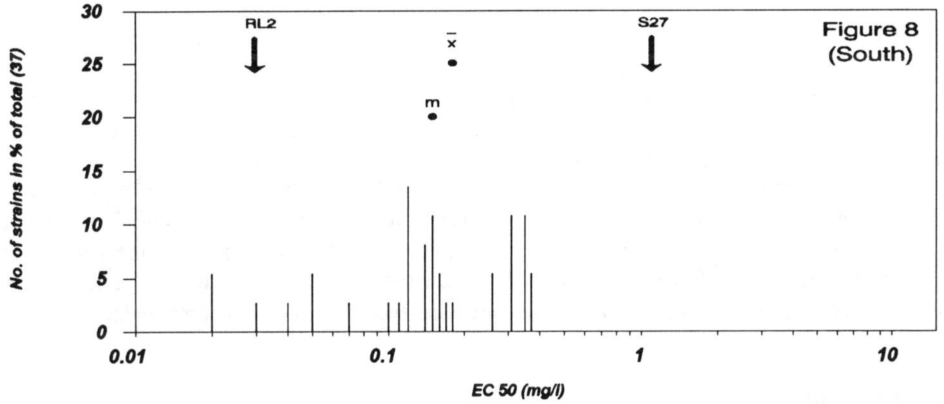

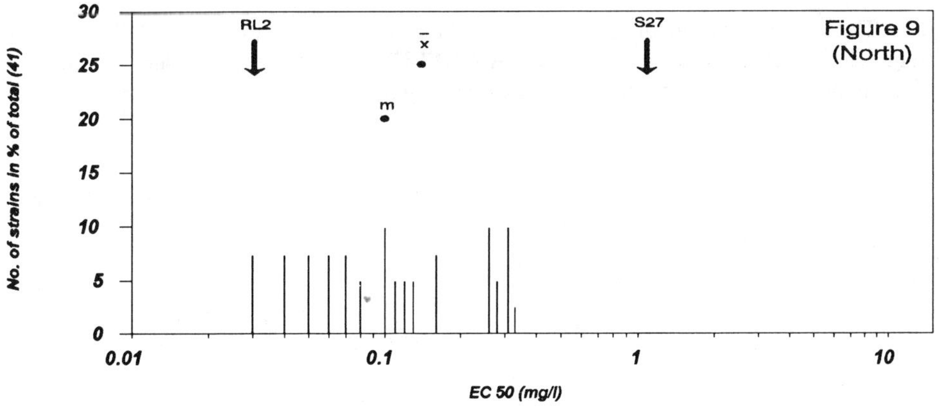

calculated to show whether certain sensitivity levels would dominate throughout the season (Fig. 10). No increase or decrease in sensitivity levels throughout the season was found, i.e. no correlation between the EC 50 values and the sampling date was observed. The highest EC values were found in samples taken in the middle of July.

FIG. 10: Sensitivity of *S. tritici* to cyproconazole, isolated from wheat fields in 1993 in East Anglia/Midlands as a function of the sampling date

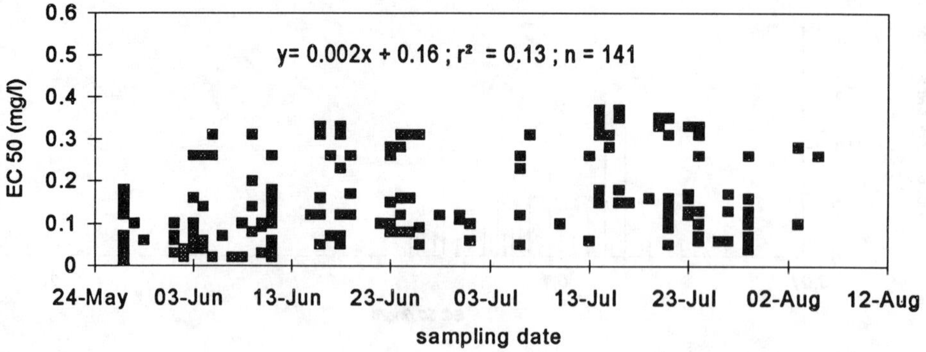

CONCLUSIONS

Comparing the two years 1992 and 1993, there is no indication of changes in sensitivity of *S. tritici* populations towards cyproconazole and flutriafol. Because there are no base line data available from years prior to the extensive use of triazole fungicides, it is difficult to conclude whether a sensitivity shift has already occurred and stabilized on a lower sensitivity level (as in powdery mildew of cereals) or, more likely, whether such a shift has not yet occurred, as with rusts of wheat (Ohl and Gisi, this volume). The years to come will reveal whether a sensitivity shift will occur with this pathogen. So far, effective triazole fungicides control the disease caused by *S. tritici* if they are used according to the recommendations.

ACKNOWLEDGEMENT

The authors wish to thank U. Hugelshofer and C. Steiner for the skillful sensitivity testing. Thanks are also extended to the SANDOZ-UK team for the excellent cooperation during sampling and to D. Hollomon for providing us with the reference isolates.

RESISTANCE OF *UNCINULA NECATOR* TO DMI FUNGICIDES IN CALIFORNIA VINES

W.D. GUBLER[1], H.L. YPEMA[1], D.G. OUIMETTE[2], L.J. BETTIGA[3]

[1]Department of Plant Pathology, University of California, Davis, CA, USA. [2]Dow Elanco, Indianapolis, IN, USA. [3]University of California Cooperative Extension, Salinas, CA, USA.

ABSTRACT

This paper discusses interactions between environmental conditions, pathogen life-cycle and application strategy, and their effects on disease severity and development of Demethylation Inhibitor (DMI) resistance in *Uncinula necator* in Californian vines.

A programme of screening for DMI resistance in isolates from 19 vineyards in different regions of California was initiated in 1990. Isolates were tested for sensitivity to triadimefon, myclobutanil and fenarimol using a leaf disc test. Frequency distribution curves indicated higher levels of resistance to triadimefon and to lesser extent to myclobutanil in most regions. Increased levels of resistance found to fenarimol were not as evident compared to wild-type sensitivity.

A time-course experiment was conducted to follow the development of resistance to the three chemicals in a vineyard subjected to multiple applications of triadimefon. Resistance levels increased during the growing season. Resistance levels to myclobutanil and fenarimol in ascospore populations collected prior to and after the growing season were similar, but resistance levels to triadimefon were found to be increased in the latter ascospore population. Cross-resistance relationships are discussed.

INTRODUCTION

Grape production in California utilizes 300,000 ha of land in virtually every agricultural production area. Coastal regions, particularly Central and South are prone to severe powdery mildew pressure annually, while some North Coast regions, the Sacramento Valley and most of the San Joaquin Valley deal with heavy pressure from *Uncinula necator* less frequently. Though varietal differences exist between these areas and certainly can contribute to disease development, differences in optimum and minimum temperatures are more important. Under optimum temperature conditions, *U. necator* can complete its life-cycle and reproduce in as little as 5 days. At 33°C, reproduction may require up to 15 days and colonies may be killed when temperatures exceed 37°C for only a few hours.

It is in this context that resistance to DMI fungicides has been somewhat ignored in *U. necator* populations and quite possibly in other pathogen systems where single site action fungicide resistance occurs.

The effect of genetic recombination in fungicide-resistant populations should be addressed. Though cleistothecia are produced and ascospores function in the disease cycle in nearly all California vineyards, there are production areas in which cleistothecia are produced in much higher numbers, thus increasing the risk of increased resistance due to genetic recombination. In California, the greatest number of cleistothecia are produced in the cool coastal areas and in the Delta area and southernmost areas of the San Joaquin Valley. Germination and infection efficiency of ascospores in California are very high. In the aforementioned production areas, viable ascospores have been released from cleistothecia in late summer and fall, as well as from overwintered cleistothecia from the bark in early spring. We suspect that two ascospore-derived populations per year can occur in these areas.

Resistance to the DMI fungicides in California was suspected in 1985, three years after the introduction of triadimefon. In 1985, control failures were reported from nearly every production area, but were more severe in vineyards along the coast. In 1986, disease pressure was extremely high statewide and control failures were prevalent. Isolates collected in 1986 ranged in LC50 values from 2.5 to 39 mg/l. Because no baseline sensitivity data existed for triadimefon for any *U. necator* population, attempts were made to show resistance by investigating whether isolates with differing LC50 values could differentially attack leaves of seedling grapevines after they had been treated with 154 mg/l of triadimefon. Though these tests were successful, in that isolates with the highest LC50 values were able to infect and colonize leaves seven to ten days earlier than isolates with lower LC50 values, poor performance of triadimefon in California was assigned to coverage problems, rate problems or fast tractor speeds during application, rather than resistance. In 1987 and 1988, temperatures statewide were above normal and no problems with powdery mildew spray programmes occurred except in some coastal vineyards. In 1989 and again in 1990, large scale losses occurred in many of the production areas.

MATERIALS AND METHODS

Pathogen sampling was initiated in 1990 from 19 vineyards in California, including one, 15 year old, isolated vineyard (Renaissance) which had never been treated with DMI fungicides. Because the latter vineyard was treated with sulfur only, we designated isolates from this vineyard as wild types.

Thirty mass isolates were collected from 30 individual vines in each vineyard. The isolates were transferred to individual Carignan grapevine seedlings and maintained in isolation tubes held at 23-25°C. Isolates were screened for resistance, using leaf discs treated with a concentration range of each of the three DMI fungicides; triadimefon, myclobutanil and fenarimol. Discs were inoculated with conidia brushed from colonized leaves at the top of a 1.5 m settling tower, then incubated at 24°C for 10 days. Discs were rated for percent surface area covered by the fungus using a Horsefall-Barratt system and LC50 values were calculated by probit transformation of the percentages of inhibition and regression against the logarithm of the concentrations used.

RESULTS AND DISCUSSION

Results of the 1990 survey showed that all nineteen vineyards contained isolates of *U. necator* resistant to triadimefon (ie. exceeding maximum LC50 values obtained from Renaissance vineyard), while resistance to myclobutanil and fenarimol was observed in ten and three vineyards, respectively. For the purposes of this paper, isolates from these 19 vineyards have been regionalized and Table 1 shows the patterns and levels of resistance found in the previously discussed regions of California, i.e. North Coast, Central Coast, Central Valley, South San Joaquin Valley and Renaissance vineyard, where the *U. necator* population has been designated as representing wild type.

TABLE 1. Mean, minimum and maximum LC50-values of *U. necator* populations to triadimefon, myclobutanil and fenarimol in different regions of California, 1990.

Region in California	Fungicide	Number of isolates	Mean LC50 (mg/l)	Minimum LC50 (mg/l)	Maximum LC50 (mg/l)
Renaissance	triadimefon	61	1.40	0.70	3.91
	myclobutanil	62	0.15	0.01	0.80
	fenarimol	62	0.13	0.01	0.59
North Coast	triadimefon	120	4.33	0.62	16.70
	myclobutanil	121	0.67	0.01	6.80
	fenarimol	121	0.22	0.01	1.29
Central Coast	triadimefon	103	16.48	0.49	65.51
	myclobutanil	102	2.83	0.05	6.44
	fenarimol	101	0.44	0.06	1.19
Central Valley	triadimefon	131	5.32	0.10	43.12
	myclobutanil	133	0.58	0.01	1.82
	fenarimol	133	0.19	0.01	1.14
South San Joaquin Valley	triadimefon	105	6.02	0.29	45.83
	myclobutanil	108	0.69	0.01	3.54
	fenarimol	106	0.20	0.01	0.64

The mass isolates collected from Renaissance vineyard demonstrated LC50 values not exceeding values of 3.91 mg/l (mean 1.40 mg/l), 0.80 mg/l (mean 0.15 mg/l) and 0.59 mg/l (mean 0.13 mg/l) for triadimefon, myclobutanil and fenarimol, respectively.

The highest mean levels of resistance (expressed as LC50 values), to all three fungicides were observed in vineyards in the Central Coast, and also the widest ranges of sensitivity levels were found in this area. For triadimefon, and to a lesser extent for myclobutanil, frequency distribution curves range far into levels of higher resistance. This effect is much less pronounced for fenarimol. The frequency distribution curves for the Central Valley, South San Joaquin Valley and North Coast isolates showed a less pronounced occurrence of higher levels of resistance. However occurrence of higher levels of resistance to triadimefon and to a lesser extent myclobutanil is evident. Mean and highest levels of resistance to fenarimol in the latter regions were similar to those found in Renaissance vineyard.

Overwintering of resistance has been documented to occur in ascospore populations which reside in cleistothecia on the bark of cordons, spurs, canes and trunks of grapevines. This phenomenon occurs prevalently in coastal production areas and it is not uncommon to find 600-700 cleistothecia in 10 grams of bark. In addition, ascospore germination and infection efficiency is approximately 50% in these same areas. This phenomenon, coupled with heavy disease pressure starting early in the growing season and continuing through the season, partially explains the high levels of resistance observed in these production areas. Table 2 lists the results of a time course sampling experiment from one vineyard in Monterey County. Ascospores were released from cleistothecia collected in January 1990 from cordon bark of cv. Chardonnay. Thirty ascospore-derived colonies were screened for sensitivity to triadimefon, myclobutanil and fenarimol and mean LC50 values were 21.35 mg/l, 3.16 mg/l and 0.41 mg/l, respectively. After one application of triadimefon (Bayleton 50WP, 285g/ha), thirty mass conidial isolates were collected on May 4 and screened. The average LC50 values were found to have increased with respect to the ascospore-derived population to 33.35, 3.27 and 0.69 mg/l for triadimefon, myclobutanil and fenarimol, respectively. The mean level of sensitivity was found to have decreased further in conidial samples collected on June 25, July 24 and September 3, after the second, third and fourth application of triadimefon, respectively (Table 2). On October 23, a second ascospore population of thirty isolates was collected from cleistothecia residing on leaves. These overwintering resistant isolates appeared to be quite fit and survived well in nature. The above mentioned ascospore samples were also examined for sensitivity to myclobutanil and fenarimol but development of resistance was not as severe for these fungicides and the average LC50 values found in October were not significantly different from those of January. Distributions for the sampled conidial populations showed bimodal curves for both triadimefon and myclobutanil, whereas population curves for fenarimol were normal with only slight movement towards higher levels of resistance from May to September.

TABLE 2. Mean, minimum and maximum LC50-values of 30 samples of a *U. necator* population in Lone Oak vineyard, Monterey County to triadimefon, myclobutanil and fenarimol between January 22, 1990 and October 23, 1990.

Fungicide	Sampling Date	Population derived from:	Mean LC50 (mg/l)	Minimum LC50 (mg/l)	Maximum LC50 (mg/l)
triadimefon	January 22	ascospores	21.35 A	2.27	49.34
	May 4	conidia	33.35 B	8.04	124.15
	June 25	conidia	34.01 B	15.65	66.21
	July 24	conidia	42.93 BC	9.22	66.61
	September 3	conidia	54.41 C	7.21	86.33
	October 23	ascospores	32.20 B	7.22	61.92
myclobutanil	January 22	ascospores	3.16 A	0.44	15.77
	May 4	conidia	3.27 A	0.21	12.08
	June 25	conidia	3.77 A	0.13	13.83
	July 24	conidia	4.51 AB	0.75	9.69
	September 3	conidia	7.94 B	1.33	36.92
	October 23	ascospores	3.52 A	0.38	11.58
fenarimol	January 22	ascospores	0.41 A	0.16	1.65
	May 4	conidia	0.69 A	0.13	2.51
	June 25	conidia	0.73 AB	0.12	2.09
	July 24	conidia	0.85 AB	0.22	2.23
	September 3	conidia	0.94 B	0.25	2.61
	October 23	ascospores	0.43 A	0.13	0.93

Mean values followed by different letters indicate significant differences at the 5% probability level.

The effect of temperature on pathogen increase and development of resistance to DMI-fungicides has largely been ignored. When resistance to DMI fungicides first appeared in *U. necator* populations, the common thinking was that economic loss due to resistance was unlikely to occur because of the gradual movement of the population towards resistance. However, our data indicate that directional shifts in resistance can mimic disruptive shifts which can result in large scale disease control failures. This is primarily due to population increase as effected by environmental conditions. For example, when a *U. necator* population with sufficiently high levels of resistance to triadimefon is subjected to optimal environmental conditions, selection for these high levels of resistance occurs rapidly, due to maximum reproduction rate of the pathogen and selection by the fungicide. This leads to the increased occurrence of members of the population capable of attacking leaves and fruit on the same day the spray is applied. Under these conditions, economic losses have occurred in vineyards throughout California. On the other hand, a resistant population can be controlled easily when temperatures are high and reproduction rates are decreased, because selection for and increase of the resistant population do not take place at a rate fast enough to cause economic losses.

From the presented data, it is evident that resistance of *U. necator* to fenarimol in California vineyards has remained relatively low, indicating that cross resistance between fenarimol and other DMIs may not be wholly representative of situations with other pathogens on other crops. The fact that population curves for DMI fungicide-resistant populations of *U. necator* do not exhibit the same distribution or shape could indicate that development of resistance to fenarimol is at an early stage in California. However, more recent monitoring over a three year period in all production areas has not demonstrated increased levels of resistance to fenarimol even in years conducive to disease development. If no further increase in resistance is observed in future years, fenarimol could possibly be used in integrated control programs to prevent an increase in strains resistant to the other DMI fungicides.

CONCLUSIONS

This research documents the overwintering of resistance to the DMIs, triadimefon, myclobutanil and, to a lesser extent, fenarimol, in ascospores of *U. necator*.

In vineyards where this occurred a time course sampling experiment demonstrated that the mean sensitivity of conidial populations to triadimefon and myclobutanil declined through the growing season as successive applications of triadimefon were made to control disease. Changes in sensitivity to fenarimol were only slight and not significant.

From other studies carried out since 1986 but not reported here, we would like to pass on the following conclusions or observations:

- Resistance to DMI fungicides has occurred more prevalently in California vineyards where triadimefon was used exclusively for disease control between 1982 and 1985.

- Vineyards having the highest levels of resistance in 1986 have the same ranking in 1993, indicating that while control programmes based on sulfur and DMI fungicides appeared to have stabilized the development of resistance, they have had little impact on reverting populations to wild type.

- Resistance has been consistently more severe in vineyards in which cleistothecia are produced. In addition there appears to be a correlation between numbers of cleistothecia produced and levels of resistance in a particular vineyard or region.

- Environmental conditions, primarily temperature, can influence the rate of development of resistance by influencing the reproduction rate of a population of *U. necator*. Under conditions of rapid increase of a population in which high levels of resistance occur, DMI fungicides remove sensitive isolates and allow for a rapid increase in the frequency of the most resistant isolates. Under these conditions, directional shifts in resistance can mimic disruptive shifts with respect to disease control failure.

SENSITIVITY OF EYESPOT TO PROCHLORAZ

R J BIRCHMORE, P E RUSSELL

AgrEvo UK Limited, Chesterford Park, Saffron Walden, Essex, CB10 1XL

H BUSCHHAUS

AgrEvo, Postfach 65 0331, D-1000, Berlin 65, Germany

ABSTRACT

Monitoring carried out between 1985 and 1991 revealed no evidence of changes in sensitivity to prochloraz of French populations of Pseudocercosporella herpotrichoides. However, sampling during 1992 and 1993 identified a number of less-sensitive, slow-growing, R-type isolates from sites in Northern France. The level of reduced sensitivity was relatively low, in common with other reported instances of altered sensitivity to sterol biosynthesis inhibitor fungicides. Prochloraz efficacy was reduced at some, but not all sites where less-sensitive strains were found. However, yield increases were obtained at the majority of sites studied. There was no consistent evidence of prochloraz applications selecting for the R-type at sites studied in detail.

INTRODUCTION

Prochloraz has been the leading fungicide for control of the cereal eyespot pathogen, Pseudocercosporella herpotrichoides, since the mid-1980s, when resistance to carbendazim became widespread in European populations. One of the strengths of prochloraz is its equal activity against both W and R strains of the pathogen under normal circumstances.

The sensitivity of European populations of P. herpotrichoides to prochloraz has remained remarkably stable since the introduction of the compound in the early 1980s. However, there have been several recent reports of the occurrence of isolates with reduced sensitivity to prochloraz in the North of France (Leroux and Marchegay, 1991, Cavelier et al, 1992, Migeon et al, 1992). Initially these authors reported the detection of a reduction in sensitivity under laboratory conditions only, but Leroux and Migeon (1993) concluded that in some instances there was a correlation between in vitro insensitivity and reduced field efficacy.

As part of a programme of work to investigate these findings a large number of trials were carried out at sites in France during 1992 and 1993. This paper will present the results of these trials as well as reviewing recent literature on this subject.

MATERIALS AND METHODS

Field trials were established across N. France, all in winter wheat with known cropping histories. The sites were chosen as being representative of prochloraz use, but included ones identified as having less sensitive eyespot populations which were studied in detail. Trials were arranged as random block designs and prochloraz applied in 200 l water/ha at decimal growth stage 31 to 32 (Tottman, 1987). Levels of eyespot were assessed regularly throughout the season and samples of infected stems were removed at intervals for isolation of the pathogen at Chesterford Park.

Sampling, and subsequent sensitivity testing of the isolates, were carried out according to the published method (Birchmore, 1991), using a range of concentrations of prochloraz from 0.0075 mg/l to 1.0 mg/l. The colony diameters obtained after 14 days growth at 20°C were used to calculate an IG_{50} value for each isolate, by applying linear regression. The IG_{50} value is defined as the concentration of prochloraz which is required to give 50% inhibition of growth (IG) of each isolate on potato dextrose agar (PDA, Oxoid).

RESULTS AND DISCUSSION

The results of sensitivity-monitoring carried out at Chesterford Park between 1985 and 1991 are shown in Fig 1. During these years the monitoring process consisted of a 0.5 mg prochloraz/l single dose-rate test for all isolates, followed by a dose-response test carried out on a random-selection of approximately 100 isolates per year from each country. The data show that no changes were detected in the populations sampled during these seven years.

Figure 1 Cumulative frequency distributions of prochloraz IG_{50} values from dose-response testing of *P. herpotrichoides* isolates from French trials between 1985 and 1991.

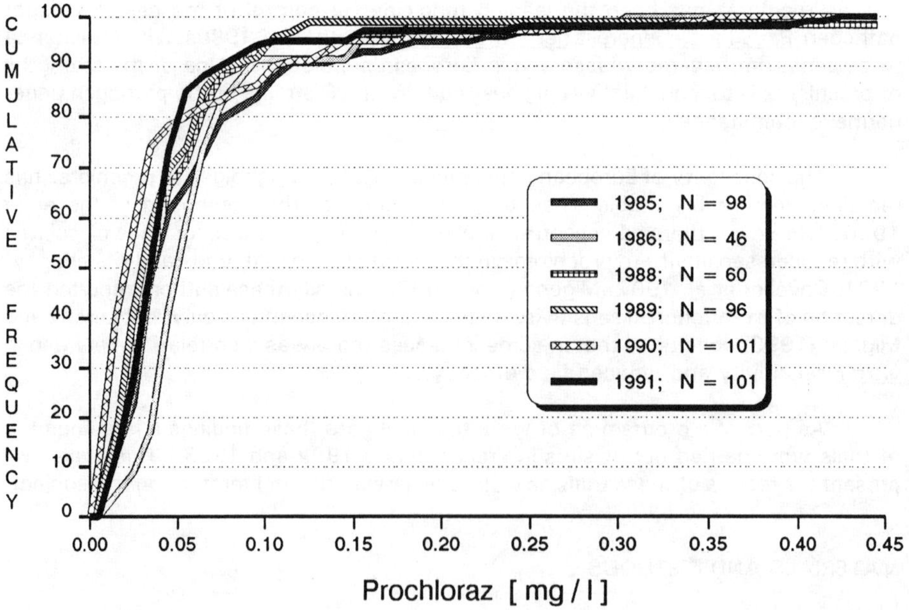

This conclusion is confirmed by the single dose-rate data which have also been published (Birchmore et al, 1992). However, during 1991 the detection of isolates of P. herpotrichoides with reduced sensitivity to prochloraz was reported (Leroux and Marchegay, 1991). These isolates were obtained during 1990 from two sites north of Paris, in the Oise department, and one site further south in Essonne. All of the isolates were of the type II, which can be loosely correlated with the R, or slow-growing type and were designated type IIp. At that time there were no indications of any effect on the efficacy of prochloraz. However, a later report by Leroux and Migeon (1993) indicated that there were instances of reduced efficacy at the end of the season at sites where isolates with reduced sensitivity to prochloraz were found.

During 1992 and 1993 the monitoring programme at Chesterford Park was expanded so that dose-response testing was carried out on all isolates. Sampling during 1992 was concentrated on a relatively small number of sites in Oise, including Jamericourt and Ressons, the two locations reported by Leroux and Marchegay (1991) to have less-sensitive populations. In addition, a more random selection of sites was sampled, spread over a wider area. An increased total of 378 isolates was fully tested in 1992, in comparison with 101 in 1991. Substantial numbers of the R type or slow growing isolates were found to give IG_{50} values above the previous limit of approximately 0.5 mg prochloraz/l. However, the majority of these less-sensitive isolates were derived from a small number of sites which were intensively sampled, including those at Jamericourt and Ressons (Fig 2). Even at these sites, 76.3% of isolates were found to have IG_{50} values below the 0.5 mg prochloraz/l threshold level. Fig 3 shows the population derived from the sites where less-sensitive isolates were not found.

Figure 2 Results of dose-response testing of 329 eyespot isolates obtained during 1992 from French "less-sensitive" sites out of a total of 378 isolates

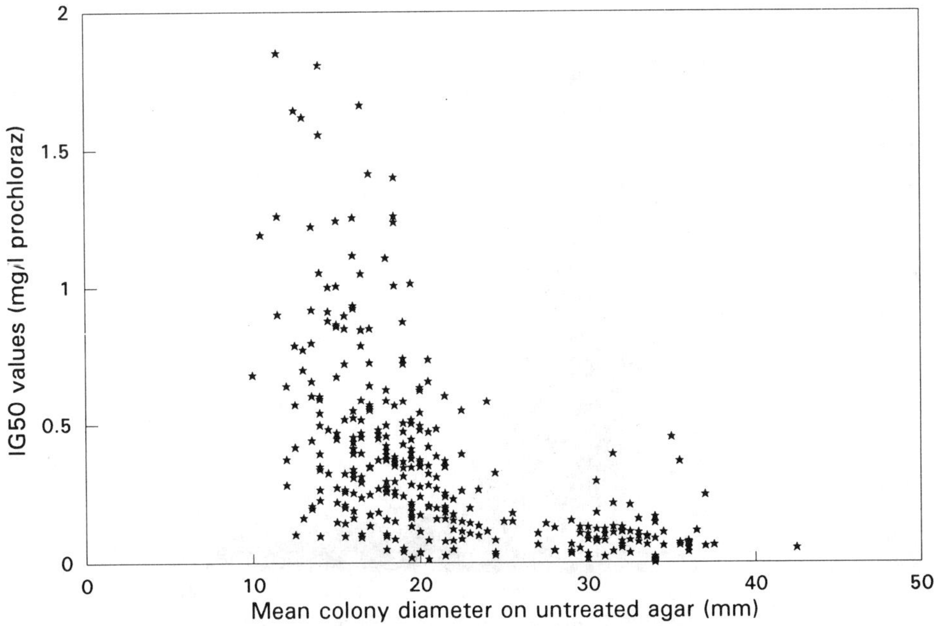

Figure 3 Results of dose-response testing of 49 eyespot isolates from "sensitive" sites

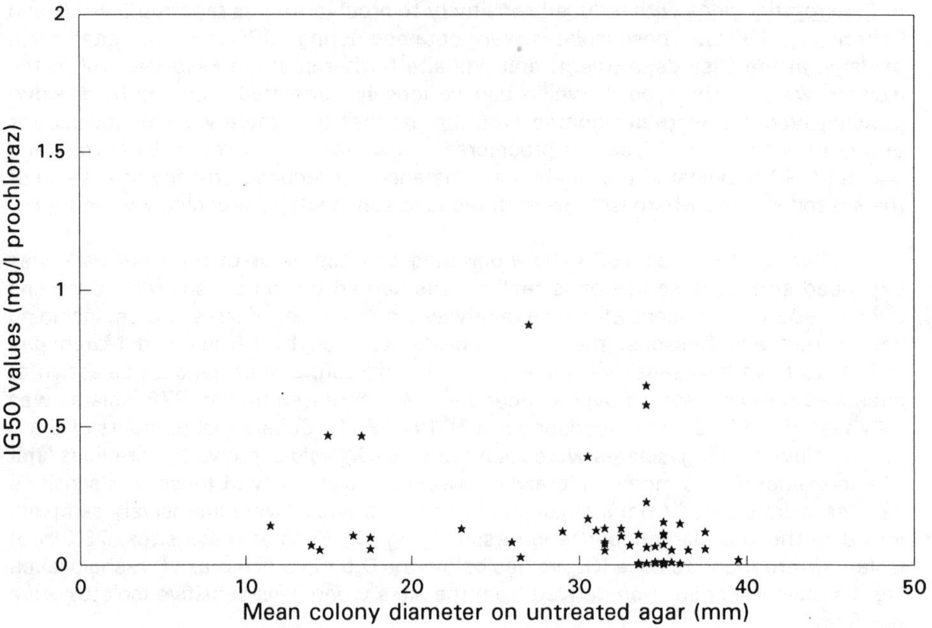

Figure 4 Results of dose-response testing 2000 eyespot isolates obtained during monitoring of French sites in 1993

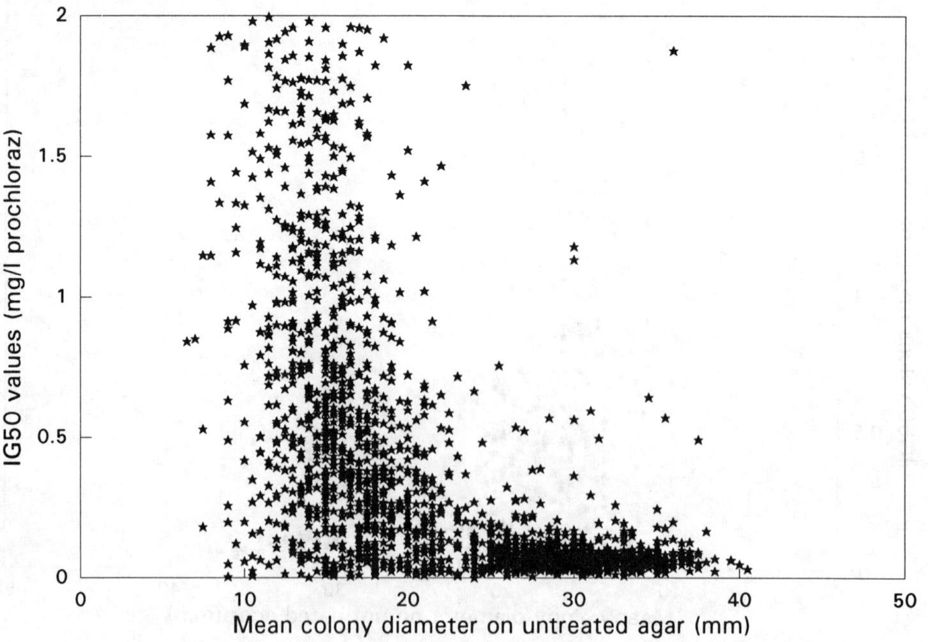

A similar pattern to 1992 (Fig 2) was seen in 1993 (Fig 4) and again the emphasis was placed on studying sites where less sensitive isolates were thought to be present.

An even larger number of samples was monitored than in 1992, with 2000 isolates being tested. The results showed that, while a high proportion (25%) of less-sensitive, R-type isolates were detected, nevertheless, the majority (75%) of the overall population was prochloraz sensitive, that is, had an IG_{50} value of less than 0.5 mg/l.

A very small number of fast-growing isolates were also found to be less-sensitive (Fig 4). These are currently under investigation, as the possibility exists that these could be fast-growing sectors of R types, as reported by Julian (1990).

Our results suggest that the margin between sensitive and less-sensitive isolates was relatively modest and that there still appears to be a continuous distribution. This may indicate a limited, step-wise movement towards reduced sensitivity, as seen with other SBI inhibitors, rather than selection of a mutation with high levels of resistance.

Occurrence of less-sensitive isolates

Migeon et al (1994) reported the results of a large scale monitoring programme, carried out during 1993. This covered 25 departments in the north of France, with samples being taken from over 230 plots and 6,000 isolates of P. herpotrichoides being evaluated for sensitivity to prochloraz. They concluded that W or fast-growing types were in the majority and were not implicated in prochloraz "resistance". However, the proportion of W type isolates which were resistant to triazoles had risen over recent years, to become more common than the sensitive portion of the population. In agreement with our findings in 1992 and 1993, the type IIp isolates were found to be "limited to a few North Western Departments where their distribution is irregular". In the absence of a field-by-field mapping exercise it is not possible to say with precision where and why these isolates occur.

Efficacy and Yield

Trials were carried out in France during 1993 with the objective of examining the relationships between the occurrence of less-sensitive isolates, the efficacy of prochloraz applications and the subsequent yield.

In each of the trials 450g AI/ha of prochloraz was applied during early April at GS 31-32 to winter wheat crops. The efficacy of this application was assessed during the first two weeks of June, when the crops were at GS 74-75. The yields from treated and untreated plots were measured at the time of harvesting and the increase in yield given by the treatment calculated.

Despite variability in efficacy at the end of the season, cost-effective yield increases, above 1.0 decitonne/ha, were obtained at 85% of the 48 sites examined. It was possible to identify 6 sites which had a higher than normal proportion of less-sensitive isolates. The frequencies of occurence of these, together with disease control and yield are shown in Table 1.

Table 1 Effect of 450g AI/ha prochloraz applied at GS 31 on eyespot levels, assessed 9/6/93 at GS 74-75.

Site location (department)	Mean Eyespot Index			% Control	Yield % increase over untreated	% of R types classed as insensitive at harvest
	Untreated	Treated	SE ±			
1 Oise (60)	0.56	0.10	0.13	81.5	2.6 NS	40
2 Oise (60)	2.21	1.27	0.21	42.5	7.9 •	33
3 Oise (60)	1.16	1.68	0.13	0	0	37
4 Oise (60)	1.54	0.76	0.14	50.7	8.8 •	48
5 Oise (60)	0.65	0.89	0.08	0	6.7 •	30
6 Somme (80)	1.55	1.21	0.10	41.0	3.7 •	40

The maximum eyespot index = 4.0

•, significantly different from untreated at P = 0.05

NS, not significantly different from untreated at P = 0.05

The six sites were all located in the north of France, in the Oise and Somme departments. Levels of eyespot at the beginning of June varied from low at two of the sites (1 and 5) to moderate at sites 3, 4 and 6 and high at site 2. The levels of control given by prochloraz at this time also varied, from 81.5% at site 1, to moderate efficacy at sites 2, 4 and 6 and no apparent control at sites 3 and 5.

Yield increases were seen at 5 of the 6 sites and all but one of these increases was significant. The single case, site 1, where the increase was found not to be significant, despite excellent efficacy, could have been due to the levels of disease being so low that there was only limited potential for improvement.

These wide variations in levels of infection, efficacy and yield were seen despite the relative uniformity of the proportions of less-sensitive R type isolates, which ranged from 30% to 48% of the eyespot populations obtained from the sites. The presence of these isolates did not, therefore, appear to be the major factor determining yield and efficacy at these sites.

Similarly, Migeon et al (1994) concluded that at certain sites, prochloraz gave good efficacy, despite the presence of significant proportions of less sensitive isolates.

Effects of prochloraz application on eyespot populations

Bateman et al (1986) reported that prochloraz selected for R-types when applied to trial plots artificially inoculated with mixtures of W and R-types. In vitro assessments of sensitivity, however, have indicated that the R-type is slightly more sensitive to prochloraz than the W-type, on average (Birchmore et al, 1986, Gallimore et al, 1987).

Analysis of the data from the trials conducted during 1993 indicates that there is no clear relationship between prochloraz application and the proportions of R-types (Table 2).

Table 2 Effects of an application of 450g prochloraz/ha on proportions of R-types isolated from stem samples

	% R types isolated		
Department	April Pre-treatment	June	
		Untreated	Prochloraz treated
Eure (27)	70	75	97
Seine-et-Marne (77)	15	20	0
Marne (51)	15	60	15
Marne (51)	13	20	25
Oise (60)	30	30	50
Somme (80)	44	70	70
Oise (60)	67	58	74
Oise (60)	84	57	61

Any differences seen may not be due to one type being more susceptible than another but to differences in the timing and speed of infection of the cereal stem. W-types have been found to infect and develop in stem-bases much more rapidly than R-types and may, therefore, be more susceptible to fungicide applied early in the season than the later, slower developing R-type (Goulds and Fitt, 1990).

CONCLUSIONS

Extensive investigations carried out by both independent researchers and ourselves have confirmed the presence of isolates with decreased sensitivity to prochloraz at a limited number of sites in northern France. Research has shown that at individual trial sites, such isolates can form a significant proportion of the R-type population. However, prochloraz can provide useful efficacy and yield benefits even in their presence.

REFERENCES

Bateman, G.L., Fitt, B.D.L., Creighton, N.F. & Hollomon D.W. (1986) Seasonal changes in populations of *Pseudocercosporella herpotrichoides* (eyespot) in wheat crops. Proceedings of the 1986 British Crop Protection Conference - Pests and Diseases, 1:441-446.

Birchmore, R.J., (1991) Monitoring populations of *Pseudocercosporella herpotrichoides* for sensitivity to prochloraz. Bulletin OEPP/EPPO. 21:313-315.

Birchmore, R.J., Russell, P.E., Ashman, P.I., Stanley, S. & Buschhaus, H. (1992) Sensitivity to prochloraz of *Pseudocercosporella herpotrichoides* isolates from France. Mededelingen van de Faculteit Landbouwwetenschappen Universiteit Gent. 57 : 205-212.

Birchmore, R.J., Vernié, P., Russell, P.E. & Buschhaus, H. (1986) Sensitivity of *Pseudocercosporella herpotrichoides* populations to prochloraz. Proceedings of the 1986 British Crop Protection Conference - Pests and Diseases, 3:787-792.

Cavelier, N., Lorée, F. & Prunier, M., (1992) Resistance of the eyespot fungus, *Pseudocercosporella herpotrichoides*, to DMI fungicides. Proceedings of the Brighton Crop Protection Conference - Pests and Diseases - 1992, 1:189-194.

Gallimore, K., Knights, I.K. & Barnes, G. (1987) Sensitivity of *Pseudocercosporella herpotrichoides* to the fungicide prochloraz. Plant Pathology 36:290-296.

Goulds, A. & Fitt, B.D.L. (1990) Effects of fungicide timing on the severity of winter wheat or winter barley inoculated with W-type or R-type isolates of *Pseudocercosporella herpotrichoides*. Crop Protection 9:265-270.

Julian, A.M., (1990) Variation and fungicide sensitivity in the cereal eyespot pathogen *Pseudocercosporella herpotrichoides*. PhD thesis, University of Nottingham.

Leroux, P. & Marchegay, P. (1991) Caractérisation des souches de *Pseudocercosporella herpotrichoides*, agent du piétin-verse des céréales, résistantes au prochloraze, isolées en France sur blé tendre d'hiver. Agronomie 11:767-776.

Leroux, P. & Migeon, J-L (1993) La resistance en pratique du piétin-verse des céréales vis-a-vis du prochloraze. Une certitude en 1992. Phytoma 448 : 20-24.

Migeon, J-L., Mathon, J-P. & Chudzicki, A-M. (1992) Piétin-verse de blé (*Pseudocercosporella herpotrichoides*) : Etude de la sensibilité au prochloraze dans le nord de la France. Mededelingen van de Faculteit Landbouwwetenschappen Universiteit Gent 57:205-212.

Migeon, J-L., Mathon, M-P., Chudzicki, A-M., Leroy, J-P. (1994) Piétin-verse des céréales : cartographie 1993. Phytoma 458 24-26.

Tottman, D.R. (1987) The decimal code for the growth stages of cereals, with illustrations. Annals of Applied Biology 110, 441-454.

SENSITIVITY OF ERYSIPHE GRAMINIS F.SP. TRITICI TO DEMETHYLATION INHIBITING FUNGICIDES IN EUROPE

F.G. FELSENSTEIN

Institute of Agronomy and Plant Breeding, Technical University of Munich, D-85350 Freising-Weihenstephan

ABSTRACT

To study the sensitivity of the wheat mildew pathogen to demethylation inhibiting fungicides (DMIs) on a European scale, a monitoring programme has been followed since 1986. Single-colony isolates from different regions were analysed. In relation to standard isolates with wild-type sensitivity, the resistance factor (RF) and for samples the mean/median RF (MRF) were calculated. The current results show evident regional differences in the DMI sensitivity of the pathogen, as well as differences in its resistance level to several DMI compounds. The sensitivity distribution within Europe is devided into three large areas. There is the north-west with the highest MRF level, the east with lower MRFs, and the south with a sensitivity level close to that of the standard isolates. Results are postulated to be due to selection pressures and wind dissemination of the pathogen. In recent years the DMI sensitivity of wheat mildew has partly stabilized at a reduced level in North-western Europe. Genetic recombination of the pathogen is considered to be the main reason.

INTRODUCTION

Powdery mildew on wheat, caused by Erysiphe graminis f.sp. tritici, appears periodically from year to year in all European wheat-growing areas. Because of its frequently epidemic occurrence, wheat powdery mildew is often one of the main target pathogens for chemical disease control. In the early 1980s, new fungicides with modern active ingredients, namely the demethylation inhibitors (DMIs) triadimenol and propiconazole, promised highly successful mildew control, and this led to their wide spread and common use. However, it was soon realised that the pathogen showed unexpectedly good adaptability towards DMIs. First reports of a decrease in DMI-sensitivity (Bennet & van Kints, 1982; Buchenauer, 1983), as well as different regional observations and experiences, led to intense discussions at the beginning of this specific gradual evolution of fungicide resistance. Up to now, reduced DMI sensitivity of the wheat mildew pathogen has not, in general, led to disease control failures, but rather a change in efficacy of DMIs under some field conditions.

To determine the sensitivity situation of wheat powdery mildew towards different active compounds and to study the changes in sensitivity of pathogen populations with time, due to selection and wind dispersal of the pathogen, a European-wide monitoring programme, based at Weihenstephan, was started in 1986. A survey of current data on sensitivity to triadimenol, tebuconazole, cyproconazole and propiconazole is presented in this paper, as well as comparison with data from previous years.

MATERIALS AND METHODS

To produce representative data from different regional mildew populations, random samples were taken from the air above areas of interest. Conidio-spores were collected with a jet spore trap (Schwarzbach, 1979) mounted on the roof of a car. While passing through the regions, trapped spores fall onto segments of primary leaves of a highly susceptible wheat variety placed in Petri dishes on water agar (0.6 % agar, 45 mg/l benzimidazole). The trapping distance within an area was approximately 100 km on average.

In the laboratory, the sampled spores grew up to single colony isolates (climate chamber: 18 °C, 10 µEinstein/m²s continuous light). They were transferred onto fresh leaf segments (water agar: 0,6 % agar, 35 mg/l benzimidazole) for storage and multiplication before testing.

For propiconazole, current sensitivity results were obtained from field samples. Mildew-infected leaves were sent from different sites in Ireland, UK, France and Germany by co-workers of Ciba. In the laboratory, conidio-spores of freshly-sporulating colonies were transferred onto leaf segments, and their sensitivities assayed as described below.

The sensitivity of each single colony progeny to triadimenol was determined on a test set of 3 cm long leaf segments. These were cut from the first leaf of ten day old seedlings grown from Baytan treated seed (200 µE/m²s continuous light, 20 °C). Only the middle sections of the primary leaves were used, because of uneven distribution of the fungicide in the leaf. Fungicide treatment was graded logarithmically by a factor of 2. Each test set was inoculated with one isolate using a mini inoculation tower. After 10 days' incubation (18 °C, 10 µE/m²s continuous light), disease coverage was scored relative to the untreated control, and the highest dose allowing $\geq$ 50 % sporulation was determined for each isolate. If available at least 30 isolates per sample (region) were analysed in this way.

In order to analyse sensitivity to the other active ingredients, seedlings were sprayed with different fungicide solutions, containing concentrations graded as above, one day before cutting and inoculation of the test sets. To avoid gas phase interactions among differently treated leaf segments, separate disposable Petri dishes of 6 cm diameter were used for each concentration for every single test set; each Petri dish contained leaf segments of 5 replicates. Thus a test set for analysing one isolate, involving e.g. 10 fungicide concentrations (including untreated control) consisted of 10 Petri dishes. Only during inoculation, the dishes of a test set were placed next to each other under a settling tower, and the leaves exposed to conidia for about 60 seconds. In this way, 10 isolates per sample were tested. After 10 days' incubation, each test set was scored for sporulating diseased area, and the LD50 of each test isolate was calculated by probit analysis.

Standard (wild-type) isolates were included in the sensitivity tests. They were obtained from the field in the 1970s, before the fungicides in question were commercialized, and therefore represent the sensitivity of the fungus in original, unselected populations. If the sensitivity of each test isolate is related to that of the standard isolates, a resistance factor, RF, can be calculated. In order to characterize each random sample, the median resistance factor was determined for triadimenol (seed treatment) and the mean (geometrical mean) resistance factor for the other DMIs (leaf treatment). The abbreviation MRF will be used below for both.

RESULTS

Since 1986, an extensive monitoring programme has been carried out to measure triadimenol sensitivity. When investigations were started, selection pressure had been present for some years. For North-western Europe, results from 1986 showed MRF values of about 15 for the most part, with a range of about 5 to 30 (Felsenstein, 1991; Felsenstein *et al.*, 1991). Only in the southern European regions investigated (South of France, North of Italy) MRFs still remained close to 1. In the following years, the evolution of resistance continued, mainly in North-western Europe, and led to the 1989 situation shown in Figure 1, with MRF values up to 90 and enormous regional differences in triadimenol sensitivity within Europe. Populations with sensitivities close to the level of the wild-type isolates could only be found south of the Pyrenees and the Alps. Apparent differences were also obtained between North-western and Eastern Europe, where MRFs remained at a level of about 10. The current sensitivity situation is presented in Figure 2. On the whole, only few striking

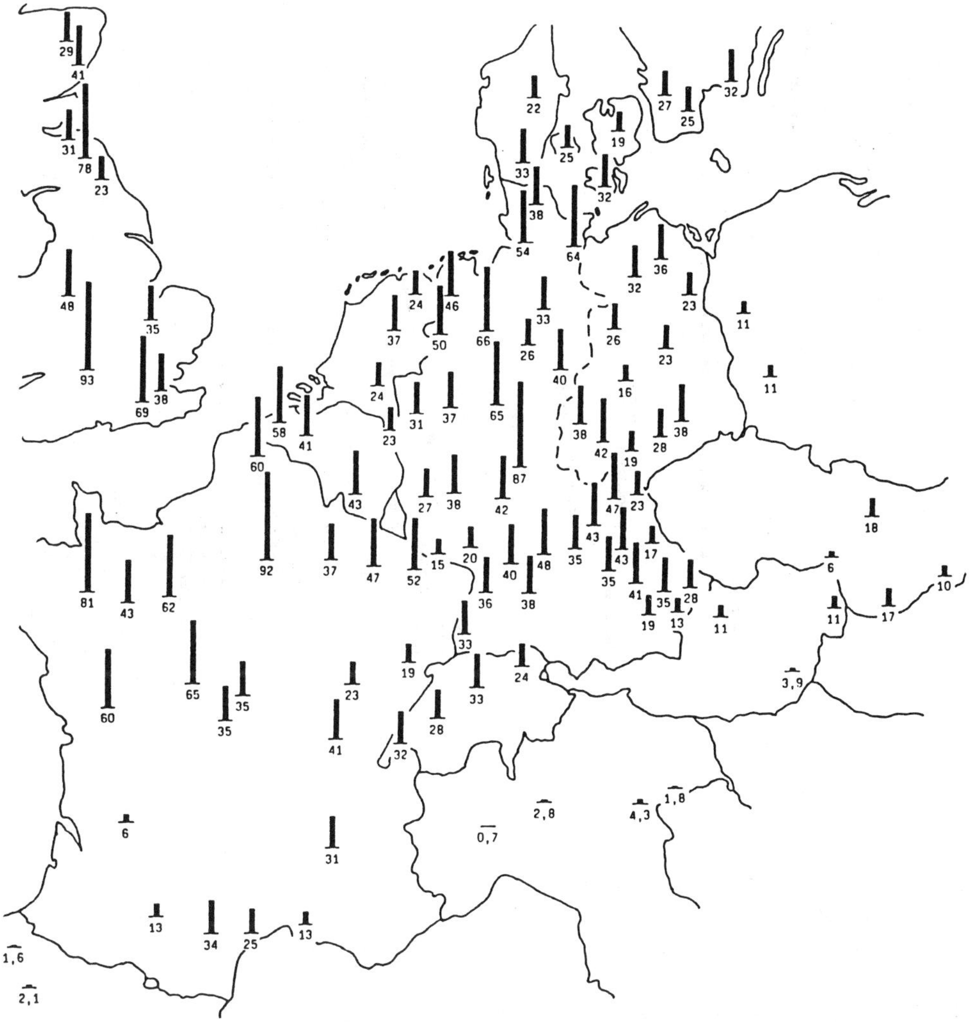

FIGURE 1. Regional differentiation of MRFs of wheat powdery mildew towards triadimenol, 1989

sensitivity changes have occurred on a European scale since 1989. Only in some areas of North-western Europe, namely Denmark and Eastern Germany, the resistance levels detected increased up to a MRF range from 30 to 70. Up to 1993 there were remarkable differences in resistance level between the north-west, east and south of Europe.

For tebuconazole MRF values of random samples from different regional wheat mildew populations are shown in Table 1. Investigations were started in 1990 as tebuconazole was launched commercially in Europe. At this time, MRF values were between approximately 10 and 15 in North-western Europe. MRF values in the east and south were distinctly lower. In the following years up to 1993, a slow shift in tebuconazole sensitivity has occurred for a number of populations. At present the MRFs of most investigated populations in North-west Europe vary around 20 and are still in contrast to those of the east and

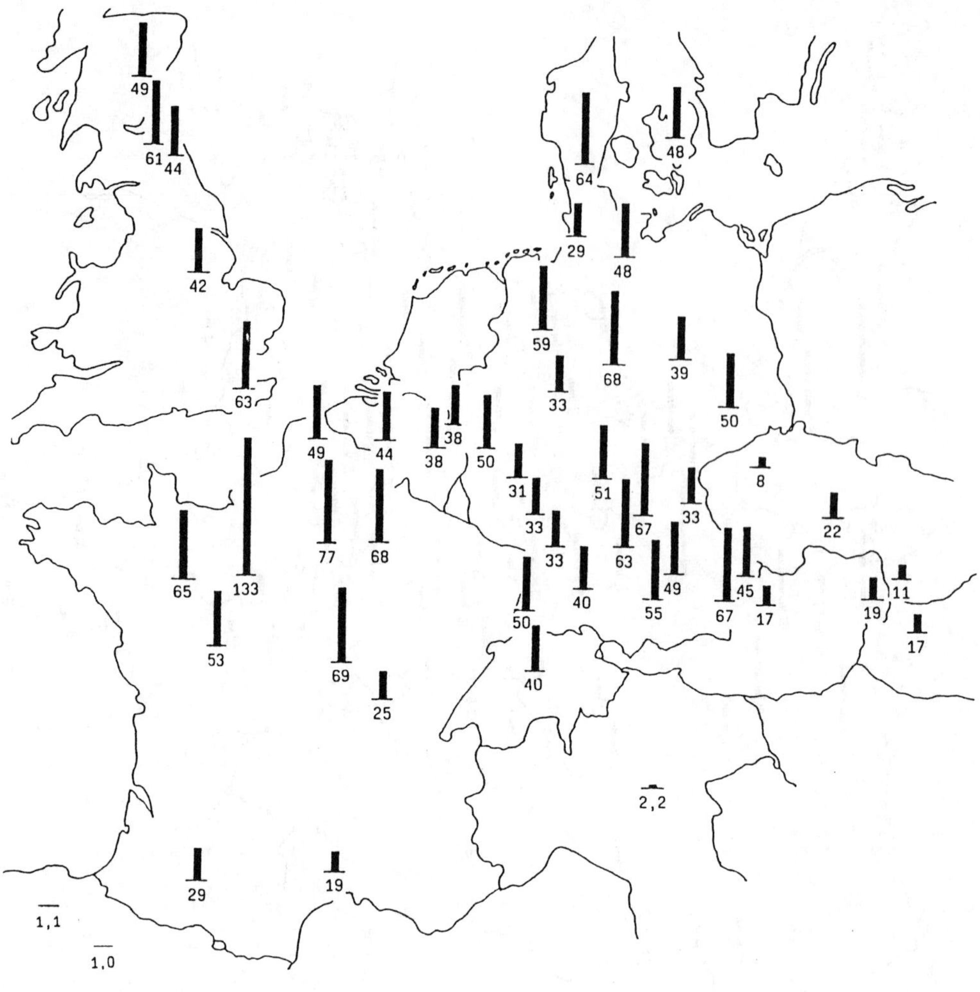

FIGURE 2. Regional differentiation of MRFs of wheat powdery mildew towards triadimenol, 1993 (values in northern Spain from 1992)

south. In comparison with the triadimenol-sensitivity data (see above), tebuconazole resistance factors are, in general, lower.

Cyproconazole is also an active compound which was introduced on the European market more recently than triadimenol. The MRF values of random samples from 5 different European wheat mildew populations between 1990 and 1993 are listed in Table 2. As with the results obtained with tebuconazole, MRF values had a level of about 10 in North-western Europe before the fungicide was launched. There is a clear difference between the north-west and the south of Europe as well. From 1990 to 1993, only slight changes in the populations investigated were observed.

Like triadimenol, propiconazole was first sold in Europe in the early 80s, and it has also been widely used. In Table 3, current data obtained from field samples are presented for North-west Europe, including the average MRF values of each country. Comparison between the four countries shows a relatively homogeneous sensitivity situation, with MRFs

TABLE 1. Mean resistance factors (MRFs) of random samples out of regional wheat mildew populations towards tebuconazole within Europe, 1990-1993

Region	1990	1991	1992	1993
GB:				
Edinburgh-Grantsh.	-	10.4	24.0	31.3
Cambridge-Dover	14.5	9.0	15.3	22.7
F:				
Calais-Mons/Lille	13.8	-	-	21.3
Paris-Reims	16.0	10.4	20.0	22.5
Bourges-Nevers	16.5	11.6	20.0	23.7
Narb./Auch-Toulouse	-	8.2	10.2	15.2
DK:				
Nyborg-Kopenhagen	12.8	8.2	12.3	14.0
D:				
Hamburg-Neustadt	16.0	15.2	20.0	23.2
Hannover-Kassel	-	22.0	17.0	21.0
Magdeburg-Halle	8.0	13.4	15.5	18.3
Nürnberg-Freising	11.8	9.2	11.3	17.2
A:				
Marchfeld/b. Wien	3.8	3.4	5.3	5.8
I:				
Verona-Venedig	1.5	1.4	2.7	3.8

TABLE 2. Mean resistance factors (MRFs) of random samples out of regional wheat mildew populations towards cyproconazole within Europe, 1990-1993

Region	1990	1991	1992	1993
GB:				
Edinburgh-Grantsh.	-	11.5	11.9	10.5
F				
Paris-Reims	10.9	9.8	10.9	12.5
D				
Hamburg-Neustadt	12.0	13.6	12.1	10.2
CH				
Baden-Bern	6.0	8.8	9.4	8.0
I				
Verona-Venedig	1.9	2.4	2.7	1.6

TABLE 3. Mean resistance factors (MRFs) of field samples of wheat mildew from different sites of Ireland, Great Britain, France and Germany towards propiconazole, 1993

Country	No of samples	MRFs (min. - max.)	average-MRF
IRE	10	10.1 - 23.1	18
GB	10	9.9 - 23.8	17
F	10	10.4 - 25.8	18
D	10	10.1 - 23.1	16

varying in each country between nearly 10 and 25. Results from air-borne spore samples, collected mainly in Southern Germany, confirm those obtained with field samples. Their MRF values seldom reached a level of 30 (Felsenstein, unpublished). Thus, despite selection pressure for more than one decade, European wheat mildew populations reached only a level of resistance to propiconazole similar or a little higher than to other fungicides which were introduced on the market recently. Comparison with triadimenol-sensitivity data shows that there is still a difference in resistance behaviour of the pathogen towards these two triazoles. However, for both compounds, only a few sensitivity changes have occurred in the populations of North-west Europe in the last few years, and resistance evolution has stopped in some areas.

DISCUSSION

The results show a clear picture of the current resistance situation of the wheat mildew pathogen towards DMI fungicides on a European scale. It has to be emphasized that within Europe there are populations with a relatively high level of resistance, as well as populations with wild-type DMI sensitivity. Data show a distinct division of Europe into 3 large areas. One of them is the north-west. In this area a more or less evident shift in sensitivity took place during the 1980s. The reason was partly a high selection pressure due to regionally extensive use of DMI fungicides. Until 1993, the highest median/mean resistance factors were obtained mainly in important wheat growing areas, even though it seems that wind dispersal of the pathogen causes increased mixing of neighbouring populations. Furthermore, comparison of data between the active ingredients investigated shows that the wheat mildew pathogen has reached different resistance levels towards the different compounds. It is also true for azoles which have been used on a similar scale and over a similar time period, like triadimenol and propiconazole. In contrast to reduction of sensitivity mainly caused by selection pressure, MRF values towards tebuconazole and cyproconazole, found at the beginning of the 1990s, can only be explained by positive cross resistance of the pathogen towards azoles. These relationships have been described for cereal mildews in several reports (Butters *et al.*, 1984; Buchenauer & Hellwald, 1985; Gisi *et al.*, 1986; de Waard *et al.*, 1986), and their findings are confirmed in this report. Moreover, data presented indicate that since the end of the 1980s, there is an evident reduction in the rate of sensitivity change of most wheat mildew populations in North-western Europe. For some regions it seems that a balance is now reached at a reduced sensitivity level, between forces which promote and impede resistance development.

The second one of the 3 large areas in Europe mentioned above is the east, where clearly lower MRF values (still) predominate. A relatively low fungicide input in the past is responsible. Whilst in Austria, varieties carrying effective mildew resistance were commonly used, the other Eastern countries did not use fungicide treatments because of a lack of foreign currency. A normally broad spectrum of sensitivity in each population (Felsenstein, 1991) infers not only a lower level of fungicide use, but also an influence by wind spread of the pathogen out of western neighbouring regions. In particular, if economic situations change in the future, there might also be a change in the sensitivity level, equivalent to that seen in North-western Europe. The wheat mildew populations in the area of the former GDR provide a clear example.

In contrast to the two large areas described above, there is a third one in Southern Europe, south of the Alps and Pyrenees. There, sensitivity of the isolates tested is predominently the same as the wild-type (and the standards), and the MRFs still remain around one. There are two reasons for the wide-spread unchanged levels of sensitivity. Firstly, low selection pressure because of low fungicide input due to low priority in controlling the pathogen has to be taken into account. Secondly, there are the high mountains, namely the Alps and Pyrenees, which seem to protect the populations in the south of Europe from an influx of spores from the north and north-west, respectively. It is evident that the high mountains act as an epidemiological barrier and that they allow little pathogen exchange between neighbouring populations on either side. This observation is confirmed by investigations into the virulence situation of wheat and barley powdery mildew in Europe (Felsenstein, 1991; Limpert *et al.*, 1991). Therefore, if local selection pressure does not increase, the sensitivity of wheat powdery mildew to DMIs is expected to change more moderately and over a longer time scale in Southern Europe beyond the Alps and Pyrenees.

Finally, if attention is concentrated again on the north-west of Europe, the question arises as to why in the recent past (since 1989) sensitivity levels have regionally stabilised and resistance evolution has been relatively moderate. There are three main points which have to be discussed: Firstly, there is increasing diversification in the use of azole derivates. With regard to the single DMI compounds, there is now less specific selection pressure. Indeed, there is positive cross resistance of the pathogen towards DMIs on the one hand, but on the other hand, there are also variations concerning its expression (see above). Treatment with different azoles is supplemented by an increased use of fungicide mixtures and morpholines, partly within the implementation of appropriate anti-resistance strategies. Also there is no cross-sensitivity between azoles and morpholines /piperidine compounds, which show a different mode of action from the DMIs.

Secondly, there is a possible reduction in fitness of the pathogen, due to increased DMI resistance. In the 1980s, some reports suggested to this relationship (Buchenauer, 1983; Buchenauer & Hellwald, 1985), leading to the opinion that evolution of resistance in the pathogen might have no practical effect in the field. As other investigations could not confirm these results and conclusions (Butters *et al.*, 1984; Wolfe, 1985; Porras *et al.*, 1990), this question was much discussed. Today, on consideration of all available information, it seems that the phenomenon of reduced fitness based on an increase of DMI resistance plays only a subordinate role in the stabilized sensitivity levels observed in North-western Europe.

Thus, a third factor seems to be mainly responsible: Poly-(oligo-)genic control of the gradual (quantitative) resistance evolution towards SBIs (Butters *et al.*, 1984; Hollomon *et al.*, 1984; Skylakakis, 1985), combined with evident genetic recombination of the pathogen. It has to be considered that the polygenic background of DMI resistance encompasses an effect which acts against unimpeded, continued decrease in sensitivity. The more resistant the pathogen becomes, the more genetic changes are necessary. But the polygenic model does not automatically guarantee limited resistance development, as is illustrated e.g. by the barley powdery mildew pathogen, E. graminis f.sp. hordei, where, in general, much higher resistance factors have been obtained (Limpert, 1991; Felsenstein, unpublished). Thus, in the specific case of wheat mildew, the polygenic control of resistance evolution, combined with genetic recombination due to the sexual stage (ascospores) of the pathogen has to be taken into account. The latter yields a large number of isolates with different sensitivities (Hollomon *et al.*, 1984; Butters *et al.*, 1986). In particular, data on virulence and pathotypes indicate that within the yearly reproduction cycle, ascospores are a much more important factor in the life cycle of wheat powdery mildew (Felsenstein, 1991) than they are for barley powdery mildew (Welz & Kranz, 1987; Brown & Wolfe, 1990). Because of more intensive yearly redistribution of genes responsible for DMI resistance, maintenance and in particular multiplication of pathotypes with an exceptionally low DMI sensitivity (high RFs) are probably not possible to the same extent as reported for barley powdery mildew (Welz & Kranz, 1987; Brown & Wolfe, 1990). Thus, the phenomenon of genetic recombination in the wheat mildew pathogen is considered to be the main reason for the observed reduction in DMI sensitivity changes in North-west Europe.

ACKNOWLEDGEMENT

The author thanks Maria Goldbrunner, Mariele Felsenstein, Johanna Stieber and Hannelore Meckl for excellent technical assistance.

REFERENCES

Bennett, F.G.A.; Kints, T.M.C. van (1982) Mildew of wheat. *U.K. Cereal Pathogen Virulence Survey, 1981 Annual Report*, 3-17.

Brown, J.K.M.; Wolfe, M.S. (1990) Structure and evolution of a population of Erysiphe graminis f.sp. hordei. *Plant Pathology*, **39**, 376-390.

Buchenauer, H. (1983) Studies on resistance of Ustilago avenae, Erysiphe graminis f.sp. tritici and E. graminis f.sp. hordei against the ergosterol synthesis inhibitors. *4th International Congress of Plant Pathology*, Melbourne, Abstr.

Buchenauer, H.; Hellwald, K.-H. (1985) Resistance of Erysiphe graminis on barley and wheat to sterol C-14-demethylation inhibitors. *Bulletin OEPPO/EPPO* Bulletin **15**, 459-466.

Butters, J.; Clark, J.; Hollomon, D.W. (1984) Resistance to inhibitors of sterol biosynthesis in barley powdery mildew. *Mededelingen van de Faculteit Landbouwwetenschappen Rijksuniversiteit Gent*, **49**, 143-151.

Butters, J.; Clark, J.; Hollomon, D.W. (1986) Recombination as a means of predicting fungicide resistance in barley powdery mildew. *British Crop Protection Conference - Pests and Diseases*, 561-565.

Felsenstein, F.G. (1991) Virulenz und Fungizidsensitivität des Weizenmehltaus, Erysiphe graminis DC f.sp. tritici Marchal, in Europa. PhD Thesis, Technical University of Munich, Weihenstephan.

Felsenstein, F.G.; Limpert, E.; Fischbeck, G. (1991) Wheat mildew populations in the FRG and neighbouring regions, 1986-1988 - some aspects of their change. In: *Integrated Control of Cereal Mildews: Virulence Patterns and Their Change*, J.H. Jorgensen (Ed), Riso National Laboratory, Roskilde, Denmark, pp. 1-7.

Gisi, U.; Rimbach, E.; Binder, H.; Altweg, P.; Hugelsdorfer, U. (1986) Biological profile of SAN 619 F and related EBI-fungicides. *Brighton Crop Protection Conference - Pests and Diseases*, 857-864.

Hollomon, D.W.; Butters, J.A.; Clark, J. (1984) Genetic control of triadimenol resistance in barley powdery mildew. *British Crop Protection Conference - Pests and Diseases*, **2**, 477-482.

Limpert, E. (1991) Fungicide resistance in populations of plant pathogens: Determination, correlations with virulence, and recent evolution of resistance to triadimenol in the barley mildew pathogen in Europe. In: *Integrated Control of Cereal Mildews: Virulence Patterns and Their Change*, J.H. Jorgensen (Ed), Riso National Laboratory, Roskilde, Denmark, pp. 177-185.

Limpert, E.; Andrivon, D.; Knittel, R.; Fischbeck, G. (1991) Barley mildew in Europe: Patterns of composition of the pathogen population during the period 1985-1988. In: *Integrated Control of Cereal Mildews: Virulence Patterns and Their Change*, J.H. Jorgensen (Ed), Riso National Laboratory, Roskilde, Denmark, pp. 87-103.

Porras, L.; Gisi, U.; Stähle-Csech, U. (1990) Selection dynamics in triazole treated populations of Erysiphe graminis on barley. *Brighton Crop Protection Conference - Pests and Diseases*, 1163-1168.

Schwarzbach, E. (1979) A high throughput jet trap for collecting mildew spores on living leaves. *Phytopathologische Zeitschrift*, **94**, 165-171.

Skylakakis, G. (1985) Two different processes for the selection of fungicide-resistant subpopulations. *EPPO Bulletin* **15**, 519-525.

de Waard, M.A.; Kipp, E.M.C.; Horn, N.M.; Nistelrooy, J.G.M. van (1986) Variation in sensitivity to fungicides which inhibit ergosterol biosynthesis in wheat powdery mildew. *Netherland Journal of Plant Pathology*, **92**, 21-32.

Welz, G.; Kranz, J. (1987) Effects of recombination on races of a barley powdery mildew population. *Plant Paphology*, **36**, 107-113.

Wolfe, M.S. (1985) Dynamics of the response of barley mildew to the use of sterol synthesis inhibitors. *EPPO Bulletin* **15**, 451-457.

EVALUATION OF ANTI-RESISTANCE STRATEGIES

K.H. Kuck,

Bayer AG, Agrochemicals Division, Research Fungicides

ABSTRACT

The feasibility and the success of any anti-resistance strategy depend not only on the anti-resistance strategy itself, but on several additional factors.
One of these is the availability of rapid and reliable monitoring methods which allow control of the efficacy of a certain strategy. Another crucial point is the availability of companion partners. As a result of increased development costs and political and legislative hurdles, the number of active ingredients in plant protection has decreased in a dramatic manner. Under these circumstances, existing anti-resistance strategies are mostly based on the use of preventive fungicides as companion partners of DMIs.
In addition to this narrow range of options concerning the available fungicide partners anti-resistance strategies have to fit economical, ecological and legislative requirements.

INTRODUCTION

The first DMI-fungicides were introduced nearly twenty years ago. They rapidly became the most important group of fungicides, representing a new standard of modern specific fungicides. Although the risk of resistance for DMIs was initially considered to be low to moderate, first reports on decreased sensitivity of powdery mildew fungi against DMIs were published in the early eighties.
Today, growers, agronomists and scientists have learned to deal with resistance towards DMIs and a lot of experience with resistance development against DMI fungicides has been gained.

In spite of these favourable assumption, there are still many unsolved questions concerning the most appropriate anti-resistance strategies for DMIs. Most of the difficulties have their origin in the fact that DMIs have an extremely broad spectrum of fungicidal activity in a wide variety of crops. This results in a wide variation in

a) pathogens or pathogen complexes to be controlled in a given crop.
b) the number of treatments with DMIs and with fungicides from other (non cross resistant) chemical classes in one season in a crop.

Many biochemical studies have elucidated the biochemical mode of action of DMIs in several different fungi. However, studies on the mechanisms of DMI resistance in fungi are much rarer. It is still unclear whether changes in target sensitivity, in the uptake of fungicide into the fungus or in other compensating biochemical alterations are the main cause of resistance development in those pathogens which have meanwhile shown potential to develop more or less pronounced resistance to DMIs.

Some indications suggest that resistance of fungi to DMIs has a multigenic basis. This means that only an accumulation of several independent mutational changes would allow the development of a high degree of resistance. Moreover, if this is true, it is probable that different combinations of different resistance mechanisms exist in different fungal species. For that reason, it is unlikely that only one theoretical model describing the molecular basis of resistance to DMIs in all fungi will be determined in the near future.

The following analysis aims, therefore, to show
- the difficulties in evaluating the success of anti-resistance strategies
- the practical limitations which hinder the use of an optimal anti-resistance strategy
- the availability of suitable companion partners from non-cross resistant fungicides
- the status of practically existing anti-resistance strategies in several crops and countries.

DIFFICULTIES IN EVALUATING THE SUCCESS OF STRATEGIES

Under ideal circumstances, the implementation of an anti-resistance strategy has to be accompanied by monitoring methods which allow rapid and reliable feed-back of the efficacy of a given strategy.

With DMIs, there are several hurdles which usually make it very expensive to optimise the success of a given anti-resistance strategy with the aid of simultaneous sensitivity monitoring.

Many of the target fungi are either obligate parasites (e.g. powdery mildews or rusts) or they are very slow growing fungi in ordinary *in vitro* cultures (e.g. *Mycosphaerella fijiensis*, *Venturia inaequalis*). This necessitates the use of more time consuming and/or more expensive monitoring methods compared to those which can be used for monitoring sensitivity in other fungicide classes.

The fact that resistance of fungi to DMIs is typically characterised by a relatively slow and continuous selection process (shifting type), requires the determination of sensitivity profiles of fungal populations. In order to guarantee statistically sound results, large sample numbers are the consequence. Additionally, it is usually not sufficient to differentiate the sensitivities of a fungal population using only one discriminatory concentration. The use of several concentrations covering a wide range is necessary to determine LC_{50} values or equivalent data.

ECONOMICAL LIMITATIONS

Every anti-resistance strategy must overcome a basic problem. As long as no actual resistance problems are obvious, an anti-resistance management is of secondary priority for the farmer who is primarily concerned with economical problems. Why should the farmer not use the fungicide or the fungicide class with the best price / efficacy ratio all the time? Educational efforts from officials and industry can only in part open the mind to the fact that anti-resistance strategies are mostly a profitable investment in the long term. The implementation of an anti-resistance strategy has, therefore, to be a compromise between the technical need and the narrow economical and legislative framework.

Usually, every anti-resistance management strategy is more expensive than conventional farming methods. For example, the use of two-way mixtures at full rates, which is often recommended by researchers, is clearly more expensive than alternation or the use of mixtures at reduced rates.

Economical reasons are also one of the main causes of numerous efforts to reduce the cost of chemical input by using reduced and split dosages. In some regions, practical farming has, generally, adopted this habit (Jørgensen and Nielsen, 1992; Bosse *et al.*, 1991). Although a potentially negative influence on the effectiveness of anti-resistance strategies was presumed from the beginning, evaluation of the effect of split and reduced rates on the selection of less sensitive fungal strains has been carried out only recently. Experiments by FRAC members gave clear indications that the use of reduced and split doses of fungicides may increase the selection of less sensitive fungal strains (Anonymous, 1994).

LEGISLATIVE LIMITATIONS

Any good anti-resistance strategy should include all factors which are usually described by terms such as "Integrated Pest Management" (IPM) or "Good Agricultural Practice". This includes, for example, the use of resistant cultivars, adequate fertilisation, crop rotation etc.

In most cases, an anti-resistance strategy based on chemicals is additionally necessary. The basis of any anti-fungicide-resistance strategy is the availability of active ingredients from non cross resistant fungicide classes which are effective against the pathogens of interest.

In recent years, the number of registered active ingredients has decreased dramatically in all countries. In Germany, for example, the number of registered active ingredients has decreased by more than 30 % since 1986. On a world-wide basis, the number of newly introduced active ingredients has decreased in a similar way.

The reasons for this decline are multifarious:

- dramatically increasing costs for development of new pesticides and for the re-registration of older compounds have seriously narrowed the variety of options especially in minor and midsize crops.
- public and political pressure to reduce pesticide usage.

AVAILABILITY OF COMPANION PARTNERS

The benefits of the use of companion products for DMIs do not only include anti-resistance strategies. In apple scab, for example, the simultaneous use of preventive and curative products improves the performance in a significant manner. In cereals, the broadening of the activity spectrum of DMIs against several secondary diseases is a reason for the use of companion products (Urech, 1988).

In many crops, DMIs are the only available fungicide class which is highly active and which can be used curatively. Examples include economically important pathogens such as

-powdery mildew in grapevine
-apple scab
-*Septoria* diseases in cereals

If only non-systemics are available as potential companion partners, it is evident that the evaluation of possible advantages or disadvantages of the use of systemic partners (which have a resistance risk on their own) versus the use of non systemic multisite inhibitors (with a very low risk of resistance) is only of theoretical value. Curative products need curative companion partners (Urech, 1988). For that reason the use of DMIs in mixture with preventive fungicides has to be preventive. In alternation programs, the application intervals have to be adapted to take into account the lasting effect of the preventive partner.

Only in a minority of cases are systemic fungicides available which on the one hand are suitable companion partners for DMIs, and which on the other hand do not have severe resistance problems on their own. One of the most successful examples is the use of morpholines and of DMIs for the control of cereal powdery mildew in Europe. The regular use of both fungicide groups has without doubt considerably retarded the development of DMI resistance in this pathogen.

STATUS OF EXISTING ANTI-RESISTANCE STRATEGIES FOR DMIS

Generally speaking, producers, officials, and farmers have learned to treat the phenomenon of DMI resistance development in fungi in an objective and technical manner. At the basis of this was the realisation that resistance of fungi to DMIs is a continuous not a sudden event which is normally correlated with gradually decreasing efficacy in the field.

In face of the limited choice of companion partners, several general rules have found a broad acceptance. These are well reflected in the recommendations of the FRAC SBI-Working Group (Anonymous, 1994).

Generally, with "high risk" pathogens, repeated applications of DMIs alone should be avoided. The use of DMIs should be reserved for the critical parts of the season. The use of mixtures, or alternation with non cross-resistant fungicides are equally recommended.

In crops where several pathogens have to be controlled simultaneously, such as in cereals, the use of mixtures with morpholines or with preventive fungicides is well established. Mixtures are also mostly used in the control of scab on apple, an example of a crop where relatively small failures of disease control can cause severe economic losses.

Alternation programs with tridemorph and other (mostly preventive) fungicides are normally used in Sigatoka control in bananas.

In grapevine, DMIs are preferably used in the most critical period of the season (around flowering), whereas preventive fungicides dominate the early and the later season treatments.

REFERENCES

Anonymous (1994): Summaries and recommendations from the 1993 meeting of the FRAC-SBI working group. *GIFAP Resistance Newsletter*, in press

Bosse, W., Schönberger, H., Dölger, D., and Kropf, U. (1991). Bekämpfung von Getreidekrankheiten mit reduzierten Aufwandmengen. *Feldwirtschaft*, **32,** 276-280.

Jørgensen, L.N. and Nielsen, B.J. (1992). Reduced dosages of fungicides for controlling wheat diseases in Denmark. Proceedings Brighton Crop Protection Conference 1992, 609-614.

Urech, P.E. (1988) Companion products for use in fungicide resistance strategies. In: Fungicide Resistance in North America. Ed.: Ch. J. Delp, APS Press pp. 95-97

THE EFFECT OF REDUCED DOSE ON THE EVOLUTION OF FUNGICIDE RESISTANCE IN *SEPTORIA TRITICI*

M. W. SHAW, C. F.N. PIJLS

Department of Agricultural Botany, School of Plant Sciences, University of Reading, 2 Earley Gate, Whiteknights, Reading RG6 2AU, UK

ABSTRACT

Genetic variation in flutriafol resistance of *Septoria tritici* exists, and reduced fungicide rates produced reduced control of the pathogen population in field experiments. Despite this, no reproducible shifts in resistance appear to have occurred in response to selection by flutriafol applications at either full-rate or 1/4 rate, or to a mixture of flutriafol and chlorothalanil, or in plots sprayed with water only. Possible explanations are discussed.

INTRODUCTION

There are three hypotheses as to how reduction of the rate of a single fungicide application might affect the rate at which resistance evolves. The reduced dose might increase the rate of evolution, because partly resistant forms could survive and subsequently give rise to more resistant individuals by crossing or mutating; the reduced dose might, over the range within which fungal growth was affected at all, do nothing; or the reduced dose might reduce the rate of evolution, because more sensitive forms survive to breed in the future. In the latter case, a quantitative understanding is needed to decide the overall effect on resistance evolution when alternative treatments are, say, several small doses or one large dose.

The theoretical justification for the use of maximal doses to retard the evolution of resistance is clearest in the case of diploid sexually reproducing organisms with resistance inherited as an allele at a single locus. Here a large dose should minimise the expression of resistance in heterozygotes. Since almost all resistant genes are in heterozygotes initially, because of Hardy-Weinberg assortment, it is on the fitness of these that the rate of evolution of resistance depends. This argument has been strongly made by, for example, Mani (Mani, 1989). However, the effectiveness of this has been disputed by others, not least because a pesticide must pass through all concentrations less than the application rate as it decays, and therefore there will be stages at which there is an advantage to the heterozygote (R T Roush, pers. comm., 1991). Even if applicable to organisms like insects, there are few fungi which are diploid and sexually reproducing during the epidemic phase when selection occurs.

Shaw (Shaw, 1989) studied a model of the evolution of polygenic resistance in haploid or clonal fungi, which suggested that dose was irrelevant to the evolution of resistance. This prediction depended on the assumption in the model that intrinsic growth rate of both forms was similarly affected by fungicide application, but that of the resistant form was always greater around reasonable field doses. Shaw (Shaw, 1989) acknowledged that this could be at best an approximation, roughly true around field doses, but showed that the approximation was likely to be quite good between about 1/4 and 4 times a reference dose.

The argument that an individual reduced dose should have less effect on the genetic composition of the pathogen population than a standard dose is a common-sense one. It can be based on comparisons with plant or animal breeding, where weaker selection, not unreasonably, means slower selective progress. What is much less clear is that the reduction will be proportional to the dose: if the selective effect of a half dose is more than half that of a full dose, but two half doses are applied, the evolution of resistance will clearly be speeded up overall.

We are trying to test these hypotheses about the relative effect per application of full and reduced doses, using the wheat-*Septoria tritici* (*Mycosphaerella graminicola*) pathosystem. This seems to be well-suited to the work. The population in a field is believed to be initiated by widely dispersed and fairly abundant sexually generated ascospores (Shaw & Royle, 1989). Genetic evidence based on molecular polymorphism supports this view: the populations surveyed so far are extremely variable on a very fine scale and contain many clones, and almost all the variability in a population is contained within a field (McDonald & Martinez, 1990; McDonald & Martinez, 1991). This means that the past history of a site has a very minor effect on the population within it, and that plots should be representative of the surrounding farming district and normal populations. Infection in the autumn is certain, and artificial inoculation unnecessary to augment the population. Multiplication thereafter is rapid through splash-dispersed, clonally produced spores, which disperse only over a few metres, so immigration into a field should be numerically negligible after the autumn (Shaw & Royle, 1989), and plots of 100 m^2 or so should behave as independent populations over a period of one year (Shaw & Royle, 1993). Thus, sites separated by several km, or sampled in different years, should serve as true replicates (Hurlbert, 1984), encompassing the variability in the founding populations, and allowing us to make inferences about agricultural populations of the pathogen.

We report here partial results from the first two years' work in this system.

MATERIALS AND METHODS

Experimental Design of Field trials

Field experiments have now been planted for three consecutive years at two locations, Sonning Farm (Reading University) and Jealott's Hill Experimental Station (Zeneca), about 20 km apart. However, we shall be concerned with only the first two years' crops. At Sonning the crop was on a new site, not sown to wheat the year before, every year while at Jealott's Hill plots (and treatments) were on the same site in the same position for the first two seasons (1991/92, 1992/93).

The lay-out at each site was a split-plot design, using two winter wheat cultivars (Mercia and Riband) and four different fungicide treatments. Riband is very susceptible to *S. tritici* while Mercia is moderately resistant. Each site had two main plots composed of four split plots of the same cultivar. Each split-plot was separated from all the others by 3 m wide strips of winter barley to minimise gene flow between treatments. Each split-plot was at least 12x20m. Plots were sprayed at around GS 37 (flag leaf just visible) with one of four treatments at a spray rate of 250 l/ha: water; ¼ l/ha Impact (125 g/l flutriafol); 1 l/ha Impact or 2.66 l/ha Impact Excel (47 g/l flutriafol + 300 g/l chlorothalonil). The formulations used were made up every year and the same batch was used at both locations.

Field performance of the sprays was assessed on a regular basis by taking random samples of the leaf below the flag at about 0.5m intervals on two diagonal walks across the field. Disease incidence was expressed as percentage leaves with one or more lesions bearing pycnidia.

Sampling of pathogen population

Wheat leaves bearing pycnidia were collected from each plot in May, just before spraying, The youngest leaves with pycnidia were taken, since these were most likely to infect the top leaves. Within 6 weeks after spraying, another sample was taken, this time from the leaf below the flag. This six weeks limit was set to avoid sampling the second generation of pycnidia after spraying. Depending on the amount of disease present in a plot, 75-100 leaves were taken at a time. Dirty leaves were washed before they were dried for several hours at room temperature. Subsequently, leaves were stored frozen at -20°C.

Assay of Fungicide Resistance

Details of the assay are being published (Pijls et al., 1994), so only a brief outline will be given here. Leaves were surface sterilised in a 1% NaOCl-solution for 30 s, and incubated in a sandwich box containing a thin layer of tap-water agar. The box was then covered by paper tissue and put away for 24-48 hours at 17°C after which a single dry, curly cirri could be picked off each sampled lesion, and suspended in sterile water. These spores were used in an assay of flutriafol resistance based on light absorbance in liquid medium.

Flat bottomed microtitre plates with 8 rows of 12 wells were used. Each row was filled with pycnidiospores of a single S. tritici isolate and medium containing different fungicide concentrations. The range of final fungicide concentrations was: 0, 0, 0.010, 0.0316, 0.100, 0.158, 0.251, 0.398, 0.631, 1.0, 1.78, 3.16 μg A.I. ml^{-1}. Preliminary experiments had shown that most field isolates had an EC_{50}-value between 0.056 and 0.56 μg A.I. ml^{-1}, and therefore the range was made more precise in this interval. Two rows in every plate were used for control isolates, Rl2 and S27 (D.W. Hollomon, Long Ashton Research Station) with a known sensitivity to flutriafol. After 10 days incubation in the dark at 17°C, growth was measured using absorbance of light at 405 nm. A dose response curve was then fitted to the absorbance data to estimate the fungicide concentration reducing absorbance by one-half (EC_{50}). EC_{50}-values based on this method seem to be adequately correlated with preliminary results from a bio-assay on wheat seedlings (GS12-13).

RESULTS

Field performance of sprays

The results on Riband for the first season are shown in Tables 1 and 2. These show that some multiplication of disease was possible under all fungicide treatments, and that substantially more was possible under the reduced dose than the full dose. Notwithstanding this, the fungicide very greatly reduced the incidence of disease. There was also a difference in disease severity (proportion of leaf area infected). Infected leaves from plots untreated (0) or treated with reduced (¼) dosage usually had more and bigger lesions than those from plots treated at the full rate (1) or with the mixture with chlorothalonil (1+C). This suggests that the pathogen was not only growing in leaves which had remained by chance completely free from fungicide. Findings in the second season were similar.

TABLE 1. Sonning, 1992: Incidence of disease caused by *S. tritici* on leaf 2 of wheat cv. Riband at various dates after spraying on 14 May.

		Incidence (%)		
Cultivar	Spray	23 May	2 June	9 June
Riband	0	6.1	30.3	72.1
	1/4	7.1	15.2	43.5
	1	3.3	6.9	22.0
	1+C	0	3.4	18.6

TABLE 2. Jealott's Hill, 1992: Incidence of disease caused by *S. tritici* on leaf 2 of cv Riband at various dates after spraying on 6 May.

	Incidence (%)			
Spray	17 May	23 May	2 June	9 June
0	14.3	74.6	94.2	100
1/4	5.7	42.6	89.0	98.9
1	1.4	27.3	72.2	92.5
1+C	* 5.7	27.1	48.3	52.9

*: high incidence due to uneven plant development

The results of fungicide assays conducted so far on isolates from Riband are presented in Figures 1 and 2, as 'box-and-whisker' plots of the distributions. Each plot shows the extremes, the quartiles and the median of the distribution for the plot. The Kolmogorov-Smirnoff test for differences between probability distributions was used to test for differences between the distributions before and after treatment (Siegel, 1956).

Several points are striking in these data. First, there is no systematic change in sensitivity after spraying, at any dose. Second, such significant changes as are present are as likely to be to lower levels of resistance as to higher, and occur equally with the water and fungicide sprays. Third, plots appear to differ in fungicide sensitivity distribution before spraying, certainly across years. For example, in Sonning in 1993 the distributions before spraying were all much narrower than in 1992.

FIGURE 1. 1992, cv Riband: distribution of fungicide sensitivity (EC50) in each plot before and after spraying. The sample size (N) is shown in parentheses to the right of each plot; the plot is identified to the left: 0 - water spray; 1/4 - 1/4 rate spray with flutriafol; 1 - spray with full recommended dose of flutriafol; 1+C - sprayed with a mixture of flutriafol and chlorothalanil. A scale appears in the middle of the figure. Each plot shows the extremes, quartiles and median of the distribution of the observed EC50. ★ : significant ($P \leq 0.05$) difference between before and after, according to a Kolmogorov-Smirnoff test.

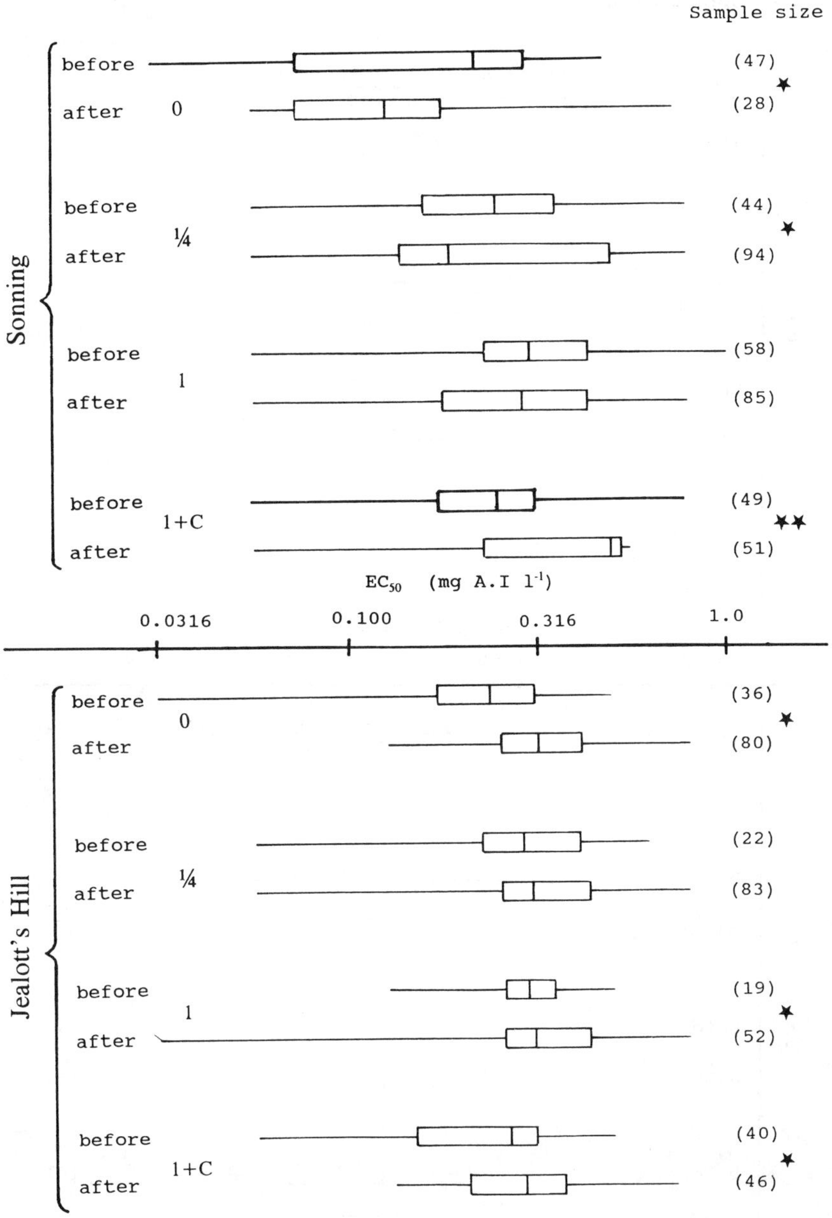

FIGURE 2. 1993, cv Riband: distribution of fungicide sensistivity (EC50) in each plot before and after spraying. The sample size (N) is shown in parentheses to the right of each plot; the plot is identified to the left: 0 - water spray; 1/4 - 1/4 rate spray with flutriafol; 1 - spray with full recommended dose of flutriafol; 1+C - sprayed with a mixture of flutriafol and chlorothalanil. A scale appears in the middle of the figure. Each plot shows the extremes, quartiles and median of the distribution of the observed EC50. ★,★★ : significant ($P \leq 0.05$ or $P \leq 0.01$) difference between before and after, according to a Kolmogorov-Smirnoff test.

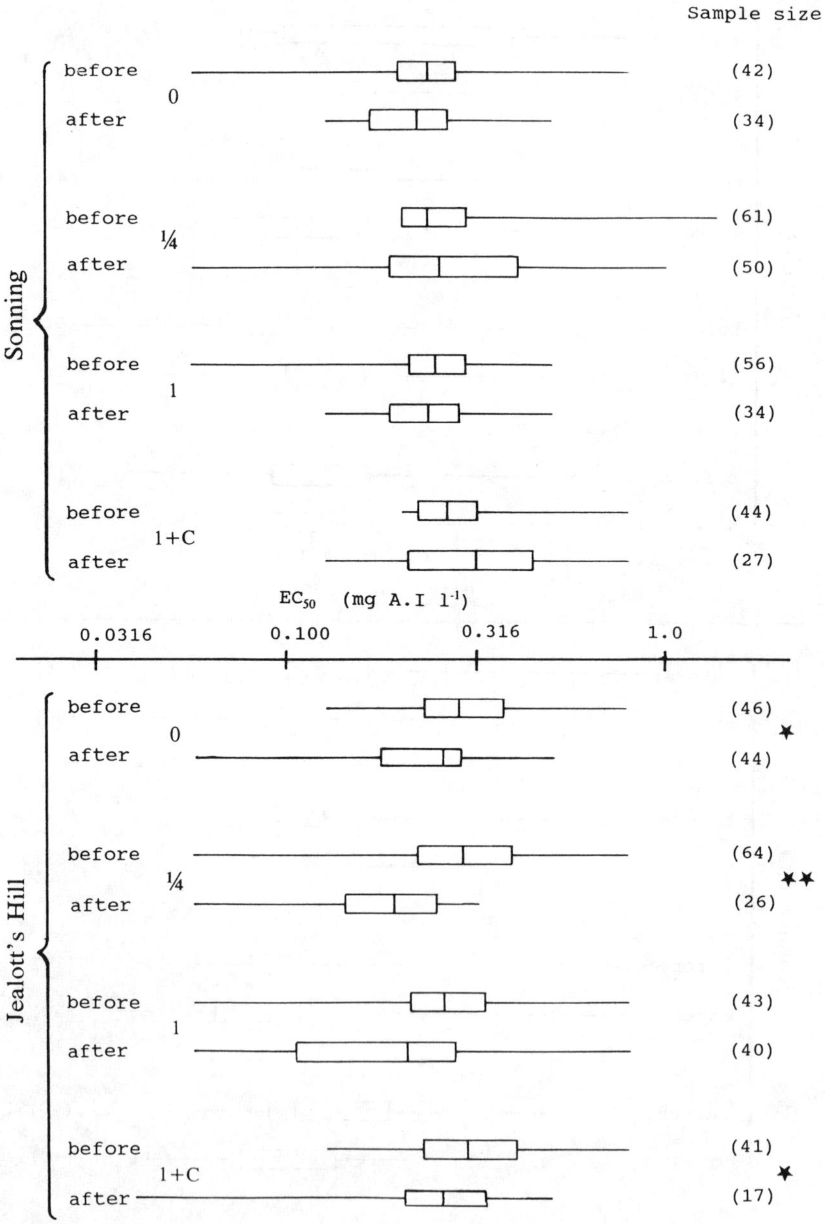

DISCUSSION

The results are more surprising than may appear at first sight. There is genetic variation in fungicide sensitivity present in the population, spanning a range of about 30-fold in EC_{50}, although most of the variation is contained within a roughly 3-fold range, form 0.2 to 0.6 mg/l. The EC_{50} of an isolate is stable following reinoculation onto plants and reisolation. Thus, the character measured is heritable and reflects the ability of the fungus to cause disease in the presence of fungicide. Selection was applied by the fungicide treatments, because the population size of the pathogen was substantially reduced. Therefore, the fungicide sensitivity distribution should have changed after spraying; the median sensitivity should have decreased. This did not happen, to the accuracy of our experiments.

There are a number of possible explanations of our observations, some of which can be excluded fairly quickly using data we already have. We shall consider some of the simplest explanations in turn.

First, selection could have been so strong that the variation we have seen was irrelevant: disease found after treatment was in leaves or parts of leaves without fungicide, and the area available to a more resistant isolate for successful disease expression would be no larger than for a more susceptible isolate. However, as we noted, lesions in the 1/4 rate and water treated plots tended to be larger than in the full rate and mixture plots, and control was better with full rate than with 1/4 rate. This explanation therefore requires that a graph of area of foliage with a given fungicide concentration against fungicide concentration have a slope of 0 near the maximum concentration at which growth is possible, but also that in the 1/4 rate plots much more disease is expressed. This implies that the change in disease incidence caused by reducing the rate to 1/4 was entirely due to an increased area of the foliage containing no fungicide at all . However, with a fully systemic and quite mobile chemical like flutriafol, this seems improbable.

Second, perhaps more resistant isolates are systematically less able to cause infection than more susceptible ones. However, in this case, resistance should have decreased in the water treated plots where no selection by fungicide occurred.

Third, perhaps the origin of the inoculum for the disease on the upper leaves was not within the plots. This is not consistent with what we know of the epidemiology of the disease, and, in any case, such immigrating inoculum should have been similar for all plots, so there should be differences in fungicide sensitivity distribution between the water and fungicide treated plots.

Fourth, perhaps selection for ability to infect the particular variety used under the particular conditions prevailing completely outweighs selection for resistance, so that resistance shifts appear random. This argument works only if the proportion of genotypes capable of infecting at any given time is actually very small, so that selection by fungicide is negligible compared to selection by the environment and the host. But this flies in the face of the excellent control achieved by fungicide and the difference between full and 1/4 rates, which demonstrate at least the potential strength of selection by fungicide.

Fifth, if only one or a few clones could best infect cv. Riband under the prevailing environmental conditions, then random association between the virulence characteristics of the isolate and its fungicide sensitivity could generate essentially random shifts in fungicide sensitivity. This requires effective population sizes in each plot to be small. In the US this seems not to be the case (McDonald & Martinez, 1990; McDonald & Martinez, 1991), and our preliminary, unpublished, evidence concurs in the UK. However, this explanation is the only one of those put forward which can explain the increase in resistance in the water treated plot at Jealott's Hill in 1992, or the decrease in the 1/4 treated plot at Jealott's Hill in 1993.

To sum up, insofar as our results provide an answer to the question we have posed, it suggests that resistance is no more likely to be selected by reduced rates than by full rates; but since the results suggest that no effective selection for resistance is happening in the system, this result needs substantial further investigation before it can be regarded as a trustworthy example.

ACKNOWLEDGEMENTS

This work is funded by the Home-Grown Cereals Authority, as part of its programme on fungicide resistance. We are very grateful to Zeneca plc for maintaining and allowing us to use plots at Jealott's Hill, for the gift of fungicides and special formulations, and for very helpful discussions.

REFERENCES

Hurlbert SH, 1984. Pseudoreplication and the design of ecological experiments. *Ecological Monographs* **54**, 187-211.

Mani GS, 1989. Evolution of resistance with sequential application of insecticides in time and space. *Proceedings of the Royal Society B* **238**, 245-276.

McDonald BA, Martinez JP, 1990. DNA restriction fragment length polymorphisms among *Mycosphaerella graminicola* (anamorph *Septoria tritici*) isolates collected from a single wheat field. *Phytopathology* **80**, 1368-1373.

McDonald BA, Martinez JP, 1991. DNA fingerprinting of the plant pathogenic fungus *Mycosphaerella graminicola. Experimental Mycology* **15**, 146-158.

Pijls CNF, Shaw MW, Parker A, 1994. A rapid test to evaluate *in vitro* sensitivity of *Septoria tritici* to flutriafol, using a microtitre plate reader. *Plant Pathology* **43**, in press.

Shaw MW, 1989. A model of the evolution of polygenically controlled fungicide resistance. *Plant Pathology* **38**, 44-55.

Shaw MW, Royle DJ, 1989. Airborne inoculum as a major source of *Septoria tritici* (*Mycosphaerella graminicola*) infections in winter wheat crops in the UK. *Plant Pathology* **38**,35-43.

Shaw MW, Royle DJ, 1993. Factors determining the severity of *Mycosphaerella graminicola* (*Septoria tritici*) on winter wheat in the UK. *Plant Pathology* **42**, 882-900.

Siegel S, 1956. *Nonparametric Statistics for the Behavioural Sciences*. McGraw-Hill.

EVALUATING ANTI-RESISTANCE STRATEGIES FOR CONTROL OF *ERYSIPHE GRAMINIS* F.SP. *TRITICI*

U. SCHULZ

BAYER AG, PF-F/F, Pflanzenschutzzentrum Monheim, D-51368 Leverkusen

ABSTRACT

The present article, which deals with the problem of resistance to fungicides, looks specifically at powdery mildew in cereals. Different approaches to the solution of this practical problem are discussed and evaluated with respect to their effect on control of powdery mildew and its sensitivity to various fungicides. The most appropriate strategy, and the one advocated by the SBI Working Group of FRAC (Fungicide Resistance Action Committee), is to mix fungicides with different resistance mechanisms. A reduction of the field application rates is inadvisable. Resistant cereal varieties should be used.

INTRODUCTION

Problems arising from the loss of efficacy of azole fungicides against powdery mildew in cereals emerged towards the end of the 1970s in England and Scotland and concerned the seed treatment of spring barley. The first reports of problems with azoles in wheat appeared later, around the early 1980s, and were to do with winter wheat crops in The Netherlands and Northern Germany.

Identifying the start of resistance development in mildew has proved as difficult in wheat as in barley. Initially, the situation in practice was also masked by years with very wet spring months, e.g.1983 in Germany, when the fields were hardly workable and late spraying made the fungicides appear less effective.

As the good field performance of many of the azoles continues to indicate, resistance in the sense of 'complete loss of performance' could not be proven for powdery mildew (Beer, 1985), but regionally fields from which mildew with reduced sensitivity to azoles could be isolated were encountered repeatedly (Beer, 1987). Recent observations indicate that in regions in which the fungicides used were systematically alternated (from azoles to morpholins), azoles showed distinctly better efficacy when they were reintroduced some time later (e.g. in barley in Scotland).

Experience gained in the last 10 years also shows that the situation in practice is not determined by the direct interaction of the fungus and fungicide alone. We now know just how much factors such as the choice of variety, use of fertilizer, and climatic conditions affect the entire situation in the field.

THOUGHT PATTERN FOR STRATEGIES

So how does one tackle the problem of resistance? Different conceptual models would be conceivable, involving both the use of fungicides and the choice of the correct variety of the crop in question. The following options will be discussed which, of course, cannot solely be looked at from the aspect of powdery mildew:

a. mixing or alternating an azole with a fungicide from another chemical class, e.g. morpholine
b. splitting and reducing the field application rates of a fungicide
c. choosing and/or mixing cereal varieties with resistance genes that are effective against mildew

EXPERIENCE FROM PRACTICE

Mixing or alternating fungicides

A long-term field trial has been conducted by BASF with different spray regimes (mixtures and alternations) of fenpropimorph and triadimenol, in order to obtain information for possible resistance strategies (Lorenz et al., 1992). "Data .... indicate that triazole/morpholine combinations and alternations are equally well suited to avoid a shift in fenpropimorph sensitivity."

In the year 1988 field trials were performed in 'Caribo' winter wheat with triadimenol, tridemorph and a mixture of triadimenol&tridemorph. The 'triadimenol&tridemorph' mixture was much more effective in controlling disease than the alternation 'tridemorph/triadimenol/tridemorph', which in turn was more effective than triadimenol. There was hardly any difference between the sensitivity to triadimenol of mildew isolates from the mixture plot and mildew isolates from the triadimenol treated plot. Mildew from the alternating spray sequence tended to be more sensitive to triadimenol, as the last treatment was performed with tridemorph. However, the individual plots showed different levels of residual infection.

In 1989, field trials were performed in 'Caribo' winter wheat with tebuconazole and a mixture of tebuconazole&tridemorph. After 2 treatments, the efficacy of the mixture was far superior to that of tebuconazole. Mildew isolates from all plots, including the control, showed only minimal differences in their sensitivity to tebuconazole.

In 1992, field trials were carried out in 'Kanzler' and 'Apollo' winter wheat with tebuconazole, fenpropimorph and a mixture of tebuconazole+fenpropimorph (tank mix).
In Kanzler, no clear differences in efficacy were detectable on account of heavy infection pressure although disease control of the mixture was slightly better. At the end of the season, sensitivity to fenpropimorph was greater in isolates from the mixture plot than in isolates from the fenpropimorph treated plot.
In Apollo, the mixture was much more effective in disease control than the single compounds. Sensitivity to tebuconazole was greater in isolates from the mixture plot than in isolates from the tebuconazole treated plot.

In 1993, field trials were performed in 'Kanzler' winter wheat with tebuconazole, fenpropimorph and a mixture of tebuconazole+ fenpropimorph. The mixture was slightly more effective in disease control than tebuconazole and fenpropimorph. At the end of the season, and with similar performance of the different fungicides, mildew isolates from the mixture plot were as sensitive to tebuconazole and much more sensitive to fenpropimorph than isolates from the tebuconazole and fenpropimorph treated plots respectively.

Our conclusions from the studies mentioned above are that mixtures of azoles and morpholines can positively influence both field performance and changes in sensitivity with respect to the individual mixture components.

Splitting and reducing application rates

A proposed control strategy is to reduce the rate at which a fungicide is applied and to apply it at more frequent intervals (splitting doses).

The next step would then be to dispense with one or more of the applications in order to reduce the amount of fungicide used. In Denmark, spray models of this kind are sought and supported by both the public and politicians. Jørgensen and Nielsen (1992) have examined this topic thoroughly, and pointed out both the benefits of such programmes and the problems and risks of incorrect use. Spray programmes using lower application rates were also advocated in Germany (Bosse et al., 1991; Schönberger et al., 1993).

In 1992, field trials were carried out in 'Kanzler' and 'Apollo' winter wheat. A strict spraying regime using 2x1/1, 4x1/2 and 8x1/4 of the application rates of tebuconazole+ fenpropimorph, tebuconazole and fenpropimorph was demanded. The quantities of fungicide applied were thus split, but not reduced.
The available data demonstrate no negative influence of a split application rate on the fungicide performance. The crucial point is the correct timing of the sprays.
No general trend to reduced sensitivities could be proven when reduced rates were applied. The frequency distributions of the sensitivities seemed to have been mainly influenced by the fungicide performance. Good performance left a small amount of less sensitive mildew isolates surviving. Poor performance left a mixture of sensitive and less sensitive isolates.

In 1993, field trials were performed in 'Kanzler' winter wheat. No strict spraying regime was followed. The fungicides tebuconazole+fenpropimorph, tebuconazole or fenpropimorph at rates of 1/1, 1/2, 1/4 and 1/8 were applied whenever the field evaluation found 3-5% living mildew. The application rates were thus split and at the same time reduced, if by the end of the spraying period the total amount of active ingredient used did not reach the amount used at the full application rate.

Reduced rates, applied with adequate timing, were able to control the powdery mildew as well as full rates. This holds for all the three fungicides: tebuconazole+fenpropimorph, tebuconazole, fenpropimorph. However, there is a limit for the splitting and reduction of rates. Below this limit, sprays are not feasible due to 'no acceptable efficacy'.

In general, shifts in sensitivity to tebuconazole and fenpropimorph were surprisingly limited. The performance of a fungicide programme, independent of the amount of AI applied, seems to be the critical factor determining sensitivity changes in a population. The better the performance the higher the proportion of less sensitive strains which survived. If performance decreased after long spray intervals, even with high dose rates, populations could have a higher proportion of more sensitive strains at the end of the programme. The mixture of azole and morpholine influenced the sensitivity to both compounds positively, i.e. the mildew populations became more sensitive.

In conclusion, where frequent application of reduced rates leads to good performance, a high and continuous selection pressure is maintained, which may negatively influence the sensitivty of mildew to the fungicide in question.

Choosing and mixing varieties

It has been known for quite some time that not only the choice of the mildew fungicide but also the choice of the crop variety contributes decisively to the preservation of crop health (Wolfe, 1984; Schaffner et al., 1992). The idea that the efficacy of the mildew fungicides can and should be supported by the aid of cereal varieties with effective resistance genes has been discussed in various countries, and proposals for its practical realization have been made (Wolfe, 1985; Frahm, 1986).

One should expect neither "permanently effective" fungicides nor "permanently effective" varietal resistances. A mildew population reacts to each and every selection pressure to which it is exposed. Fungicides and crop varieties select from the population those strains which withstand the selection pressure. It is therefore likely that a small fraction of the population will always survive.

For agricultural practice, then, it is important to be able to choose relatively resistant crop varieties which, together with the necessary fungicidal applications, would give optimal yields (Schulz and Lein, 1993). This would also be in line with the Integrated Crop Management model, which takes account of various cultivation parameters, combining them in an appropriate way.

CONCLUSIONS

All experience from research and practice with mildew fungicides in cereals can be summarized in the following recommendations, formulated by FRAC's SBI Working group in 1993 (FRAC, 1994):

"Repeated applications of DMI or 'morpholine' fungicides alone should not be used on the same crop in one season against a high risk pathogen (eg. cereal powdery mildew) in areas of high disease pressure for that particular pathogen. Split/reduced rate programmes using repeated applications which provide continuous selection pressure should be avoided."
"For control of cereal powdery mildews, mixtures or alternation of a DMI with a 'morpholine' fungicide represent the best currently available none cross-resistant combination."
"Fungicide input is only one part of crop management. Fungicide use does not replace the need for resistant crop varieties, good agronomic practice, plant hygiene/sanitation, etc..."

Breeders and manufacturers of crop protection products are here jointly called upon to provide concepts for agricultural practice which take account of all the important parameters and combine them in an appropriate way within the context of Integrated Crop Management. The aim is to put a system at the grower's disposal which is as stable as possible, and the more complex its composition the more stable it will be.

REFERENCES

Beer, E. (1985) Blattkrankheiten an Winterweizen und Winterroggen - Keine Resistenz des Mehltaus gegen Azolderivate. *Landwirtschaftsblatt Weser-Ems*, **18**, 12-19

Beer, E. (1987) Zur Wirksamkeit von Fungiziden gegen Echten Mehltau (*Erysiphe graminis* D.C.) an Wintergerste und Winterweizen. *Gesunde Pflanzen*, **39**/10, 402-408

Bosse, W.; Schönberger, H.; Dölger, D.; Kropf, U. (1991) Bekämpfung von Getreidekrankheiten mit reduzierten Aufwandmengen. *Feldwirtschaft* **32**, 276-280

Frahm, J. (1986) Resistente Sorte oder Fungizid? Vielfältige Wechselwirkungen-sinnvolle Ergänzung. *DLG-Mitteilungen*, 7, 382-386

FRAC (1994) Summaries and recommendations from the 1993 meeting of the FRAC-SBI working group. *GIFAP Resistance Newsletter*, in press

Jørgensen, L.N.; Nielsen, B.J. (1992) Reduced dosages of fungicides for controlling wheat diseases in Denmark. *Proc. British Crop Protection Conference - Pests and Diseases*, 609-614

Lorenz, G.; Saur, R.; Schelberger, K.; Forster, B.; Küng, R.; Zobrist, P. (1992) Long term monitoring results of wheat powdery mildew sensitivity towards fenpropimorph and strategies to avoid the development of resistance. *Proc. British Crop Protection Conference - Pests and Diseases*, 171-176

Schaffner, D.; Koller, B.; Müller, K.; Wolfe, M.S. (1992) Response of populations of *Erysiphe graminis* f.sp.*hordei* to large-scale use of variety mixtures. *Vortr. Pflanzenzüchtg.*, **24**, 317-319

Schönberger, H.; Bosse, W.; Kropf, U. (1993) Krankheiten mit kleinen Mengen gezielt bekämpfen. *Der Pflanzenarzt Sonderteil Pflanzenschutz-Praxis*, 2, 20-22

Schulz, U.; Lein, K.A. (1993) Use and significance of resistance genes against powdery mildew (*Erysiphe graminis*) in German barley varieties in the years 1980 to 1992. *Pflanzenschutz-Nachrichten Bayer-English Edition*, **46**/2, 183-224

Wolfe, M.S. (1984) Progress with variety mixtures. *Vortr. Pflanzenzüchtg.*, **6,** 58-68

Wolfe, M.S. (1985) Integration of host resistance and fungicide use. *Bulletin OEPP/EPPO Bulletin*, **15**, 563-570

EVALUATING ANTI-RESISTANCE STRATEGIES FOR CONTROL OF *UNCINULA NECATOR*

H. STEVA

BIORIZON S.A., Centre de Ressources, Bordeaux-Montesquieu, F-33651 Martillac Cedex, France

ABSTRACT

Since 1989, different strategies of treatment (mixtures, alternations) have been compared to cope with *Uncinula necator* resistance to sterol demethylation inhibitors in French vineyards. After treatments with DMI fungicides, populations show at first a progressive evolution with an increase in the proportion of less sensitive phenotypes, followed eventually by the appearance of resistant phenotypes. Low application rate of triadimenol (18,75 g/ha) results in rapid reduction in the population sensitivity compared to the recommended rate (37,5 g/ha). Mixtures containing triadimenol and sulphur do not slow down the evolution of resistant spores in natural populations. Reduction of the number of treatments, and use of sulphur are the only strategies helpful in slowing down the evolution of resistant phenotypes and keeping the disease under control.

INTRODUCTION

Resistance of grape powdery mildew to fungicides that inhibit the C-14 demethylation of sterols (DMI) was discovered for the first time in 1988 near Lisbon, Portugal (Steva et al, 1988) after reduction in efficacy of these fungicides in the vineyard. Since then, resistant strains have been identified in France (Steva et al, 1989) and in Italy (Aloi et al, 1990). In spite of this fact, DMI fungicides were still effective in the majority of situations in recent years. This normal performance should not lead one to believe there was no risk that resistant strains would develop in the field. To define this risk and develop treatment strategies that will limit the emergence and multiplication of resistant strains, it was necessary to compare different programs of treatment.

The purpose of our research was to observe any changes within a mildew population (*Uncinula necator* (Schw.) Burr.) as a result of different fungicide pressures during four years (1989 to 1992). Exclusive sprayings of DMI fungicide during the season were compared to an alternation or a mixture of DMI and sulphur and DMI and dinocap.

METHOD TO ASSESS THE SENSITIVITY OF *U. NECATOR* TO DMI FUNGICIDES

Since *U. necator* is an obligate fungus, it has to be grown on plant material. Our method is based on the use of leaf discs maintained on agar medium and observation of an individual hyphal growth .

Leaf disc test

The sensitivity of conidia of *U. necator* to triadimenol was evaluated using a test on leaf discs kept alive on a water agar medium (20 g/l) amended with benzimidazole (30 mg/l). The discs (18 mm diameter) were punched out from grape leaves (cv. Cinsaut) and disinfected for 10 minutes in a solution of calcium hypochoride (50 g/l). The upper surface of the leaf disks was placed in contact with filter paper imbibed with 3 ml of triadimenol (Baytan 5, 50 g/l provided by Bayer France, Paris). The range of fungicide concentrations was as follows : 0.01, 0.03, 0.1, 0.3, 1, 3 and 10 mg/l. After 24 hours of incubation at 20°C, the discs were transferred onto water agar medium. The upper surfaces of the leaf discs were powdered with spores by placing the Petri dishes at the base of a settling tower (0.09 m^2 and 0.6 m high). The Petri dishes were placed in a culture chamber at 21±1°C with 16 hours light per day at an intensity of 25 $\mu E.m^{-2}.s^{-1}$.

Observation of hyphal growth

Since *U. necator* is an ectoparasite, hyphae were removed after 3 days of incubation by touching the upper surface of the inoculated leaf disc with Scotch tape. The tape was then stuck to a microscope slide on a drop of cotton blue stain. Hyphae were observed and measured using a light microscope at 100 magnification with a binocular micrometer.

Determination of population sensitivity

Preliminary tests with sensitive and resistant strains and different concentrations of triadimenol showed that measurement of hyphal length allows discrimination between sensitive and resistant spores at a given concentration of fungicide. Therefore, under our experimental conditions we determine that:

- Conidia are sensitive to the concentration of triadimenol when hyphal length is less than 250 µm (hyphae will not form conidiophores and secondary conidia);
- Conidia are resistant to the same concentration when hyphal length is more than 250 µm.

Using this simple measurement, it is possible to calculate, for a given population of spores exposed to a range of concentrations of triadimenol, the percentage of sensitive and resistant spores at each concentration.

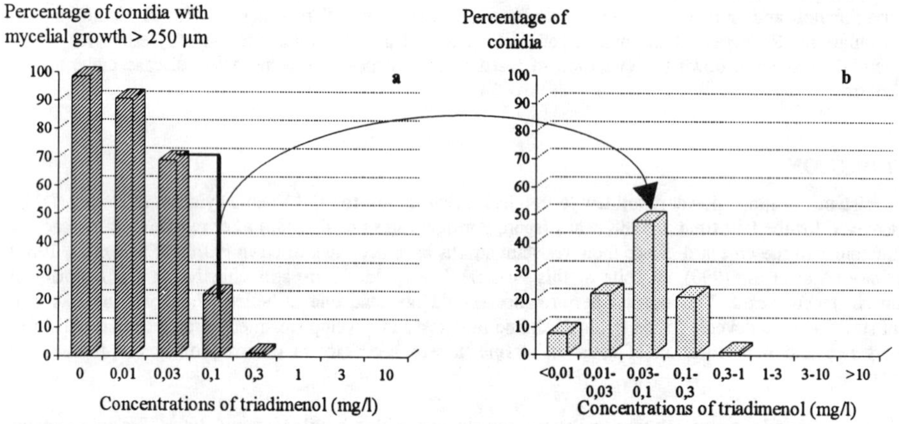

Figure 1. Representation in the case of *Uncinula necator* resistance to DMI fungicides of population sensitivity. **a** : percentage of conidia in each dose witch mycelial growth is superior to 250 µm; **b** : percentage of conidia witch sensitivity is comprise between two concentrations.

Figure 1 illustrates the process used to evaluate the population sensitivity. This population was isolated in 1989 in the south of France, from a DMI-untreated vineyard plot. Figure 1a shows for each concentration the percentage of conidia with hyphal length greater than 250 µm. For example, this value is 70% for the 0.03 mg/l concentration and 20% for the 0.1mg/l concentration. The difference between the percentages obtained for two consecutive concentrations of triadimenol is showed in figure 1b. Each bar represents the percentage of spores that grows on the lower concentration but are killed on the upper dose.

This example illustrates the variation in sensitivity that exists among a wild population of conidia of *U. necator*.

FIELD TRIALS

Triadimenol sensitivity evolution of sulphur (10 000 g AI/ha) or DMI (triadimenol, 37.5 g AI/ha) sprayed populations was studied using different experimental designs. Our goal was to define the optimal plot size to avoid natural external contamination. These results are described as follows.

Experimental design

Experiments were conducted on cv. Carignane vines planted with 1.5 m between plants in the row and 2 m between rows. Sulphur and triadimenol were each sprayed 6 times per year for 4 years on a 500 m² plot of 5 rows, each with 40 vine plants. There was no replication and rows were planted in the prevailing wind direction.

Sampling

Infected material (30 leaves or bunches) was harvested only on the central row. This material was wrapped in healthy leaves and stored at 20-25°C. When received at the laboratory, samples were directly inoculated onto leaf discs treated with a range of triadimenol concentrations. Population sensitivity was then determined as described previously.

Population sensitivity evolution

Figure 2 shows evolution of the population sensitivity in 1992 on plots located in Narbonne (France), and sprayed since 1989 with sulphur and DMI. Three samplings were conducted during the season (1 per month starting in May).

On the sulphur sprayed plot, the population was still sensitive and no evolution during the season was observed (figure 2a). At the last sampling, only a slight increase was observed in the percentage of spores with sensitivities between 0.1 and 0.3 mg/l of triadimenol.

On the DMI sprayed plot, two less-sensitive phenotypes were identified at the first date of sampling (figure 2b). These grew on leaf discs treated with 0.3 mg/l of triadimenol but were killed by 3 mg/l. At the last observation, a resistant sub-population appeared. Sixty percent of spores had a sensitivity within the 0.1 and 0.3 mg/l range. Few spores (< 2%) survived at the 10 mg/l concentration.

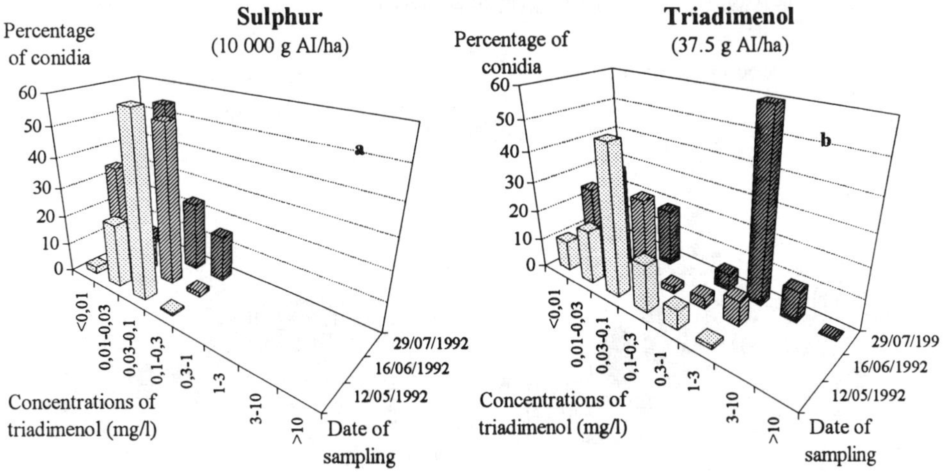

Figure 2. Evolution of *Uncinula necator* population sensitivity to triadimenol during the 1992 season in a vineyard located in France (Narbonne) after 4 years of treatments (6 per year) with : **a**. sulphur (10 000 g AI/ha); **b**. triadimenol (37.5 g AI/ha)

This example demonstrates that such a design allows the precise survey of the selective effect of a DMI fungicide at the vineyard level. Large scale field trials and determination of population sensitivity as described comprise an accurate approach for evaluating the selective effect of different anti-resistance strategies for the control of *U. necator*.

EVALUATION OF ANTI-RESISTANCE STRATEGIES

There are two main rules when applying an anti-resistance strategy :

- the selective effect should be as low as possible;
- the efficacy should be as high as possible, even on the resistant populations.

To define such optimal conditions, the effects of dose, mixtures and reduction of the number of DMI application have been studied in different field trials.

Selective effects

Programs of treatment

Various programs were compared in practice:
- standard non selective program: sulphur (10 000 g AI/ha)
- DMI program: triadimenol (37.5 g AI/ha);
- half rate DMI: triadimenol (18.75 g AI/ha);
- mixtures: triadimenol+sulphur (25+4000 g AI/ha) or triadimenol+dinocap (25+105 g AI/ha);
- alternation of DMI with non selective compound: triadimenol (37.5 gAI/ha) / sulphur (10000 g AI/ha) or triadimenol (37.5 g AI/ha) / dinocap (210 g AI/ha).

Field experiments

Two experiments have been carried out in the south of France (Narbonne and Perpignan) on Carignane vineyard. Each year, a high frequency of flag shoots was observed after bud burst. Programs were compared according to the previously described experimental design (500 m² plot, sampling on central row). A total of 6 sprayis was applied each year for four consecutive years on the same plot. Disease pressure was assessed each year at the end of the season and 3 samples were taken : first before spraying, second at the flowering stage, third after the last spray and before véraison.

Evolution of the sensitivity of the populations in different plots

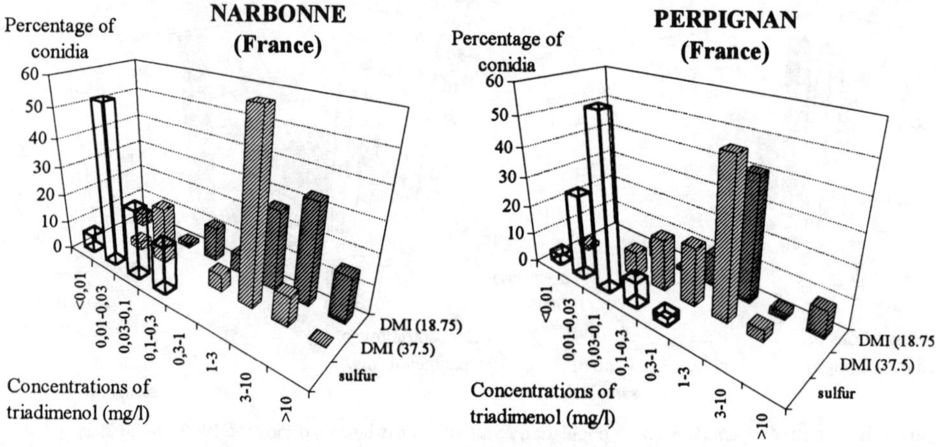

Figure 3. Sensitivity to triadimenol of *Uncinula necator* populations treated for 4 years (6 treatments per year) with sulfur (10 000 g AI/ha); a registered rate of triadimenol (37.5 g AI/ha) and a half rate of triadimenol (18.75 g AI/ha) in two french vineyards in Narbonne and Perpignan, populations sampled after the last spray

On sulphur treated plots, the population was still sensitive (figure 3). There was a slight difference between Perpignan and Narbonne locations. In Narbonne, spore growth was inhibited by 0.3 mg/l, in Perpignan this dose is higher: 1 mg/l of fungicide. There was no significant shift in comparison with the population from the beginning of the experiment before treatements wee applied.

On DMI sprayed plots (triadimenol, 37.5 g AI/ha), the most sensitive phenotypes disappeared. The populations showed a shift towards increasing levels of resistance. The MIC value in Perpignan is higher than 3 mg/l and 10 mg/l in Narbonne (figure 3).

On *half rate DMI* sprayed plots (triadimenol 18.75 g AI/ha), shift of sensitivity was greater than that observed with the population sprayed with the full concentration of triadimenol. In Narbonne (figure 3), the percentage of very resistant (> 10 mg/l) and resistant (3-10 mg/l) phenotypes reached 20% and 34% of the total respectively. In Perpignan (figure 3), phenotypes resistant to 10 mg/l were also detected but at lower frequencies.

Mixtures. Sulphur and dinocap, two surface fungicides were mixed with triadimenol at 25 g AI/ha.

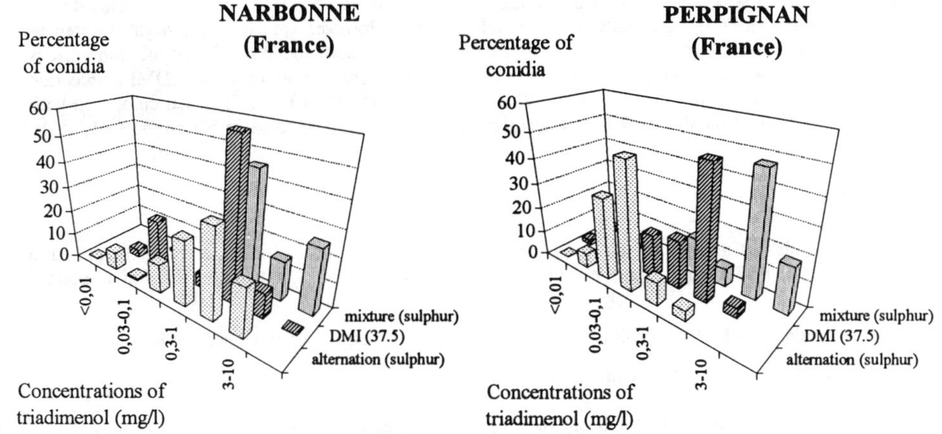

Figure 4. Sensitivity to triadimenol of *Uncinula necator* populations treated during 4 years (6 treatments per year) with triadimenol (37.5 g AI/ha), an alternation of triadimenol (37.5 g AI/ha) and sulphur (10000 g AI/ha) and a mixture of triadimenol+sulphur (25+4000 g AI/ha) in two french vineyards in Narbonne and Perpignan.

With sulphur (figure 4), we notice a strong selective effect of the mixture. The shift in sensitivity was greater than with triadimenol alone at its registered dose (37.5 g AI/ha).

The percentage of spores growing on leaf discs treated with 3 mg/l was 50% in Narbonne and 70% in Perpignan. In both cases, the most sensitive part of the population disappeared.

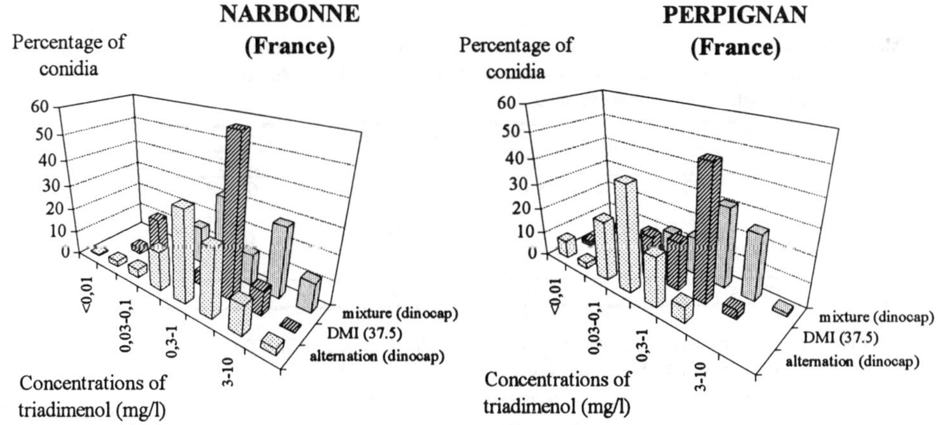

Figure 5. Sensitivity to triadimenol of *Uncinula necator* populations treated for 4 years (6 treatments per year) with triadimenol (37.5 g AI/ha), an alternation of triadimenol (37.5 g AI/ha) and dinocap (210 g AI/ha) and a mixture of triadimenol+dinocap (25+105 g AI/ha) in two french vineyards in Narbonne and Perpignan.

The selective effect of dinocap was stronger than triadimenol alone but remained lower than that of the sulphur mixture. In Narbonne (figure 5), there was a bimodal distribution of two (sensitive and resistant) subpopulations. In Perpignan, spores tolerant to 10 mg/l were detected at low frequency only in DMI+dinocap sprayed plots (figure 5).

Alternation. We tested an alternation spray progrmme where DMI and contact fungicides alternated at each application. The first spraying was a DMI and there were no more than 3 triadimenol treatments during the season. In this case, there was a generally slower population shift. However, we can notice slight differences within plots and compounds. With sulphur alternations (figure 4), in both locations the population shift is lower than with the DMI alone. In Narbonne, the MIC values for the alternation and exclusive DMI application were respectively lower and higher than 10 mg/l. With dinocap alternations (figure 5), when compared with DMI used alone, the population shift was less in Perpignan (figure 5) but the same in Narbonne (MIC > 10 mg/l).

Evaluation of efficacy

Our contention is that an anti-resistance strategy has to maintain constant efficacy, even when applied to resistant populations and after many years of application. We tried to highlight two specific and important points:

- efficacy should be as high as possible whatever the population sensitivity and disease pressure ;
- efficacy should be preserved over time.

To address these two points, two kinds of experiments have been carried out.

Similar trials were set up in different countries (France, Portugal, Italy) and on different sensitive vine cultivars (Cardinal, Carignane, Gros Manseing). The experimental design was the block type with three or four replicates and contiguous untreated control. Disease pressure on bunches was assessed after the last treatment.

Table 1 shows rates of diseases for three experiments in the south of France. In each location three application schedules were compared. High level of disease occured when triadimenol was exclusively used. This result illustrated the loss of efficacy of DMI fungicides in practice. Disease pressure was still strong when triadimenol was reduced to three consecutive sprays around the flowering stage. The most efficient results were obtained when two triadimenol applications were alternated with two applications of powdered sulphur. Statistical analysis between strategies was the same for all the locations. However, the efficacy level varies with location, probably because of different initial population sensitivity and disease pressure.

Table 1. Percentage of powdery mildew disease on bunches at the end of the season after treatments with different strategies in an experiment conducted by the CNPPA during three years at the same location in **Portugal** (Azambuja).

Programs of treatment										Percentages of disease in each year		
			*Flo***			*Bt*			*Vé*	1989	1990	1991
T*	T		T	T	T	T	T	T		85.4 d***	46.1 d	66.9 d
Sp			Sp			Sp				12.1 b	2.2 a	1.9 a
Sp	Sm	Sm	Sp	T	T	Sp	T	T		5.5 a	6.4 b	14.2 bc
D	D	D	D	T	T	D	T	T		22.2 c	17.8 c	21.3 c
			Untreated							***100.0***	***83.5***	***74.2***

*Treatments : **T**. triadimenol (50 g AI/ha), **Sm**. wettable sulphur (10000 g AI/ha), **Sp**. powder sulfur (30000 g AI/ha), **D**. dinocap (210 g AI/ha).

Phenological stages : ***Flo*. flowering, ***Bt***. berry touch, ***Vé***. véraison.

***Means followed by the same letter are not significantly different according to the Newman and Keuls test after analysis of variance.

Table 2 illustrates the evolution of disease severity during three years at the same location (Azambuja, Portugal). When triadimenol alone was exclusively sprayed, disease severity was the same as the untreated control every year, except in 1990. When two sets of two consecutive triadimenol spraying are alternated with powder sulphur the damage increases between 1990 and 1991. Data from population sensitivity (STEVA, 1992) demonstrated that was the result of the selective effect of this strategy.

Table 2. Percentage of powdery mildew disease on bunches at the end of the season after treatments with different strategies in field trials conducted by VIVADOUR during two years (1991 and 1992) in three locations in the South West of **France**.

Programs of treatment							Percentages of disease in each year		
*Flo***			*Bt*			*Vé*	**Viella**	**Labarthète**	**Monpezat**
T*	T	T	T	T	T	T	59.9 c***	90.3 c	62.3 c
Sm	T	T	T	Sm	Sm	Sm	42.2 b	78.9 b	43.7 b
Sp	T	Sp	T	Sp	Sm	Sm	15.5 a	37.1 a	14.9 a
Untreated							***63.2***	***94.9***	***93.5***

*Treatments : **T**. triadimenol (37.5 g AI/ha), **Sm**. wettable sulphur (10000 g AI/ha), **Sp**. powdered sulphur (30000 g AI/ha).

Phenological stages : ***Flo* . flowering, ***Bt*** . berry touch, ***Vé*** . véraison.

*** Means followed by the same letter in the same column are not significantly different according to the Newman and Keuls test after analysis of variance.

CONCLUSION

The objectives of these studies were to describe the steps we took to assess the effect of different strategies of coping with DMI fungicide resistance in *U. necator*.

Considering the aerial dispersal characteristic of this obligate fungus, and the progressive nature of resistance, it was absolutely necessary to produce :

- a precise method for analyzing population sensitivity.
- an experimental design that takes into account the epidemiology of the grape powdery mildew.

Once such parameters have been defined, we could study the advantages of different strategies in various French vineyards. Four years of successive treatments led us to these conclusions.

On the vineyard scale, DMI fungicides had a strong selective effect on natural populations. As a first step, this does not necessarily lead to loss of field efficacy. Not only the number of sprayis but the DMI rate per hectare can induce a selective effect. Half-rates were more effective in selecting less sensitive populations than full registered doses.

Fungicide mixtures of DMIs with sulphur or dinocap were not an effective anti-resistance strategy in the case of *U. necator*. With sulphur, such results can be explained by an antagonism with DMI fungicides (STEVA, 1992).

When compared with exclusive DMI spraying, application schedules where DMI and classical compounds (sulphur and dinocap) alternate, showed good ability to slow down the resistance evolution process. This result can be easily explained by the reduction of DMI treatments.

One of the main questions that still needs to be answered concerns the acceptable maximum number of DMI applications and their period of spraying during the season, in relation to the existing sensistivity structure of the population to be treated.

ACKNOWLEDGEMENTS

We are most grateful to Mrs Carole Cazenave for her technical assistance in laboratory analysis. Many people have also helped with field trials and we thank in particular : Gérard Brarda, Maria Teresa Gomes da Silva, Henri Guillemont, Philippe Mauranx and Daniel Novoa.

REFERENCES

Aloi, C.; Gullino, M.L.; Garibaldi, A. (1990). Reduced sensitivity to fenarimol and triadimefon in field population of *Uncinula necator* in Italy. *Seventh International Congress of Pesticide Chemistry*, Ed. Frehse H., Hamburg, Vol1; 441.

Steva, H. (1992). Résistance de l'oïdium de la vigne (*Uncinula necator* (Schw. et Burr.) aux fongicides Inhibiteurs de la Biosynthèse des Stérols. *Thèse de Docteur de l'Université de Bordeaux II*, 237pp.

Steva, H.; Cartolaro, P.; Clerjeau, M.; Lafon, R.; Gomes da Silva, M.T. (1988). Une résistance de l'oïdium au Portugal ? *Phytoma*, **402**, 49-50.

Steva, H.; Cartolaro, P.; Clerjeau, M.; Lafon, R. (1989). Premier cas de résistance de l'oïdium à un traitement fongicide. *Viti*, **137**, 124-125.

CHARACTERISTICS OF *PSEUDOCERCOSPORELLA HERPOTRICHOIDES* ISOLATES RESISTANT TO PROCHLORAZ

N. CAVELIER, C. PINEAU, M. PRUNIER

INRA, SRIV, Domaine de la Motte, BP 29, 35650 LE RHEU FRANCE

ABSTRACT

Prochloraz-resistant isolates of *P. herpotrichoides* were isolated from wheat stems in fields in different regions of France. Some of them of fast and slow growing type were used to inoculate plants in greenhouse experiments. Pathogenicity and competitive ability of sensitive and resistant strains are similar. The resistant strains seem to be as competitive or less competitive than the sensitive ones.

INTRODUCTION

Prochloraz and flusilazole are the most commonly used fungicides for controlling eyespot caused by *Pseudocercosporella herpotrichoides* in France. Prochloraz-resistant strains were isolated for the first time in 1990 on winter wheat (Leroux & Marchegay, 1991). Populations of *P. herpotrichoides* were monitored for sensitivity to prochloraz in 1990, 1991, 1992 and 1993. Few isolates were resistant to prochloraz in both the fast and the slow growing type (Cavelier *et al.*, 1992). In 1993, the percentage of isolates growing on 0.5 mg /l, 1 mg/l and 2 mg/l was 60, 27 and 9 % respectively (56 plots tested).

The knowledge of the fitness of the resistant strains may give some information about the use of prochloraz.

This paper presents the results of glasshouse experiments which compare two fitness parameters of resistant and sensitive isolates: pathogenicity and competition in the absence of fungicide.

MATERIALS AND METHODS

The isolates selected for this study were isolated from winter wheat in 1991 and 1992 in different regions of France. To test the sensitivity to prochloraz, each isolate was transferred to Petri dishes amended with 2 mg/l of the formulated compound (Cavelier *et al.* 1992). Experiments were carried out in a glasshouse heated only to give frost protection. Pots were arranged in a randomised design, with five blocks, and five plants per pot. Plants, cv. "Camp-Remy", were inoculated at the three leaf growth stage. The mean frequency of plants with symptoms was calculated at GS 60 (Zadoks *et al.* 1974). The severity of eyespot was assessed by the method of Cavelier & Le Page (1985). Plants were scored on a 0-2 scale: 0- healthy plants, 1- sheaths with lesions, 2- stems with lesions. Mean scores were calculated. On the stem, when eyespot lesions were present, the mean infected proportion of a cross-section of the stem was calculated.

Pathogenicity

Inoculation was carried out using a suspension of finely chopped mycelium. Four isolates were of the fast growing type, two were sensitive to prochloraz and two were resistant. Six isolates were of the slow growing type, three were sensitive to prochloraz, two were resistant. Isolates are designated resistant if they grow on 2 mg/l of prochloraz.

Competition

Plants were inoculated by spore suspension (100000/ml). Five mixtures of one resistant and one sensitive strain were made in different proportions: sensitive/resistant 100/0, 90/10, 50/50, 10/90, 0/100.

1: fast growing sensitive/ fast growing resistant
2: slow growing sensitive/ slow growing resistant
3: slow growing sensitive/ slow growing resistant
4: slow growing sensitive/ slow growing resistant
5: fast growing sensitive/ slow growing resistant

The strain mixtures were made with isolates of the same fields.

Isolations were made from stems with lesions to identify the strain originally involved.

RESULTS

Pathogenicity

As in previous work (Cavelier *et al.*, 1992), there was no difference between the pathogenicity of resistant and sensitive isolates (Table 1).

TABLE 1. Pathogenicity of prochloraz-sensitive or resistant isolates of *P.herpotrichoides*. (mean of 4 experiments)

Isolate type	mean frequency plant lesions	mean disease score 0-2	mean proportion stem cross-section infected
fast growing sensitive	82	1.6	58.9
fast growing resistant	78	1.6	57.6
slow growing sensitive	61	1.0	21.6
slow growing resistant	59	1.1	27.9

Competition

The pathogenicity of the mixture is higher than that of the single strains for the mixtures 2 and 5. (Table 2)

TABLE 2. Pathogenicity of mixtures of prochloraz-sensitive and resistant strains of *P. herpotrichoides* in known proportion. Mean disease score 0-2

	strains	proportion of sensitive/resistant strains 100/0	90.10	50/50	10/90	0/100
1	fast growing	1.4	1.5	1.2	1.4	0.6
2	slow growing	0.6	1.0	1.4	0.9	0.8
3	slow growing	0.8	1.2	1.2	0.7	1.2
4	slow growing	0.8	1.1	0.1	0.5	0.7
5	fast growing/ slow growing	0.6	1.2	1.6	0.7	0.8

The frequency of strains re-isolated from plants inoculated with a fast growing strain in the mixture was about the same for both strains. But, with a mixture of slow growing strains, the strains re-isolated were mainly sensitive ones.(Table 3)

TABLE 3. Mean frequency of prochloraz-resistant strains(%) re-isolated from plants inoculated with mixtures of prochloraz-sensitive and resistant strains of *P. herpotrichoides*, in known proportions.

Inoculated strains	Proportions of resistant strains % 10	50	90
fast growing	26	65	-
slow growing	0	0	10
fast growing/ slow growing	0	11	89

CONCLUSIONS

The similar pathogenicities of prochloraz-sensitive and resistant strains suggest that this parameter would have no influence in the evolution of populations of *P. herpotrichoides*. On the other hand, it was surprising that some mixtures of strains appeared more pathogenic than the single strains. These results differ from preceding observations, were similar mixtures of fast growing type/slow growing type were compared (Cavelier *et al.*, 1987).

The re- isolations from plants suggest that the resistant strains are equally competitive (fast-growing strains) or less competitive (slow-growing strains) than the sensitive ones. These results are in contrast to these seen with MBC resistant and sensitive strains, where the resistant strains were more competitive (Cavelier & Le Page, 1985), MBC resistance is now widely developed. This suggests that the evolution of prochloraz resistance could differ from that of MBC resistance.

REFERENCES

Cavelier, N.; Le Page, D. (1985) Caractéritiques de souches de *P. herpotrichoides* (Fron) Deighton (agent du piétin-verse des céréales) résistantes aux fongicides benzimidazoles et thiophanates: pouvoir pathogène, capacité de développement. *ANPP, premières jounées d'études sur les maladies des plantes*, **1**, 49-56.

Cavelier, N.; Lorée, F.; Prunier, M. (1992) Resistance of the eyespot fungus, *Pseudocercosporella herpotrichoides,* to DMI fungicides. *British crop protection conference-Pests and diseases*, **1**, 189-194.

Cavelier, N.; Rousseau, M.; Le Page, D. (1987) Variabilité de *Pseudocercosporella herpotrichoides,* agent du piétin-verse des céréales: comportement *in vivo* de deux types d'isolats et d'une population en mélange. *Zeitschrift fuer Pflanzenkrankheiten und Pflanzenschutz*, **94**, 590-599.

Klein, U. (1991) Zur Resistenz bei *Pseudocercosporella herpotrichoides var. herpotrichoides* gegenueber Prochloraz. *Dissertation Hannover.*

Leroux, P.; Marchegay, P. (1991) Caractérisation des souches de *Pseudocercosporella herpotrichoides*, agent du piétin-verse des céréales résistantes au prochloraze, isolées en France sur blé tendre d'hiver. *Agronomie*, **11**, 767-776.

Zadoks, J.C.; Chang, T.T.; Konzak, C.F. (1974) A decimal code for the growth stages of cereals. *Weed research*, **14**, 415-421.

REDUCED SENSITIVITY TO THE EBI FUNGICIDE PROCHLORAZ IN THE CEREAL EYESPOT FUNGUS *PSEUDOCERCOSPORELLA HERPOTRICHOIDES*

J.E. HARDY, J.A. LUCAS and J.F. PEBERDY

Department of Life Science, University Park, Nottingham, NG7 2RD

A.M. JULIAN

Natural Resources Institute, c/o Department of Agriculture, University of Reading, Earley Gate, Reading, RG6 2AT.

ABSTRACT

Isolates of *Pseudocercosporella herpotrichoides* reduced in sensitivity to prochloraz were derived from ultra-violet (UV) mutation and spontaneous selection on fungicide-amended media. Frequencies of resistance were found to be similar to those obtained for other ergosterol biosynthesis inhibiting (EBI) compounds. Although many of these putative mutants reverted rapidly to wild-type sensitivity, those generated from successive rounds of mutation or from spontaneous selection were found to be more stable. Mutants reduced in sensitivity to prochloraz were compared with wild-type isolates in terms of response to prochloraz *in vitro* and *in vivo*, growth rate, sporulation ability, pathogenicity and competitive ability on a range of cereal hosts. The observations made from this study support a multi-factorial basis for prochloraz resistance, suggesting directional selection for field resistance.

INTRODUCTION

Eyespot, caused by the fungus *Pseudocercosporella herpotrichoides* (Fron.) Deighton, is an important disease of winter cereals with annual losses averaging £20 - 30 million in the UK alone. In the 1970's disease control was provided by application of the methyl-benzimidazol-2-yl carbamate (MBC) fungicides. However, wide-spread resistance had developed to this group by the early 1980's (King and Griffin, 1985) and de-methylation inhibitors such as prochloraz are currently recommended for eyespot control in the U.K.

Prior to 1990, there were no published cases of reduced control of *P. herpotrichoides* by prochloraz. Strains with reduced sensitivity had been recovered from field surveys (Gallimore *et al.*, 1987) but tolerance levels were below the recommended application rate. In 1990, strains showing reduced-sensitivity to prochloraz at levels of up to 10 mg/l agar were isolated from northern France (Leroux and Marchegay, 1991). This discovery emphasised the need for further research into the development of resistance in the field.

In the current study, strains of *P. herpotrichoides* with reduced sensitivity to prochloraz were mass-selected on fungicide-amended media or induced by UV-irradiation. The stability of resulting putative mutants was assessed through serial transfer on amended and unamended media, sporulation steps and passage through plants. Stable isolates were further characterised in terms of fitness by measuring growth rate in culture, pathogenicity and ability to compete with the sensitive wild-type after co-inoculation on cereal hosts.

MATERIALS AND METHODS

Production of mutant strains

Three field isolates of *P. herpotrichoides* were used in this study: R-types 22-12 and 22-1011 and a W-type 22-20. Isolates reduced in sensitivity to prochloraz were induced by UV irradiation of conidia or spontaneous selection on prochloraz-amended media using the method described by Julian *et al.* (1994 a). Putative mutants were purified on malt-yeast-glucose (MYG) agar (5 g malt extract, 2.5 g yeast extract, 10g glucose, 20 g agar l^{-1} distilled water) amended with 5 μM prochloraz (1 $\mu M = 0.377$ mg/l) and then serially sub-cultured three times on amended and unamended media to confirm stability. A further two rounds of UV mutagenesis were carried out on selected mutants derived from 22-12, resulting in the production of resistant strains classified as intermediate or high level mutants.

Characterisation of mutants

In vitro plate assays were used to assess the dose response to prochloraz of the mutant strains. MYG plates amended with a range of concentrations (0.1 - 200 μM) of technical grade prochloraz were inoculated with 3 mm plugs cut from the growing edge of a colony. For each treatment 6 replicate colonies were measured after 14 days at 19°C, and the percent inhibition calculated from the unamended control. This data was used to determine the LC_{50} and MIC values for each isolate. The dose response was re-tested after a 1 year period during which time isolates were sequentially sub-cultured on amended and un-amended medium. The response of strains was also tested following sporulation steps and passage through wheat cv. Avalon.

Pathogenicity was tested by inoculating 21 day old seedlings of Avalon with conidial suspensions of the test isolates (2 x 10^5 ml^{-1}) (Daniels *et al.*, 1991). For all treatments 45 plants in 9 replicate pots were assessed. Plants were maintained in a growth room (16 h daylength, 17 ($\pm$ 2)°C) for 6 - 7 weeks after which symptoms were visually scored (Scott, 1971). For fungicide treatments plants were sprayed with Sportak 40, at a rate equivalent to 200 g A.I./ha, 24 hours prior to inoculation. Re-isolations taken from infected stem-base material were tested to confirm level of sensitivity to prochloraz. For competition studies 10 day old seedlings of Avalon were co-inoculated as previously described with mixtures of the sensitive wild-type parent 22-12, and either the spontaneously selected strain 22-1103, or the UV-induced mutant 22-1128. Conidia were mixed in the following ratios: 100:0, 75:25, 50:50, 25:75, 0:100. Prochloraz was applied, at a rate equivalent to 400 g A.I./ha, 4 days after inoculation. Plants were visually assessed after 12 weeks, and 1 cm stem sections surface-sterilised prior to re-isolation to determine the proportion of sensitive and resistant fungal outgrowths.

RESULTS

Frequencies of spontaneous resistance to prochloraz were within the range of 10^{-6} to

10^{-7}. UV-irradiation increased this frequency to approximately 10^{-5} (2 - 20% survival). Although many putative mutants reverted rapidly to wild-type sensitivity, those produced after several rounds of mutagenesis proved to be relatively stable in culture (Figure 1). Sporulation steps or passage through cereal hosts did result in reduced levels of resistance in some UV-induced isolates, however, resistance levels of spontaneously selected mutants were less affected by these treatments (Julian *et al.*, 1994a).

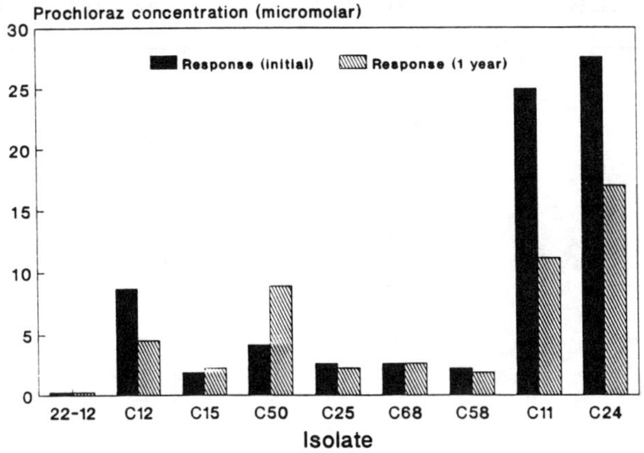

Fig. 1. The LC_{50} to prochloraz (μM) of UV-induced mutants after 1 year of continuous serial sub-culture, compared with the LC_{50} response immediately after isolation and purification.

Growth rates in culture and sporulation ability of mutant isolates were variable but there was no correlation between resistance level and growth rate (R-squared = 0.051), or conidia production (R-squared = 0.006). No reductions in pathogenicity were associated with spontaneous mutants. However, UV-induced mutants were more variable (Figure 2).

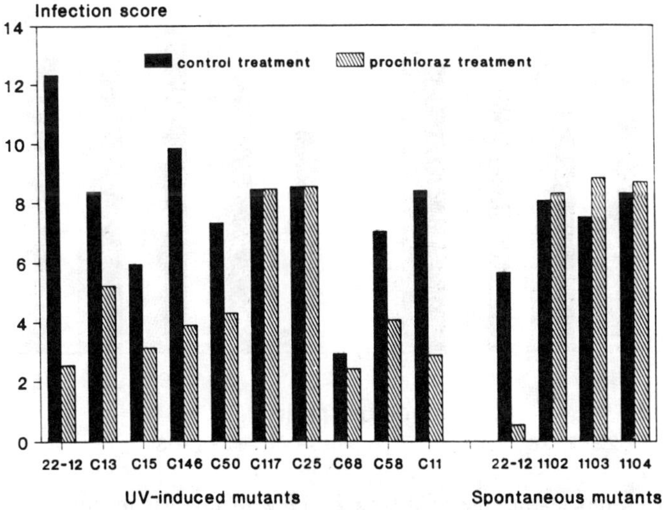

Fig. 2. Pathogenicity of UV-induced and spontaneous mutants from the R-type 22-12, on wheat cv. Avalon, in the presence and absence of a prochloraz treatment.

TABLE 1. The sensitivity of colonies re-isolated from Avalon stem bases inoculated with mixtures of the wild type 22-12, and the spontaneously selected strain 22-1103, or the UV-induced resistant strain 22-1128 to prochloraz.

Inoculum mix	Number of colonies re-isolated			
	Fungicide treated		Untreated	
	Sensitive	Resistant*	Sensitive	Resistant*
22-12:22-1103				
100:0	0	1	11	0
75:25	2	110	1	31
50:50	3	90	3	28
25:75	4	28	0	3
0:100	6	87	0	29
22-12:22-1128				
100:0	0	1	11	0
75:25	2	32	3	9
50:50	0	62	8	48
25:75	0	29	1	0
0:100	0	25	0	26

* Ability to grow on MYG amended with 5 μM prochloraz

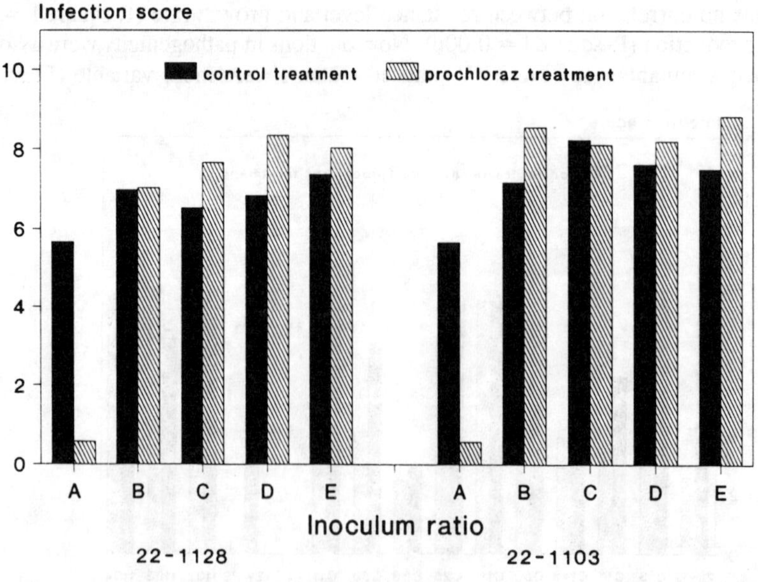

Fig. 3. The effect of co-inoculating the wild-type 22-12 with insensitive mutants 22-1128 (UV-induced) and 22-1103 (spontaneous) on disease levels on wheat cv. Avalon. Inoculation ratios (22-12:mutant): A = 100:0, B= 75:25, C = 50:50, D = 25:75, E = 0:100.

Re-isolations from stems infected during the competition experiments comprised a mixture of sensitive and resistant material (Table 1). There were no significant differences between infection scores on co-inoculated plants but plants inoculated solely with the sensitive wild-type 22-12 had significantly reduced infection scores ($P < 0.001$) particularly where a prochloraz treatment had been applied. Disease scores were often higher on prochloraz treated plants inoculated with the insensitive isolates (Figure 3).

DISCUSSION

Assessment of the characteristics of isolates of *P. herpotrichoides* resistant to prochloraz is vital when evaluating the likelihood of establishment of such strains in field populations. In order for resistance to become established, resistant strains must arise and be able to persist in the field. Factors such as rate of mutation to the resistant phenotype, selection pressure exerted by the fungicide, stability and competitive ability of the mutants and mode of dissemination will all contribute to this. The recent occurrence of prochloraz resistant strains in France suggests that these criteria can be satisfied under field conditions (Leroux and Marchegay, 1991). Although it is well-established that conidia are distributed by rain-splash (Fitt and Bainbridge, 1983), the role played by ascospores is still unclear. However, apothecia were found to occur at numerous set-aside sites throughout England during 1993 (Dyer, pers. comm.).

In this study, frequency of resistance to prochloraz after UV mutation was found to be similar to that identified with other EBI compounds. For example, induction of triadimenol resistance in W-type isolates of *P. herpotrichoides* gave frequencies of 5 x 10^{-5} for a 90% kill (Leroux *et al.*, 1988). Levels of resistance observed in the prochloraz-resistant mutants were similar to those derived from studies with imazilil, an imidazole closely related to prochloraz, with *Aspergillus nidulans* (van Tuyl, 1977). Generation of increasing numbers of mutants with successive rounds of UV mutation, along with higher resistance levels of third round mutants, may suggest a multigenic trait involving progressive alteration of more than one gene.

Most UV mutants were found to be unstable, rapidly reverting to wild-type sensitivity after mutation. The resistance initially expressed may have been of a transient nature, possibly from physiological adaptation to the fungicide. Third-round UV mutants and low level spontaneous mutants were found to be stable, even after one year in culture, possibly indicating a different resistance mechanism. Previous studies of cultural instability in the pathogen, showed that morphologically altered sectors had increased sensitivity to prochloraz compared to the parent colony. This may have important implications for resistance testing and may be indicative of a degree of instability of this type of resistance in culture (Julian *et al.*, (1994b).

Differences in pathogenicity levels on cereal hosts between classes of mutant may also indicate different resistance mechanisms, with the possibility that the mutation procedure had pleiotropic effects causing a reduction in infective ability. Additional studies found that field isolates showing reduced sensitivity to prochloraz gave higher infection scores in the presence of the fungicide than on untreated control plants (Hardy, unpublished). Co-inoculation with sensitive and resistant isolates indicated that resistant isolates were able to compete effectively with the susceptible parent.

Evidence from this study supports the hypothesis that development of resistance to prochloraz will be directional in nature, with the possibility that small incremental changes in resistance level, reflecting a multi-factorial basis for the trait, will occur. Establishment of the genetic nature of prochloraz resistance could lead to the identification of molecular markers. These could provide improved diagnostic procedures for field surveys and enable the development and spread of resistance to this fungicide to be more readily and accurately followed.

ACKNOWLEDGEMENTS

The authors would like to thank the Ministry of Agriculture, Fisheries and Food and Schering Agrochemicals for funding of this project through postgraduate studentships. The help and advice of Dr. M.J. Hocart and Dr. A. Daniels is gratefully acknowledged.

REFERENCES

Daniels A.; Lucas J.A. and Peberdy J.F. (1991) Morphology and ultrastructure of W and R pathotypes of *Pseudocercosporella herpotrichoides* on winter wheat seedlings. *Mycological Research* **95** (4) 385 - 397.

Fitt B.D.L. and Bainbridge G.L. (1983) Dispersal of *Pseudocercosporella herpotrichoides* spores from infected wheat straw. *Phytopathologische Zeitschrift* **106** 214 - 225

Gallimore K.; Knights I.K. and Barnes G. (1987) Sensitivity of *Pseudocercosporella herpotrichoides* to the fungicide prochloraz. *Plant Pathology* **36** 290 - 296.

Julian A.M.; Hardy J.E. and Lucas J.A. (1994a) The Induction and characterisation of isolates of *Pseudocercosporella herpotrichoides* with altered sensitivity to the fungicide prochloraz. *Pesticide Science* (in press).

Julian A.M.; Hardy J.E. and Lucas J.A. (1994b) Cultural variability in the cereal eyespot pathogen *Pseudocercosporella herpotrichoides*. *Mycological Research* (in press).

King J.E. and Griffin M.J. (1985) Survey of benomyl resistance in *Pseudocercosporella herpotrichoides* on winter wheat and barley in England and Wales in 1983. *Plant Pathology* **34** 272 - 283.

Leroux P.; Gredt M. and Boeda P. (1988) Resistance to inhibitors of sterol biosynthesis in field isolates or laboratory strains of the eyespot pathogen *Pseudocercosporella herpotrichoides*. *Pesticide Science* **23** 119 - 129.

Leroux P. and Marchegay P. (1991) Caractérisation des souches de *Pseudocercosporella herpotrichoides* agent du piétin verse des céréales, résistantes au prochloraze, isolées en France sur blé tendre d'hiver. *Agronomie* **11** 767 - 776

Scott P.R. (1971) The effect of temperature on eyespot (*Cercosporella herpotrichoides*) in wheat seedlings. *Annals of Applied Biology* **68** 169 - 175.

Van Tuyl J.M. (1977) Genetic aspects of resistance to imazalil in *Aspergillus nidulans*. *Netherlands Journal of Plant Pathology* **83** 169 - 176.

EVALUATION OF HYPHAL ELONGATION AS A BASIS FOR MONITORING THE SENSITIVITY OF *VENTURIA INAEQUALIS* AND *MYCOSPHAERELLA FIJIENSIS* TO FLUSILAZOLE

R.A. Hamlen, A.E. Trivellas, C.M Smith, L.E.B. Johnson, C.A. Shillingford

E. I. DuPont de Nemours and Company, Agricultural Products, Stine-Haskell Research Center, Newark, Delaware 19714-0030 USA

ABSTRACT

An *in vitro* procedure, based on elongation of primary hyphae, was developed to monitor the sensitivity to flusilazole of *Venturia inaequalis* and *Mycosphaerella fijiensis*. The lengths of primary hyphae from 50 pathogen spores, transferred from infected plant tissue, were compared after 2±0.5 days of growth on flusilazole-amended and unamended agar media. Evaluations were made when hyphae on unamended media had elongated significantly without obscuring the original spores and were not overrun by contamination. For each flusilazole concentration, isolates were considered inhibited if the lengths of their primary hyphae were ≤50% of the hyphal lengths on unamended media. Minimum inhibitory concentrations determined by hyphal elongation were equivalent to those obtained by colony radial growth. These results demonstrate that the hyphal elongation assay is a simple, rapid, and accurate method for monitoring the sensitivity of these pathogens to flusilazole.

INTRODUCTION

Flusilazole is a demethylation-inhibiting (DMI) fungicide developed by DuPont, which is effective against a broad spectrum of plant diseases (Fort & Moberg, 1984). Essential to our stewardship of flusilazole is the development and use of techniques to accurately assess the sensitivities of pathogens to this fungicide. This paper describes a novel *in vitro* procedure, based on the determination of primary hyphal elongation, to monitor the sensitivity to flusilazole of *Venturia inaequalis*, the cause of apple scab, and *Mycosphaerella fijiensis*, incitant of black sigatoka of bananas.

MATERIALS AND METHODS

Sample collection

Apple leaves exhibiting young, well-defined sporulating lesions of *V. inaequalis* were collected in 1988 in a randomized pattern from 3 U.S. orchards with and without DMI fungicide treatments, refrigerated in paper bags, and processed within 5 days of collection. Banana leaves with *M. fijiensis*-infected necrotic tissue containing ripe perithecia were obtained in 1990 from 3 commercial plantations in Central America, 2 weeks following the last DMI application; as well as from 1 location untreated with DMI fungicides. Leaf tissue was refrigerated dry in plastic bags until testing within 2 weeks of collection.

Media preparation

Technical flusilazole dissolved in acetone or formulated product in water was added at the appropriate concentrations to cooled, molten media (45°C) after autoclaving, mixed thoroughly, and poured into petri plates. Solvent was added to unamended media.

Hyphal elongation assay

V. inaequalis spores were transferred from 1 lesion per leaf as 5 short streaks onto the surface of potato dextrose agar (PDA) that was unamended or amended with 0.005, 0.01, or 0.05 mg/l AI flusilazole in each petri plate. Petri plates were incubated for 2±0.5 days at 21°C with 12-h day.

Banana leaves were first incubated for 48 hours at 20-25°C in a plastic bag with a moist towel to allow ascospore maturation. Five leaf sections (1-2 cm^2) from different locations were stapled to each 9-cm filter paper disc, immersed in water for 5 minutes, and placed inside a petri plate cover over 2% water agar that was unamended or amended with 0.001, 0.01, 0.1, 1.0, or 2.5 mg/l AI flusilazole for 1 hour to allow discharge of ascospores. Petri plates were incubated for 2±0.5 days at 26°C with 12-h day.

The lengths of primary hyphae from 50 spore streaks (*V. inaequalis*) or spores (*M. fijiensis*) were determined on amended and unamended media at 100x magnification using either an ocular micrometer or as multiples of spore length (Smith, Johnson *et al.*, 1991; Smith, Trivellas *et al.*, 1991). Evaluations were made when the primary hyphae on unamended media had grown a significant amount without obscuring the original spores and before saprophytic contamination overran pathogen hyphae. Daily examination of petri plates was necessary since the optimal time for assessment varied slightly (±0.5 day) depending on the growth rates of the pathogens and saprophytes. For each flusilazole concentration, isolates were considered inhibited if the lengths of their primary hyphae were ≤50% of the hyphal lengths on unamended media. For *V. inaequalis*, the response of the majority of spores in a streak was used to describe the entire streak due to the uniformity in response. For *M. fijiensis*, the percentage of reduction in hyphal length on flusilazole-amended media compared to the untreated controls was determined. Minimum inhibitory concentrations (MIC) for both pathogens were established.

Colony radial growth assay

Responses in the hyphal elongation assay were determined for 40 spore streaks or individual spores of wild-type *V. inaequalis* or *M. fijiensis* isolates, respectively. These propagules were then transferred to PDA (*V. inaequalis*) or Mycophyll® Agar (*M. fijiensis*) media that were unamended or amended with the flusilazole concentrations corresponding to those in the hyphal elongation assay. Colony diameters were determined after 2 weeks and compared to hyphal elongation responses.

RESULTS

Hyphal elongation assay

Spores of *V. inaequalis* and *M. fijiensis* isolates unexposed or exposed to DMI fungicides in the field initially germinated and transverse cell walls formed in germ tubes on media that were unamended or amended with flusilazole at all concentrations evaluated. However, after approximately 2 days, elongation of primary hyphae ceased on higher concentrations of flusilazole, whereas hyphal growth continued on unamended media and lower flusilazole concentrations (Table 1a,b).

At the time of evaluation, hyphal elongation in all spore streaks of the *V. inaequalis* isolates unexposed to DMI fungicides in the field was significantly inhibited at 0.05 mg/l AI flusilazole compared to unamended controls, whereas growth inhibition occurred in a decreasing number of spore streaks at 0.01 and 0.005 mg/l (Table 1a). An MIC between 0.01 and 0.05 mg/l AI flusilazole was determined. Isolates that were unexposed or exposed to DMI fungicides in the field provided similar dosage responses and MIC values.

Hyphal elongation in the *M. fijiensis* isolate unexposed to DMI fungicides in the field was completely inhibited at ≥0.1 mg/l AI flusilazole concentrations but not inhibited at ≤0.01 mg/l flusilazole (Table 1b). An MIC between 0.01 and 0.1 mg/l AI flusilazole was established. Isolates with and without exposure to DMI fungicides in the field had similar dosage responses and MIC values.

TABLE 1. Primary hyphal elongation of isolates of *Venturia inaequalis* (a) and *Mycosphaerella fijiensis* (b) on agar media amended with flusilazole compared to unamended controls.

a)

Isolate origin	Orchard treated with DMI Fungicides	% spore streaks with primary hyphal lengths ≤50% of unamended controls		
		0.005*	0.01	0.05
Penna.	No	5	58	100
New York	No	0	67	100
Delaware	Yes	3	90	100

*mg/l AI flusilazole

b)

Isolate origin	Plantation Treated with DMI fungicides	% spores with primary hyphal lengths ≤50% of unamended controls				
		0.001*	0.01	0.1	1.0	2.5
S. Catalina, HN	No	0	0	100	100	100
S. Alberto, CR	Yes	0	0	100	100	100
S. Barbara, HN	Yes	0	0	100	100	100
Tacamiche, HN	Yes	0	0	100	100	100

*mg/l AI flusilazole

Colony radial growth assay

For *V. inaequalis* and *M. fijiensis*, the respective levels of hyphal elongation and colony radial growth inhibition relative to untreated controls were comparable at all concentrations of flusilazole tested (Table 2a,b). In both the hyphal elongation and radial growth assays, the *V. inaequalis* isolate was completely inhibited at 0.05 mg/l AI flusilazole and partially inhibited at 0.005 and 0.01 mg/l (Table 2a). An MIC between 0.01 and 0.05 mg/l AI flusilazole was established for both assays.

Primary hyphal elongation and colony growth of the *M. fijiensis* isolate on 0.1 and 1.0 mg/l AI flusilazole were completely inhibited, whereas spores on 0.001 and 0.01 mg/l AI flusilazole developed as those on unamended media (Table 2b). An MIC between 0.01 and 0.1 mg/l AI flusilazole was determined for both assays.

TABLE 2. Comparison between primary hyphal elongation and colony radial growth of isolates of *Venturia inaequalis* (a) and *Mycosphaerella fijiensis* (b) on flusilazole-amended media, relative to unamended controls.

a)

Flusilazole Concentration (mg/l)	% spore streaks with primary hyphal lengths ≤50% of unamended controls	% inhibition of colony radial growth compared to unamended controls
0.005	7	21
0.01	71	79
0.05	100	100

b)

Flusilazole Concentration (mg/l)	% spores with primary hyphal lengths ≤50% of unamended controls	% inhibition of colony radial growth compared to unamended controls
0.001	0	0
0.01	0	0
0.1	100	99
1.0	100	100

CONCLUSIONS

The hyphal elongation assay is based on the effects on primary hyphal growth due to inhibition of sterol biosynthesis by flusilazole. Initial germination and germ tube growth of spores are not sensitive to sterol biosynthesis inhibition and their responses cannot be used to monitor sensitivity to flusilazole.

Use of the hyphal elongation assay to monitor the sensitivity to flusilazole of a limited number of isolates of *V. inaequalis* and *M. fijiensis* from locations in the U.S. and Central America, respectively, showed no difference in responses between isolates unexposed and exposed to DMI fungicides in the field.

Based on the results presented in this paper, the hyphal elongation assay is a simple, rapid, and accurate method for monitoring sensitivity of fungal pathogens to flusilazole. Comparisons between hyphal elongation and radial growth assays show excellent agreement. The hyphal elongation assay may be generally applicable to monitoring the sensitivity of other pathogens to various DMI fungicides.

REFERENCES

Fort, T.M.; Moberg, W.K. (1984) DPX H6573, a new broad-spectrum fungicide candidate. *1984 British Crop Protection Conference - Pests and Diseases*, **2**, 413-419.

Smith, C.M.; Johnson, L.E.B.; Shillingford, C.A. (1991) *In vitro* methods for monitoring flusilazole sensitivity of *Mycosphaerella fijiensis* (black sigatoka of bananas). In: *FRAC Methods for Monitoring Fungicide Resistance, EPPO Bulletin,* **21**, 332-335.

Smith, C.M.; Trivellas, A.E.; Johnson, L.E.B. (1991) *In vitro* methods for monitoring flusilazole sensitivity in populations of *Venturia inaequalis* (apple scab) and *Cercosporidium personatum* (peanut late leaf spot). In: *FRAC Methods for Monitoring Fungicide Resistance, EPPO Bulletin,* **21**, 323-328.

MONITORING STUDY OF THE RESISTANCE OF *Erysiphe graminis* f. Sp. *tritici* TO EBI FUNGICIDES IN HUNGARY

T. DULA, T KAPTÁS

Plant Health and Soil Conservation Station of county Heves, Eger Pf. 218. H-3301 Hungary

I. APONYI

Plant Health and Soil Conservation of the Capital, Budapest, Pf.127. H-1502

J. KAMP

Bayer Hungaria, Budapest, H-1371 Budapest, Pf.445.

ABSTRACT

A programme was initiated in 1993 to monitor the sensitivity of *Erysiphe graminis* f. sp. *tritici* to ergosterol biosynthesis inhibiting fungicides (EBIs) in Hungary. In 1993, 41 isolates from 11 countries were tested for sensitivity to triadimefon and tridemorph, using a test tube assay. The tridemorph sensitivities of all the isolates were similar to that of a standard isolate. Triadimefon sensitivities were more widely distributed, with LC50 values ranging from <2mg/l to 10mg/l. The LC50 values of only two isolates exceeded 10mg/l. These results will be used for comparison in future years, as part of a long term resistance monitoring programme.

INTRODUCTION

In Hungary, cereals are grown on 26 to 30 % of the arable acreage, of which the share of wheat is 85 to 90%. Powdery mildew is one of the limiting factors affecting profitability. The importance of *Erysiphe graminis* on wheat has increased in Hungary in recent years. As described by Szunics (1988), it causes an average of 5-8% yield loss, though in extreme cases this may be 20-30%.

Based on the severity of disease, 1 or 2 chemical treatments may be required. In addition to contact fungicides, systemic fungicides (such as benzimidazoles and EBIs) are available. The EBIs are very effective, broad spectrum fungicides, but their efficacy can be influenced by decreases in sensitivity of the pathogen (Brent & Hollomon, 1988; Schulz and Scheinpflug, 1986). Resistance of *E. graminis* to DMI and morpholine fungicides in Western Europe is reported in publications by researchers in the FRAC-SBI Committee (De Waard *et al.*, 1986; Highwood, 1990; Brent, 1992). Earlier studies in Western Hungary were made by Enisz (1988, 1990) on the resistance of *E. graminis* to DMIs, but no nationwide monitoring of resistance has yet been carried out in Hungary. Thus, it was thought necessary to survey the status of DMI and morpholine sensitivity and to initiate a monitoring study on resistance, to be carried out on a regular basis. Studies started in 1991 with the assistance of a grant obtained for research and development.

MATERIALS AND METHODS

Sampling areas were chosen from wheat growing regions of Hungary, where conditions are especially conducive to mildew development. Because of the long drought, only 49 wheat samples were collected from 11 counties. The pathogen was successfully isolated from 41 samples.

Test plant

The mildew-sensitive wheat variety, GK.Öthalom, was used for isolating the pathogen and for further propagation and sensitivity studies.

Fungicides

Tests were carried out with Bayleton 25% WP (triadimefon) the product most widely used on wheat in Hungary since 1978, and with Calixin (tridemorph) registered on wheat since 1988, with the following application rates (AI (mg/l): 0.3; 1; 3; 10; 30; 100.

Sampling, isolation of the pathogen and further propagation

20-25 infected plants were sampled per field. Mildew samples were sub-cultured once on to healthy wheat plants at the 1-leaf growth stage, prior to sensitivity testing.

Fungicide sensitivity test

The "test tube method for assessment of propiconazole sensitivity in cereal powdery mildew isolates" by Sozzi *et al* (GIFAP, 1991) was selected from the methods recommended by FRAC for monitoring fungicide resistance. A culture maintained since 1986 on wheat cv. GK.Öthalom was used as the standard sensitive isolate.

RESULTS

Before analysing the results, the following should be noted:

Because of the long dry weather prevailing in Hungary in 1993, the level of *E. graminis* infection was "low" on 29%, "medium" on 58% and "high" on only 13% of the sampled area.

- *c*. 30% of the sampled area was treated with one application of fungicide to control powdery mildew.
- DMI fungicides were applied to 58% of the treated fields. Morpholines were not applied.
- 20% of the treated fields were cultivated as monoculture.
- The 41 samples studied were taken from 17 wheat varieties.

Table 1 shows the LC50 values (mg/l) and the resistance factors.

Table 1. Results of *Erysiphe graminis* f.sp. *tritici* resistance monitoring programme 1993.

ISOLATES		TRIADIMEFON		TRIDEMORPH	
No.	County of Origin	LC50	RF	LC50	RF
1.	Nógrád	0.2	0.1	0.6	0.5
2.		0.2	0.1	0.7	0.8
3.		0.6	0.4	0.7	0.8
4.		2.3	1.5	0.6	0.5
5.		1.7	1.1	1.2	1.0
6.		0.6	0.4	0.3	0.2
7.		0.7	0.5	0.9	0.7
8.		1.7	1.1	0.2	0.2
9.	Somogy	4.3	2.9	1.2	1.0
10.		0.5	0.3	0.8	0.7
11.		13.0	8.7	1.5	1.2
12.		8.0	5.3	1.5	1.2
13.		1.8	1.2	0.2	0.2
14.	Veszprém	2.6	1.7	0.3	0.2
15.		2.0	1.3	0.4	0.3
16.		3.3	2.2	0.5	0.4
17.		1.0	0.7	1.4	1.2
18.		8.0	5.3	1.6	1.3
19.		3.1	2.1	1.8	1.4

Table 1. Results of *Erysiphe graminis* f.sp. *tritici* resistance monitoring programme 1993 (continued).

ISOLATES		TRIADIMEFON		TRIDEMORPH	
No.	County of Origin	LC50	RF	LC50	RF
20.	Pest	1.5	1.0	1.5	1.2
21.		7.0	4.7	1.1	0.9
22.		1.9	1.3	1.4	1.2
23.		2.0	1.3	1.8	1.4
24.		3.0	2.0	1.7	1.3
25.	Bács-Kiskun	7.5	5.0	1.1	0.9
26.		17.0	11.3	1.6	1.3
27.		0.2	0.1	1.2	1.0
28.		1.7	1.1	1.1	0.9
29.		0.3	0.2	1.2	1.0
30.	Komárom-Esztergom	6.5	4.3	0.4	0.3
31.		1.5	1.0	0.6	0.5
32.		1.8	1.2	0.5	0.4
33.		0.4	0.3	0.3	0.2
34.		0.9	0.6	0.3	0.2
35.		2.8	1.9	2.0	1.7
36.	Szolnok Heves	1.6	1.1	1.3	1.1
37.		0.8	0.5	0.3	0.2
38.		0.6	0.4	0.9	0.7
39.		0.6	0.4	0.2	0.2
40.	Győr	1.1	0.7	1.6	1.3
41.		0.5	0.3	0.5	0.4
	STANDARD	1.5		1.2	

Tridemorph sensitivity

The LC50 values of most isolates were similar to that of the standard isolate (1.2mg/l). No isolate had an LC50 value greater than 2mg/l. The resistance factor did not exceed 2, for any of the isolates.

Triadimefon sensitivity

With triadimefon as the active ingredient, the values were more widely distributed. However 66% of the isolates had LC50 values below 2mg/l. (The LC50 value of the standard isolate was 1.5mg/l). 17% and 12% of the isolates had LC50 values in the 2-4mg/l and 4-8mg/l categories, respectively. The LC50 values of only 2 isolates exceeded 10mg/l. The value of the R-factor of only one isolate was higher than 10.

DISCUSSION

The main objective of this study was to provide data at the start of a long term programme for monitoring fungicide resistance in *E. graminis* in Hungary. The results from 1993 are published here. When evaluating these results, the dry weather and consequent reduction in number of fungicide treatments should be taken into account. However, the results provide valuable information at the start of the resistance monitoring programme.

REFERENCES

Brent, K.J.; Hollomon, D.W. (1988) Risk of resistance against sterol biosynthesis inhibitors in plant protection. in:Berg, D., Plempel, M. (Eds.) : Sterol inhibitors, Pharmaceutical and Agrochemical Aspects; VCH-Ellis Horwood Ltd., Chichester, 1988, 332-346.

Brent, K.J. (1992) Monitoring fungicide resistance : purposes, procedures and progress. in: Denholm, I.; Devonshire, A.L.; Hollomon, D.W. (Eds.) : Resistance '91 - Achievements and developments in combating pesticide resistance; Proc. SCI Symp. Elsevier Applied Science, London, 1-18.

Enisz, J. (1988) Variation in sensitivity of *Erysiphe graminis* f. sp. *tritici* to SBI fungicides in Western Hungary. Brighton Crop Protection Conference - Pests and Diseases, 4, 373-378.

Enisz, J. (1990) Növénykórokozók peszticid rezisztenciáha és annak hatása a növényvédelmi technolólgiáka (Pesticide resistance of plant pathogens and its effect on technologies. Dissertation for scientific degree, Veszprém).

Highwood, D.P. (1990) Fungicide Resistance Action Committee; Pesticide Outlook, 1/3.

Schulz, U.; Scheinpflug, H. (1986) Investigations on sensitivity and virulence dynamics of *Erysiphe graminis* f. sp. *tritici* with and without triadimenol treatment. Proc. Br. Crop Prot. Conf. - Pests and Diseases, 531-538.

Sozzi *et al*, GIFAP (1991) FRAC methods for monitoring fungicide resistance. EPPO Bulletin, 21, 291-354.

Szunics, L. (1988) A búzalisztharmat fiziológiai specializációja, virulenciája; rezisztencia nemesítés (Physiological specialization, virulence of wheat powdery mildew, breeding for resistance, Doctor Thesis, Martonvásár Research Institute of the Hungarian Academy of Sciences).

De Waard, M.A.; Kipp, E.M.C.; Horn, N.M.; van Nistrelrooy, J.G.M. (1986) Variation in sensitivity to fungicides which inhibit ergosterol biosynthesis in wheat powdery mildew. Netherlands Journal of Plant Pathology, 92, 21-32.

RESISTANCE OF GRAPE POWDERY MILDEW (*UNCINULA NECATOR*) TO TRIADIMENOL, A STEROL BIOSYNTHESIS INHIBITOR : BIOCHEMICAL CHARACTERISATION OF SENSITIVE AND RESISTANT STRAINS.

C. DÉLYE, M-F. CORIO-COSTET*

INRA, SRIV, Phytopharmacie, BP 81, 33883 Villenave d'Ornon Cédex, France.

ABSTRACT

Similar sterol compositions were found when mycelium and conidia from two sensitive and three naturally triadimenol-resistant strains of *Uncinula necator* were analyzed. Ergosta-5, 24 (24^1)-dien-3β-ol was the major sterol. Mycelium contained a higher quantity of this compound than conidia. Triadimenol treatment led to eburicol(4, 4, 14-trimethyl-ergosta-8, 24 (24^1)-dien-3β-ol) accumulation in the mycelium only, indicating that C14-demethylase would be the target of the fungicide. Moreover, triadimenol had no significant action on conidia of resistant strains, whereas those of sensitive strains were strongly affected (c. 50% quantitative reduction of all sterols).

INTRODUCTION

Grape powdery mildew, caused by *Uncinula necator* (Schw., Burr.), is a disease of major importance on grapes (*Vitis vinifera*, L.), causing loss of production and reduction of wine quality. A significant step towards better vineyard protection was made in the late 70s, when systemic compounds inhibiting sterol biosynthesis, such as the triazole triadimenol, began to replace sulphur. However, all the sterol biosynthesis inhibitors (SBIs) used against *Uncinula necator* inhibit a single target enzyme, the C14-demethylase (Buchenauer, 1977, Vanden Bossche, 1988). Development of field resistance to SBIs was reported in Portugal in 1988, and in France in 1989 (Steva & Clerjeau, 1990). The aim of this study was therefore to obtain data concerning the sterol biochemistry of *Uncinula necator* and to achieve a better understanding of its resistance to SBIs.

METHODS

The five monoconidial isolates studied were isolated from vineyards in 1992. Inhibition curves for triadimenol were established (Steva & Clerjeau, 1990), and fungal material was obtained from inoculated young vines. Plants were either untreated or sprayed at a concentration of triadimenol inhibiting 30% of growth (IC30) 12 h before inoculation, so that sufficient material could be obtained for sterol extraction. Conidia and mycelium were harvested separately from plants incubated for 14 days, by blowing and scraping, respectively. Freeze-dried fungal material was heated under reflux with methanolic KOH (6%) for 1 h. Neutral lipids were extracted with hexane, separated by tlc, acetylated prior to glc analysis and identified by gc-ms.

RESULTS

Sterol composition of conidia and mycelium of grape powdery mildew

It appeared that although the five strains investigated differed greatly in their sensitivity to triadimenol (table 1) their sterol composition was very similar, being dominated by methylene-24 cholesterol (ergosta-5, 24 (24^1)-dien-3β-ol, E) which comprises over 60% of the total sterols. The remainder was mainly made up of cholesterol (D), sitosterol (F), eburicol (B) and cycloartenol (C) (Table 2). On the basis of their molecular ion and fragmentation pattern, sterols were identified according to Loeffler *et al.* (1992), Debieu *et al.* (1992) and Rahier & Benveniste (1989).

TABLE 1. Sensitivity of *U. necator* strains to triadimenol.

Strain	Origin	IC30 (mg/l)	IC50 (mg/l)	IC100 (mg/l)
P18 (S*)	Perpignan	0.2	0.3	0.8
P17 (R)	Perpignan	5.0	9.0	14.0
M27 (S)	Madiran	0.3	0.4	0.8
M42 (R)	Madiran	1.6	2.0	4.7
M43 (R)	Madiran	10.0	14.0	>20

*: S: sensitive (IC100 < 1mg/l), R: resistant (IC100 > 1 mg/l)

In order to discriminate plant sterols from fungal sterols, healthy leaves were scraped and the material obtained was extracted in the same manner as the fungus (data not shown).

The sterol profiles of *Uncinula necator* conidia and mycelium did not vary greatly. Nevertheless, the level of methylene-24 cholesterol in the mycelium was found to be slightly higher in resistant strains. Considering the total sterol concentration, it appeared that mycelium contained more sterols than conidia (from 30 to 60% more). This was especially true for methylene-24 cholesterol and plant sterols (A, C, D, F and minor sterols) in all strains. However, the relative amount of eburicol was generally lower in mycelium.

<u>Effect of triadimenol on sterol content of conidia and mycelium</u>

In the conidia produced by the sensitive strains after triadimenol treatment, the amounts of total sterols and methylene-24-cholesterol were drastically reduced (c. 60% of control conidia, Table 2). Conversely, no effect was noted on conidia issued from resistant strains in which the quantities of total sterols and methylene-24 cholesterol remained unchanged. In both cases, although there was a slight increase of plant sterol levels (D, F and other sterols), no accumulation of 14α-methyl sterols such as eburicol was detected.

Considering the mycelium, triadimenol treatment led to an increase from 20 to 75% in total sterol content in all strains. Likewise, amounts of eburicol increased significantly in all strains except M27. A careful examination of the relative proportions of the different sterol categories showed a marked decrease (10 to 30%) in methylene-24 cholesterol. Concurrently, the level of 4,4-dimethyl sterols increased, especially in resistant strains. This increase was due to eburicol and cycloartenol in mycelium. As in the conidia, the plant 4-desmethyl sterol fraction was greater after treatment.

Moreover, in all 4-desmethyl sterol fractions of treated samples, the presence of 14α-methyl-ergosta-8-en-3β-ol was detected by gc-ms.

DISCUSSION

For the first time, we have compared sterol profiles of conidia to those of mycelium. This study provided further data concerning sterol biosynthesis and resistance to SBI fungicides in grape powdery mildew.

The major sterol of *Uncinula necator* was methylene-24 cholesterol, as found in other *Erysiphaceae* by Loeffler *et al.* (1992) and Debieu *et al.* (1994). We detected 4,4-dimethyl sterols such as lanosterol, 4,4-dimethyl fecosterol, episterol (< 0.5% total sterols) and eburicol: this indicates that *Uncinula necator* is able to synthesize its own sterols. However, as shown in Table 2, significant amounts of plant sterols such as 4,4-dimethyl-cyclopropyl sterols (A, C and most of minor sterols) and some 4-

TABLE 2: Effect of triadimenol (IC30) on sterol composition of conidia and mycelium of *U. necator* sensitive and resistant strains.

STEROLS	Percentage of total sterols (untreated(treated))											
	SENSITIVE STRAINS				RESISTANT STRAINS							
	P18		M27		P17		M42		M43			
	con.	myc.	con.	myc.	con.	myc.	con.	myc.	con.	myc.		
4,4 dimethyl sterols												
A	- * (2)	2(3)	-(-)	2(3)	-(1)	2(4)	-(1)	1(3)	-(-)	2(3)		
B	16(21)	12(17)	28(28)	10(9)	23(22)	20(20)	23(22)	7(17)	31(22)	8(19)		
C	22(8)	18(18)	4(12)	12(20)	5(8)	7(26)	1(3)	6(9)	3(7)	8(22)		
other sterols (1)	-(-)	5(6)	-(-)	5(5)	-(-)	7(6)	-(-)	8(5)	-(-)	3(5)		
4-desmethyl sterols												
D	1(7)	1(2)	-(1)	1(2)	-(3)	1(3)	-(2)	1(1)	-(1)	2(3)		
E	58(56)	56(41)	67(55)	58(48)	70(61)	61(31)	72(63)	70(59)	63(69)	69(43)		
F	1(4)	3(11)	-(-)	6(7)	-(2)	2(8)	-(2)	4(4)	1(1)	4(3)		
other sterols (2)	1(1)	1(2)	-(4)	4(3)	2(2)	-(2)	2(6)	2(-)	1(-)	2(-)		
4,4-dimethyl-	38(31)	37(44)	32(40)	29(39)	28(31)	36(56)	24(26)	22(34)	34(29)	21(49)		
4α methyl-	1(1)	2(-)	1(-)	2(1)	-(1)	-(-)	2(1)	1(2)	1(-)	2(2)		
4-desmethyl-sterols	61(68)	61(56)	67(60)	69(60)	72(68)	64(44)	74(73)	77(64)	65(71)	77(49)		
	Total sterol amount (µg/mg dry weight) untreated(treated)											
Total sterols	3.8(2.3)	6.7(11.4	6.4(3.1)	5.3(6.4)	3.8(3.7)	6.3(7.9)	3.1(3.3)	8.4(11.4	3.7(3.1)	5.2(9.2)		
methylene-24 cholesterol	2.2(1.3)	3.8(4.6)	4.3(1.7)	3.0(2.6)	2.6(2.3)	3.8(2.5)	2.2(2.1)	5.9(6.7)	2.3(2.2)	3.6(4.0)		
eburicol	0.6(0.5)	0.8(1.9)	1.8(0.9)	0.5(0.5)	0.9(0.8)	1.3(1.6)	0.7(0.7)	0.6(1.9)	1.1(0.7)	0.4(1.8)		

con.: conidia, myc.: mycelium. *: < 0.5% total sterols.

A: unidentified sterol (RRt=1.275, $[M]^+$= 480), **B**: eburicol (4,4,14α-trimethyl-ergosta-8,24(24^1)-dien-3β-ol; RRt=1.230), **C**: cyloartenol (4,4,14α-trimethyl-9β,19-cyclo-cholest-24-en-3β-ol; RRt=1.312), **D**: cholesterol (cholest-5-en-3β-ol; RRt=1.098), **E**: methylene-24 cholesterol (ergosta-5, 24 (24^1)-dien-3β-ol; RRt=1.185), **F**: sitosterol (stigmast-5-en-3β-ol; RRt=1.273).(**1**): Lanosterol (RRt=1.21), Unidentified sterol (RRt= 1.28, $[M]^+$ = 468); isomer of methylene-24-cycloartenol (RRt= 1.345, $[M]^+$= 482); methylene-24-cycloartenol (RRt= 1.359).(**2**): cholestan-3β-ol (RRt: 1.111); cholest-7-en-3β-ol (RRt= 1.141); 14α-methyl-ergost-8-en-3β-ol (RRt= 1.023), campesterol (RRt: 1.221), stigmasterol (RRt= 1.221), episterol (RRt= 1.241), stigmastan-3β-ol (RRt= 1.283) desmethyl sterols (D, F and minor sterols) appeared in sterol profiles of both mycelium and conidia. Their presence in mycelium may in part be due to scraping it from plants, as demonstrated by controls. Nevertheless, for conidia which are gently blown from leaves, a selective accumulation of plant sterols by the biotrophic fungus *Uncinula necator* might also be considered.

Indeed, by comparing only the vegetal 4-desmethyl sterol fraction between fungus (cholesterol, stigmasterol and sitosterol) and plant scrapings, we found a difference between their relative amounts, particularly for cholesterol (10 to 15% in plant scraping, 25 to 53% in conidia). This argues in favour of selective cholesterol uptake from the plant, or of potential cholesterol biosynthesis by *Uncinula necator*.

In the presence of triadimenol, the conidia from sensitive strains displayed a drastic decrease in final sterol without precursor accumulation despite the low inhibition level (IC30): this may be due to a breakdown of sterol biosynthesis. However, the accumulation of eburicol in the mycelium of all strains and the presence of 14α-methyl-ergosta-8-en-3β-ol confirms that the C14-demethylase is the target of triadimenol in *Uncinula necator* as shown in other fungi (Hollomon *et al.*, 1990; Pontzen *et al.*, 1990). Furthermore, mycelium of resistant as well as sensitive strains showed an increase in total sterol amounts partly due to the presence of plant sterols. In the conidia, plant sterol quantities were also increased. These facts favour the idea of a greater selective accumulation of plant sterols in response to the treatment.

Our data indicate that unusual sterol biosynthesis (due to C14-demethylase deficiency) does not account for resistance to triadimenol. However, other resistance mechanisms (Hollomon *et al.*, 1990) such as reduced fungicide uptake, detoxification or mutation of the target are still to be considered.

From our results, the only specific feature of resistance seems to be the stability of the amount of methylene-24 cholesterol present in the conidia. As triadimenol has a similar effect on the sterol content of mycelium of all strains, and as conidia are only an extension of mycelium, an hypothesis explaining this fact could be that in resistant strains the mycelium would "attenuate" inhibition by neutralizing triadimenol. Another possibility may be that, in sensitive strains, C14-demethylase inhibition would generate a compound repressing the sterol biosynthesis upstream from the C14-demethylation step. This may be assumed from data in the literature (Kerkenaar et *al.*, 1984; Favata *et al.*, 1987).

In order to complete this work, experiments with stronger inhibition (IC50, IC75) will be done, including additional sensitive and resistant strains from different origins. Furthermore, it would be instructive to analyze carefully the 4α-methyl sterol fraction and the squalene content of treated and untreated samples, and to look for potential 14α-methyl sterol derivatives of *Uncinula necator* after treatment.

ACKNOWLEDGEMENTS

We thank C. Malosse (INRA - Versailles) for his assistance in monitoring gc-ms.

REFERENCES

Buchenauer, H. (1977) Mode of action of triadimefon in *Ustilago avenae*. *Pesticide Biochemistry and Physiology*, **7**, 309-320.
Debieu, D.; Gall, C.; Gredt, M.; Bach, J.; Malosse, C.; Leroux, P. (1992) Ergosterol biosynthesis and its inhibition by fenpropimorph in *Fusarium* species. *Phytochemistry*, **31**, 1223-1233.
Debieu, D.; Corio-Costet, M-F.; Steva, H.; Malosse, C.; Leroux, P. (1994)

Sterol composition of the vine powdery mildew fungus: *Uncinula necator* sensitive or resistant strains to the sterol biosynthesis inhibitor triadimenol. *Phytochemistry* (submitted).

Favata, F.M.; Trzaskos, J.M.; Chen, H.W., Fisher, R.T.; Greenberg, R.S. (1987) Modulation of 3-hydroxy-3-methylglutaryl-coenzyme A reductase by azole antimycotics requires lanosterol demethylation, but not 24, 25 -epoxylanosterol formation. *Journal of biological chemistry,* **262**, 12254-12260

Hollomon, D.W.; Butters, J.A.; Hargreaves, J.A. (1990) Resistance to sterol biosynthesis inhibiting fungicides: current status and biochemical bases. *In: Managing resistance to agrochemicals*, M.B. Green, H.M. Lebaron and W.K. Moberg (Eds.), Washington DC, American Chemical Society, pp.199-214.

Kerkenaar, A.; Van Rossum, J.M.; Versluis, G.G.; Marsman, J.W. (1984) Effect of fenpropimorph and imazalil on sterol biosynthesis in *Penicillium italicum. Pesticide science*, **15**, 177-187.

Loeffler, R.S.T.; Butters, J.A.; Hollomon, D.W. (1992) The sterol composition of powdery mildews. *Phytochemistry*, **31**, 1561-1563.

Pontzen, R.; Poppe, B.;Berg, D. (1990) Mode of action of sterol biosynthesis inhibitors in obligate parasites. *Pesticide science*, **30**, 357-360.

Rahier, A.; Benveniste, P. (1989) Mass spectral identification of phytosterols. *In: Analysis of sterols and other biologically significant steroids*, W.D. Nes and E. Parish (Eds.), New-York: Academic Press, pp. 223-250.

Steva, H. & Clerjeau, M. (1990) Cross resistance to sterol biosynthesis inhibitor fungicides in strains of *Uncinula necator* isolated in France and Portugal. *Medelingen van de FaculteitLandbouwweten-schappen Rijksuniversiteit Gent*, **55** (3a), 983-988.

Vanden Bossche, H. (1988) Mode of action of pyridine, pyrimidine and azole antifungals. *In: Sterol biosynthesis inhibitors*, D. Berg and M. Plempel (Eds.), Chichester: Ellis Horwood, pp. 79-119.

REDUCED SENSITIVITY TO DMI FUNGICIDES IN POPULATIONS OF *MONILINIA FRUCTICOLA* IN NEW ZEALAND STONE-FRUIT ORCHARDS.

P.A.G. ELMER[1], M. BRAITHWAITE[2], D.J. SAVILLE[3]

[1]HortResearch, P.O. Box 51, Lincoln, Canterbury, New Zealand.

[2]Plant Protection Centre, MAF Quality Management, P.O. Box 24, Lincoln, Canterbury, New Zealand.

[3]AgResearch, P.O. Box 60, Lincoln, Canterbury, New Zealand.

ABSTRACT

Significant reductions in sensitivity to triforine (a demethylation inhibitor, DMI) were detected in *Monilinia fructicola* populations from seven of nine Hawkes Bay orchards in 1992. The mean LC50 of a wild population from unsprayed home garden sites was 3.3 mg/litre triforine (range for individual isolates, 0.5-10.8 mg/litre). In contrast, the mean LC50 for the orchard populations ranged from 4.5 to 7.9 mg/litre triforine (range for individual isolates, 0.4-28.1 mg/litre). Isolates with reduced triforine sensitivity were also significantly less sensitive to three other DMI fungicides. We discuss these findings in relation to the need to reduce DMI selection pressure on *M. fructicola* populations.

INTRODUCTION

Repeated applications of fungicide are required in New Zealand orchards to control brown rot of stone-fruit caused by *Monilinia fructicola* (Wint.) Honey. The occurrence of DMI resistance in several host/pathogen systems (Koller and Scheinpflug, 1987) and the relative ease of selection of DMI resistant strains of *M. fructicola* in laboratory studies (Nuninger-Ney *et al.*, 1989) indicated that the DMI's may be at risk of resistance development.

Current recommendations in New Zealand reflect the need to restrict DMI selection pressure on *M. fructicola* populations and specify no more than three DMI's in any one season with non-DMI's at all other times (Prince *et al*, 1989). In 1990, three orchards in the Hawkes Bay region reported inadequate disease control following DMI applications. In a preliminary investigation significant shifts in triforine sensitivity were detected at two of the three orchards sampled (Elmer *et al*, 1992). In 1992, a greater number of orchards were surveyed to determine the extent of changes in triforine sensitivity in *M. fructicola* populations and to determine the sensitivity of selected isolates to three other DMI fungicides registered for use in New Zealand.

METHODS

Up to 50 fruit infected with *M. fructicola* were sampled randomly from each of nine orchards using a stratified sampling pattern (Delp *et al.*, 1986). A wild population of *M. fructicola* was obtained by sampling unsprayed home garden sites throughout New Zealand.

Triforine sensitivity tests

The sensitivity of each isolate was expressed as an LC50 value calculated from a dose response curve of growth rate (mm/day) to $\log_{10}$ triforine concentration (mg/litre). The mean LC50 of each orchard population was compared with the mean LC50 of a home garden population using analysis of variance (ANOVA).

Other DMI sensitivity tests

Three isolates with reduced sensitivity to triforine (mean LC50 =9.3mg/litre) and five isolates classed as triforine sensitive (mean LC50 = 4.4mg/litre) were used to determine cross-resistance to bitertanol, cyproconazole and flusilazol. For each DMI, the mean LC50 of the sensitive isolates was compared to the mean LC50 of isolates with reduced triforine sensitivity using ANOVA.

RESULTS AND DISCUSSION

The triforine sensitivity of *M. fructicola* populations from seven of nine Hawkes Bay orchards was significantly (P<0.01) less than that of the home garden population (Table 1).

TABLE 1. The triforine sensitivity of *M. fructicola* populations from three representative Hawkes Bay orchards in 1992.

Source	Number of isolates	Mean log LC50	Range
Home gardens	27	0.519 (3.3)[a]	(0.5 - 10.8)[a]
Orchard 2	26	0.658 (4.5)	(0.6 - 9.8)
Orchard 8	22	0.860 (7.2)**	(3.1 - 25.3)
Orchard 9	31	0.900 (7.9)**	(2.4 - 15.6)
LSD (5%)		0.143	
LSD (1%)		0.188	

[a] Backtransformed values are presented in brackets.
**, significantly different from the home garden population at P<0.01.

The magnitude of the shift in sensitivity was small (resistance factors based on population mean values ranged from 2.2 - 2.4) and it may be debatable whether they should be called resistant sub-populations. Therefore, we have used the term "significantly reduced sensitivity".

Isolates with reduced sensitivity to triforine were also significantly ($P<0.01$) less sensitive to bitertanol, cyproconazole and flusilazol, compared to the standard home garden isolates.

Significant shifts in sensitivity to triforine and other DMI's indicates that there is a need to reduce DMI selection pressure. At one orchard, 11 DMI fungicides were applied in one season, which exceeds current recommendations.

The emergence of populations of *M. fructicola* with reduced sensitivity to DMI's does not necessarily mean that disease control failures will occur. Research to investigate the biological characteristics of isolates with reduced sensitivity to DMI's and the relationship between reduced sensitivity and disease control on host tissues *in vivo* is in progress.

ACKNOWLEDGEMENTS

The authors thank AGCARM for funding this project and the field representatives from Bayer NZ Ltd. for sample collection. We thank Dr Greg Tate for his assistance with orchard selection and Rima Herber, Stojan Ganev and Kirsty Boyd-Wilson for technical assistance.

REFERENCES

Delp, B.R.; Stowell, L.J.; Marois, J.J. (1986) Evaluation of field sampling techniques for estimation of disease incidence. *Phytopathology* **76**, 1299-1305.

Elmer, P.A.G.; Braithwaite, M.; Saville, D.J. (1992) Changes in triforine sensitivity in populations of *Monilinia fructicola* from Hawkes Bay orchards. *Proceedings of the 45th New Zealand Plant Protection Conference.* **45**, 138-140.

Koller, W.; Scheinpflug, H. (1987) Fungal resistance to sterol biosynthesis inhibitors: A new challenge. *Plant Disease* **71**, 1066-1074.

Nuninger-Ney, C.; Schwinn F.J.; Staub, T. (1989) *In vitro* selection of sterol-biosynthesis inhibitor (SBI) resistant mutants in *Monilinia fructicola* (Wint.) Honey. *Netherlands Journal of Plant Pathology* **95**, 137-150.

Prince, R.W.; Moore, M.S.; Wearing, C.H. (1989) The New Zealand Committee on Pesticide Resistance 1989 Summary. *Proceedings 42nd New Zealand Weed and Pest Control Conference:* 278-292.

The magnitude of the shift in sensitivity was small [illegible] population means ranging from 2.2 [illegible] 2.4 [illegible] called [illegible] sub-populations; therefore we have used [illegible] "shift" [illegible] reduced sensitivity.

[illegible]

[illegible] DMI [illegible]

[illegible]

ACKNOWLEDGEMENTS

[illegible]

REFERENCES

[illegible]

[illegible] New Zealand Plant Protection Conference [illegible]

[illegible] Plant Disease [illegible]

[illegible]

[illegible]

SENSITIVITY OF *UNCINULA NECATOR* TO PENCONAZOLE IN EUROPEAN COUNTRIES

C. STEDEN, B. FORSTER

CIBA-GEIGY LIMITED, Plant Protection Division, CH-4002 Basle, Switzerland

H. STEVA

BIORIZON S.A., Bordeaux Montesqieu, F-33 651 Martillac Cedex, France

ABSTRACT

Demethylation inhibiting fungicides (DMIs) play an important role in the control of *Uncinula necator*, the causal fungus of grape powdery mildew.

A survey was initiated in 1992 to investigate the sensitivity of *U. necator* to penconazole in different European countries. In one part of the study, samples were collected from ornamental vines, which had never been treated with DMIs, in order to establish a "baseline"-sensitivity and to check the prevelance of possible shifts. In the other the sensitivity of samples collected in vineyards showing both good and weak performance of the product was tested, to examine the contribution of changes in sensitivity to disease control problems.

The "baseline"-sensitivity of *U. necator* to penconazole showed only a small variation within and between different European countries. Sensitivity data for samples from treated vineyards support the assumption that many factors are responsible for the variability in the disease control level and that a decrease in sensitivity could be one of them. It was shown that aspects of the safety margin for activity become more important compared to previous years.

INTRODUCTION

Grape powdery mildew, caused by the fungus *Uncinula necator,* is one of the most serious diseases in viticulture. Traditional control measures consisted of treatments with sulfur, which was applied either as dust or as a wettable powder. Since the introduction of the DMI-fungicides (demethylation inhibitors) at the beginning of the 1980's, these compounds have been widely and intensively used in all vine-growing areas for the effective control of this disease.

DMIs are site-specific fungicides and as such changes in sensitivity of a pathogen cannot be excluded. After many years of successful use of DMIs, several authors observed a decrease in sensitivity of grape powdery mildew (Aloi *et al.*, 1991, anonymous, 1987, Steva *et al.*, 1988). All reports mention several factors which were closely related to the reduced efficacy against the pathogen. Performance problems occurred under high disease pressure, on susceptible cultivars and with the exclusive and season-long use of a DMI over a number of years. Other important aspects are related to the application: incorrect doses of the fungicide, inadequate timing and poor spray quality.

The two main objectives of the study started in 1992 were to attempt to establish a "baseline"-sensitivity for *U. necator* to penconazole, in order to detect possible shifts in areas where the compound has been used for many years, and to obtain sensitivity data from vineyards where both good and reduced efficacy of the product had occurred, so that the role of sensitivity shifts in grape powdery mildew control could be investigated (active monitoring).

MATERIALS AND METHODS

Sampling

Samples of mildew infected leaves and bunches were collected during the 1993 season in the following European countries: France, Germany, Italy, Portugal and Switzerland. Each sample consisted of 20 - 30 leaves and bunches. All samples for the active monitoring originated from trial plots or commercially treated vineyards. In 1992, one sample was collected at a trial site in Portugal where penconazole showed unexpectedly weak performance. "Baseline"- samples were taken either from ornamental vines that had never been treated, or from vineyards where biological control had been used, in various regions of the respective countries. Table 1 describes the origin of the samples.

TABLE 1. Description of the monitoring samples of *U. necator* collected in different European countries.

country	**baseline**		**active-monitoring**	
	variety	sampling period	*variety*	sampling period
France	*Cabernet S., Chasselat, Carignan, Merlot*	25.07.-03.08.93	*Carignan*	03.08.-04.08.93
Portugal	*Loureiro*	20.07.93	*Carignan, Fernandinho, Carignan*	02.08.92 21.07.93
Germany	*Riesling, Portugieser, M. Thurgau, Gutedel*	17.07.-02.08.93	*Trollinger, Kerner, Juwel*	18.07.-27.07.93
Italy	-	-	*Moscato, Cortese*	13.07.93
Switzerland	?	11.08.-22.08.93	*Gamay, Chasselas, Pinot noir, Silvaner*	26.07.93

Sensitivity tests

Sensitivity tests for all samples from France and Portugal were carried out by BIORIZON with a mycelium growth test on leafdiscs of the susceptible cultivar "*Cinsaut*". This method was originally established for triadimenol (Steva, 1992) and for the present studies it was adapted for penconazole. Based on the evaluation of the filament length after 72 hours at a range of fungicide concentrations (0-0.01-0.03-0.1-0.3-1-3-10 mg/l AI), this procedure gives a quantitative description of the sensitivity distribution of conidia within a given sample and an estimation of the MIC-value (minimum inhibitory concentration) can be made.

Tests for the samples from the other countries were carried out by CIBA. The sensitivity of *U. necator* was determined in a sporulation test on leaf discs of grapevine variety "*Portugieser*". This test was also adapted for penconazole from the method described by Steva (1992). Dose-response curves, based on the percentage of sporulating leaf area, were established for each sample at the same range of concentrations as mentioned above.The MIC was determined 12 days after inoculation. All samples were tested either directly or after one propagation. Tests at CIBA were repeated.

RESULTS

Baseline studies

Table 2 summarizes the results of the baseline studies. Sensitivity data are expressed as MIC- values of the tested populations.

TABLE 2. Sensitivity of *U. necator* to penconazole for samples collected from ornamental vines and vineyards where biological control had been used in Europe, 1993.

country	number of samples	range of MIC-values[1)] (mg/l)
France	5	0.1-1
Portugal	6	0.1-1
Germany	7	0.3-3
Switzerland	3	1-3

1) MIC for France and Portugal estimated by BIORIZON on *cv. Cinsaut* in a mycelium growth test and for Germany and Switzerland determined by CIBA on *cv. Portugieser* in a sporulation test.

Results both from CIBA and BIORIZON indicate only little variation in the sensitivity of powdery mildew samples collected within and across the European countries (range= x3). The tests revealed that wild-type populations of *U. necator* are generally controlled at 1 mg/l penconazole under the test conditions used by BIORIZON and at 3 mg/l penconazole in the sporulation test carried out by CIBA. Small differences in the amount of AI needed can be explained by the two varieties used as host plant tissue in the different test systems (*cv. Cinsaut* and *cv. Portugieser*).

Active monitoring

Samples from penconazole treated vineyards show a reduction in sensitivity compared to the baseline (Table 3). Based on the EC50-values evaluated for the populations collected in France and Portugal, a mean resistance factor of 15 was noted (data not shown). The MIC-values of DMI-treated samples were generally detected between 1 and 10 mg/l. Data for the disease levels shown in table 3 indicate that there was no correlation between the sensitivity results and product performance.

TABLE 3. Sensitivity of *U. necator* to penconazole for samples from DMI treated vineyards in Europe, 1993 (values in mg/l AI).

country	no. of samples	no. of DMI treatments[1)]	% attack on bunches	MIC-value[2)]
France	4	6-8 (trial)	0.1-1	1-3
	1	8(trial)	1	0.3-1
Portugal	1	2	1	0.3-1
	1	2	40	3-10
	1	3	25	1-3
	2	5	<1	0.3-1; 1-3
	1[4)]	7 (trial)	7	1-3
	1(1992)[4)]	7 (trial)[3)]	27	1-3
Germany	1	1	?	1-3
	1	2	1	3-10
	1	4[3)]	25	3-10
	2	5 (trial)[3)]	15-20	3-10
Italy	2	4	2-5	3-10
Switzerland	1	1	30	1-3
	1	2	1	1-3
	1	2	50	3-10

1) use rates according to local registration. 2) MIC for France and Portugal estimated by BIORIZON on *cv. Cinsaut* in a mycelium growth test and for Germany, Italy and Switzerland determined by CIBA on *cv. Portugieser* in a sporulation test. 3) samples collected under high disease pressure and with curative treatments. 4) same trial location as in 1993.

Fig. 1 demonstrates the sensitivity distribution of a "baseline" population in comparison to a sample collected from a trial plot under the selection pressure of penconazole. Samples were analyzed with the mycelium growth test described by Steva (1992). Although both samples were collected in the same region, a quantitative shift in sensitivity becomes obvious. The sensitivity of the conidia from the baseline population is comprised between 0.01-1 mg/l penconazole, whereas the population from the treated vineyard had a range between <0.01-3 mg/l.

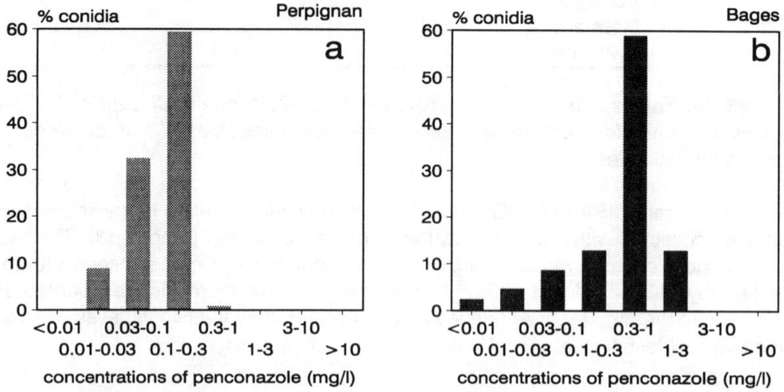

FIGURE 1. Sensitivity of *U. necator* to penconazole for populations collected in one region in France (1993) from a) a vine not previously treated ("baseline"), b) a trial plot treated 8 times with penconazole (0.2% attack on bunches).

The sensitivity distribution for samples collected in a specific trial plot in Portugal is shown in Fig. 2. In 1992 the level of attack on bunches reached 27% under a situation of high disease pressure, whereas performance was good in 1993 (7% attack), when the pressure was lower and the epidemic started later than the year before. It can be seen, that conidia in both years could be controlled with 3 mg/l penconazole in the mycelium growth test, and distributions did not differ significantly.

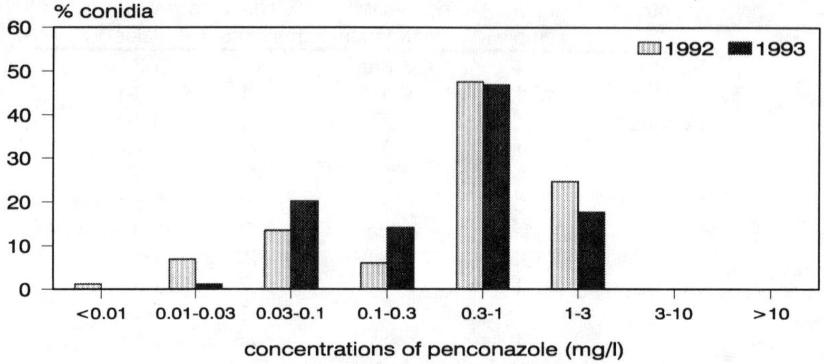

FIGURE 2. Sensitivity of *U. necator* to penconazole for 2 samples collected at the same trial site in Portugal, in a year with high (1992) and low (1993) disease pressure.

DISCUSSION

The fact that penconazole has been used for the control of *U. necator* for about a decade made it particularly difficult to find samples representing the wild-type sensitivity of the fungus to this compound. For a product already on the market this might not be a true "baseline" but it still provides useful information and should be established. Only the knowledge of the baseline sensitivity allows one to decide whether subsequently reported shifts in the pathogen sensitivity are true effects or are merely within limits of the normal variability of a population.

The "baseline"-monitoring for penconazole revealed relatively little variation in the sensitivity of *U. necator* between and within European countries. This is similar to earlier findings of Steva (1992) for triadimenol sensitivity in France and Portugal. These results now provide a basis to follow the grape powdery mildew sensitivity in future years, and to draw conclusions on any apparent significant shifts towards a decrease in sensitivity.

In this study, samples collected in vineyards where penconazole had been applied generally showed a decrease in sensitivity when compared to the "baseline". The sensitivity level of those samples collected under the selection pressure of the fungicide was reduced by a factor of 10-15. However, the number of samples tested, their origin with regard to the number of treatments and the use of the DMIs might be insufficient to draw a general conclusion about the sensitivity situation in penconazole-treated areas across Europe.

Another aspect of the influence of DMI treatments on the sensitivity of the pathogen can be seen from the data presented in table 3. Based on the sporulation test the conidia isolated from vineyards with only 1 DMI- treatment tended to show only slightly reduced sensitivity compared to those from trial plots with multiple DMI applications. This indicates an influence of the number of treatments on the sensitivity level and suggests that a limit on the number of DMI applications could help to prevent further shifts towards decreased sensitivity. However, this result should be confirmed in further trials.

As a general rule, no correlation could be shown between the sensitivity values and the level of disease control in the various vineyards. However, the two years of experience at a specific trial-site in Portugal (Fig. 2), where both the sensitivity and efficacy of penconazole were assessed, suggest that under situations of high disease pressure (1992) the safety margin for activity of the compound is no longer sufficient to provide good control at this level of reduced sensitivity, although in other cases excellent powdery mildew control was achieved.

Similar findings were published for triadimenol by Steva (1992), although he showed that the decrease in the sensitivity level for this compound was more pronounced than it could be shown for penconazole in this study. This is also concluded by Steva and Clerjeau (1990), who carried out experiments with the progeny of four single spore isolates. They grouped the DMIs into 3 classes with different sensitivity levels of *U. necator.* The isolates showed no reduced sensitivity to penconazole in comparison to triadimefon and triadimenol, where resistance factors of up to ~100 could be detected.

The present investigation demonstrates that the sensitivity of *U. necator* can decrease under selection pressure from penconazole. Therefore, it is of vital importance to strictly follow the use recommendations implemented by the Fungicide Resistance Action Committee - Sterol Biosynthesis Inhibitor working group (FRAC-SBI), in order to preserve the DMIs as an effective tool for vine growers to protect vineyards from powdery mildew.

These use recommendations are as follows (FRAC, 1993): DMIs should not be used exclusively season-long. The number of DMI sprays should be limited to a maximum of four per season, just before and after the flowering stages. It is important to use the DMIs only at the fully recommended rate in a protective, not curative manner. The use of mixtures or alternation with non-cross-resistant fungicides also might be a possible strategy. The recommended timing and volume of application must be adhered to.

REFERENCES

Aloi, M.; Gullino, M.L.; Garibaldi, A. (1991) Reduced sensitivity to fenarimol and triadimefon in field populations of *Uncinula necatrix. Pesticide Science,* **31** (1), 114-116.

anonymous, (1987) Update: Powdery mildew for '87. *California Grape Grower*, **3**, 4-9.

FRAC, (1993) Combattre les résistances aux fongicides inhibiteurs de la biosynthèse des stérols. *Phytoma*, **447**, 12-18.

Steva, H. (1992) Résistance de l'oidium de la vigne (*Uncinula necator* (Schw.)Burr.) aux fongicides inhibiteurs de la biosynthèse des stérols. *Ph.D. Thesis University of Bordeaux*, 264 pp.

Steva, H.; Cartolaro, P.; Clerjeau, M; Lafon, R.; Gomes da Silva, M.T. (1988) Une résistance de l'oidium au Portugal ? *Phytoma*, **402**, 49-50.

Steva, H.; Clerjeau, M. (1990) Cross resistance to sterol biosynthesis inhibitor fungicides in strains of *Uncinula necator* isolated in France and Portugal. *Med. Fac. Landbouww. Rijksuniv. Gent* ,**55** (3a), 983-988.

TESTING TOMATO POWDERY MILDEW (*ERYSIPHE* SP.) FOR FUNGICIDE RESISTANCE

T. M. O'NEILL

ADAS Horticulture, Brooklands Avenue, Cambridge CB2 2BL

ABSTRACT

The sensitivity of mass population samples of tomato powdery mildew (*Erysiphe* sp.) collected from commercial crops to three fungicides was determined using detached leaves and young plants. The sensitivity of one population to fenarimol seemed different to that of a reference sensitive isolate. Possible reasons for some samples failing to establish on detached leaves were investigated. A sample inoculated on to young plants established well and bupirimate and fenarimol gave control at one tenth of the label recommended rates. Further tests are required to determine if poor disease control in fenarimol-treated crops is associated with the presence of mildew showing reduced sensitivity to fenarimol.

INTRODUCTION

Powdery mildew ((*Erysiphe* sp.) was first recorded on protected tomatoes in England in 1987 (Fletcher *et al.*, 1988) and since then has affected many crops annually throughout the country. It affects leaf, stem and calyx and causes severe leaf yellowing, reduced yield and premature plant death. Fungicides applied for mildew control include the benzimidazole fungicide benomyl (Benlate), the hydroxy-pyrimidine fungicide bupirimate (Nimrod), and the sterol biosynthesis inhibitor (SBI) fenarimol (Rubigan). Resistance to all three fungicide groups has been reported in cucumber powdery mildew (*Sphaerotheca fuliginea*) (Schroeder & Provvidenti, 1969; Bent *et al.*, 1971; Schepers, 1983). Fenarimol is commonly used on tomato crops for control of mildew because, unlike benomyl and bupirimate, it does not disrupt biological pest control. However, some growers have reported poor disease control following treatment with fenarimol and have attributed this to fungicide-resistant strains of the pathogen. Work was therefore undertaken to develop techniques which could be used to determine the sensitivity of tomato powdery mildew to fenarimol and other fungicides.

MATERIALS AND METHODS

Samples of leaves affected by mildew were obtained from commercial tomato nurseries and from ADAS Reading (92/1). Sample 92/1 had not been exposed to any fungicide for at least two years. Samples 92/PV and 93/LV were both obtained as mass populations samples from fenarimol-treated crops. Various methods were used to encourage spore production from old pustules. In 1992, pustules were brushed to remove old spores and the leaves incubated in a humid chamber. In 1993, pustules were sub-cultured on untreated detached leaves.

Resistance tests were carried out using detached leaves, cvs. Gardener's Delight or Pronto, or young plants, cv. Counter. Detached leaves were dipped in fungicide, allowed to dry and inoculated with mildew spores of the test sample using a cotton bud. Pot-grown plants were sprayed to run-off with the test fungicide or water, allowed to dry and inoculated by placing small leaf discs infected with mildew in contact with treated leaves. Fungicides were tested at label recommended rates (benomyl, 500 mg/l; bupirimate 500 mg/l; fenarimol 10.8 mg/l) and at three dilutions (1 in 2, 1 in 10 and 1 in 20). Inoculated detached leaves (5 leaves/treatment; 2 inoculation sites/leaf) were incubated over water in an illuminated incubator at 23°C. Inoculated plants (3 plants/treatment; 3 inoculated leaves/plant) were covered with a clear polythene bag for 48 h before standing them in an unheated glasshouse. Leaves were assessed for the number of inoculation sites at which mildew had developed and for abundance of mildew development (0-5 scale according to proportion of inoculated area affected). Factors which might affect development of mildew were investigated including cultivar, leaf age and a comparison of abaxial and adaxial leaf surfaces.

RESULTS

Mildew developed consistently on detached leaves when spores were transferred from young, freely-sporing pustules but not when older, discoloured pustules were used. Results for sample 92/1, tested in May 1992 on leaves dipped in bupirimate or fenarimol, and sample 93/LV, tested in April 1993 on leaves dipped in fenarimol, are shown in Table 1. Sample 92/1 was also tested against benomyl and no mildew developed at 25 mg/l, the lowest concentration tested. Sample 93/LV was relatively resistant to fenarimol and developed after 12 days on most leaves treated at one tenth (1.1 mg/l) of the recommended rate.

TABLE 1. Establishment of mildew after 6 and 12 days on detached tomato leaves treated with water or fungicide.

Treatment	Rate (µg/ml)	No sites with mildew (0-10) 92/1		No sites with mildew (0-10) 93/LV		Mean disease score (0-5) 92/1		Mean disease score (0-5) 93/LV	
		6d	12d	6d	12d	6d	12d	6d	12d
Water		10	10	10	10	4.6	5.0	4.7	5.0
Uninoculated		0	0	0	0	0	0	0	0
Bupirimate	500	1	1	-	-	0.2	0.3	-	-
	250	2	3	-	-	0.3	0.6	-	-
	50	4	10	-	-	0.4	1.8	-	-
	25	5	10	-	-	0.6	2.4	-	-
Fenarimol	10.8	1	1	0	0	0.1	0.1	0	0
	5.4	0	1	0	0	0	0.2	0	0
	1.1	0	1	6	8	0	0.1	0.8	0.8
	0.5	7	9	10	10	1.9	2.5	3.7	3.7

The cultivar, age and surface of detached leaves were investigated as possible reasons for poor establishment. Mildew established more consistently on leaves of cv. Gardener's Delight (39/40 inoculation sites) than cv. Pronto (28/40 sites). The success of establishment was not affected by leaf surface (22/40 on abaxial; 23/40 on adaxial). There was slightly better establishment on young leaves of cv. Pronto (9/40) than on old leaves (3/40) of the same variety. Incubation of leaves in a damp chamber was unsuccessful in producing new pustules as leaves were rapidly rotted by grey mould (*Botrytis cinerea*) or bacterial soft rot. Transfer of spores from old pustules to untreated detached leaves was generally unsuccessful in producing new pustules.

Four weeks after inoculation of young plants with sample 92/PV, all but one of the water-treated leaves was infected with mildew. At this time both bupirimate and fenarimol were giving good control of the disease at one tenth of normal spray rates, but not at one twentieth. (Table 2).

TABLE 2. Establishment of mildew (92/PV) on young tomato plants treated with bupirimate or fenarimol.

Treatment	Rate (mg/l)	Number of inoculation sites (of 9) developing mildew		
		Weeks after inoculation		
		3	4	5
Water		6	8	8
Uninoculated		0	0	0
Bupirimate	500	0	0	3
	250	0	0	2
	50	0	0	3
	25	0	2	6
Fenarimol	10.8	0	0	0
	5.4	0	0	2
	1.1	0	0	4
	0.5	0	4	5

DISCUSSION

Development of a rapid, accurate and reproducible method for evaluating the response of tomato powdery mildew to fungicides is important because it would help to enable appropriate treatment to be selected. A detached leaf, or leaf disc technique has been used successfully for investigating *Sphaerotheca fuliginea* on cucumber (Schepers, 1983). The technique reported here for tomato mildew was relatively rapid (2-4 weeks) but was poorly reproducible. It is possible that a water film on leaves or a high light intensity (Cohen, 1993) may have affected establishment of mildew in these tests. Further work is required to define more precisely the conditions under which mildew spores will germinate and develop to produce sporing pustules on detached leaves.

When resistance to SBI fungicides occurs, it generally develops in small steps making it difficult to determine when poor control is due to fungicide resistance rather than to other factors. Reduced disease control of *S. fuliginea* by fenarimol was associated with an increase in the LD_{95} from 0.2 to 17 mg/l (Huggenberger *et al.*, 1984). In the tests described the sensitivity of the population 93/LV seems different to that of the reference sensitive isolate 92/1. However, further samples of mildew, collected from crops where different degrees of control have been observed, need to be tested to determine the levels of sensitivity that occur and to improve interpretation of tests on detached leaves. An examination of the range of sensitivities in populations using single pustule samples may be more revealing than studies on mass populations. The sensitivity of tomato powdery mildew to benomyl and bupirimate was similar to that reported by Fletcher *et al.*, (1988), who observed no mildew on tomato plants treated with benomyl at concentrations as low as 100 mg/l, and a small amount of mildew at all concentrations of bupirimate up to 370 μg/ml.

The speed of mildew development was slower on young plants (3 - 4 weeks) than on detached leaves (6 days), probably because the former tests were done at ambient temperature. It is suggested that the critical time for assessment in sensitivity tests is as soon as mildew is present at most inoculation sites on untreated leaves or plants.

ACKNOWLEDGEMENTS

Funding of this work by the Horticultural Development Council is gratefully acknowledged.

REFERENCES

Bent, K.J., Cote, A.M., Turner, J.A.W. & Woolner. (1971). Resistance of cucumber powdery mildew to dimethirimol. *Proceedings of the 6th British Insecticide and Fungicide Conference*, 274-282.

Cohen, R. (1993). A leaf disc assay for detection of resistance of melons to *Sphaerotheca fuliginea* Race 1. *Plant Disease* 77, 513-517.

Fletcher, J.T.; Smewin, B.J.; Cook, R.T.A. (1988). Tomato powdery mildew. *Plant Pathology* **37**, 594 - 598.

Huggenberger, F.; Collins, M.A.; Skylakakis, G. (1984). Decreased sensitivity of *Sphaerotheca fuliginea* to fenarimol and other ergosterol-blosynthesis inhibitors. *Crop Protection* **3,** 137 - 149.

Schepers, H.T.A.M. (1983). Decreased sensitivity of *Sphaerotheca fuliginea* to fungicides which inhibit ergosterol biosynthesis. *Netherlands Journal of Plant Pathology* **89**, 185 - 189.

Schroeder, W.T. & Provvidenti, R (1989). Resistance to benomyl in powdery mildew of cucurbits. *Plant Disease Reporter* **53**, 271-275.

VARIATION IN SENSITIVITY TO FUNGICIDES AMONG UK ISOLATES OF *VENTURIA INAEQUALIS*

A.L. ROBERTS, I.R. CRUTE[1]

Horticulture Research International, East Malling, West Malling, Kent ME19 6BJ

ABSTRACT

Dodine and myclobutanil (a triazole fungicide) are commonly used in the UK for the control of *Venturia inaequalis*, causal agent of apple scab. A total of 35 samples of *V. inaequalis* from the UK, from both fungicide treated and untreated areas, (13 single spore isolates and 22 mass spore samples) has been used in a continuing investigation of variation for sensitivity to these fungicides. The *in vitro* LD50 values for response to dodine varied from 0.85 - 2.10 mg/l. In an *in vitro* assay based on germling growth the majority of samples were as sensitive to myclobutanil as a standard isolate obtained in 1949. However samples from a site where scab control was considered to be inadequate were found to have a response to myclobutanil equivalent to a reference isolate exhibiting reduced sensitivity.

INTRODUCTION

The high level of apple scab (*Venturia inaequalis*) control demanded of UK growers is only achievable by frequent fungicide treatment. Up to fifteen applications per season of a range of fungicides is not unusual. Dodine and triazole fungicides (eg. myclobutanil) are among those most frequently used. Variation for sensitivity to dodine and triazole fungicides, associated with loss of efficacy, has been reported in North America (Sholberg *et al.*, 1989; Szkolnik & Gilpatrick, 1973) and several European countries (Fiaccadori *et al.*, 1987; Stanis & Jones, 1985; Thind *et al.*, 1986). The objective of this study is to ascertain for the first time the extent of variation for response to dodine and myclobutanil among isolates of *V. inaequalis* from UK orchards.

MATERIALS AND METHODS

Dodine

Dodine sensitivity was assessed *in vitro* on potato dextrose agar amended with 0, 0.5, 1.0 or 2.0 mg dodine/l. Percentage germination was assessed for a population of >100 spores after 24 h (Sholberg *et al.*, 1989).

[1] Present address: Horticulture Research International, Wellesbourne, Warwick CV35 9EF.

Triazoles

a) *In vitro*

Sensitivity to myclobutanil was assessed by measuring the mean length of hyphal growth from 20 germlings after incubation for 60-65 h on water agar amended with 0.05 and 0.5 mg myclobutanil/l. Data are expressed as percentage reduction in mean hyphal length compared to growth on unamended agar.

b) *In vivo*

Apple seedlings (at 3-5 leaf stage) were inoculated with conidia of *V. inaequalis* 24 h after the plants had been treated with myclobutanil (0, 10, 20 or 40 mg/l; *c.* 0.4 ml/seedling). Numbers of seedlings with sporulating lesions were recorded 14 d after inoculation.

RESULTS AND CONCLUSIONS

The results of *in vitro* assays are presented in Tables 1 and 2 and of *in vivo* assays in Table 3. Reference isolates sensitive or with reduced sensitivity to dodine were not available for inclusion in these assays. Published reports indicate that a *Venturia inaequalis* population with an LD50 >0.7 mg/l for response to dodine can result in reduced field efficacy (McKay & MacNeill, 1979). The LD50 value for sensitivity to dodine of all isolates and mass-spore samples of *V. inaequalis* tested in this study was >0.7 mg/l and a factor of approximately x3 separated the most and least sensitive samples (Table 1). Four of five single spore isolates were from trees that had not been treated with dodine and one of these had a high LD50.

A conidial germling growth assay allowed reproducible assessment of variation in sensitivity to myclobutanil (as a representative of triazole fungicides). Variation in sensitivity to myclobutanil was observed among isolates and mass-spore samples of UK origin (Table 2). The least sensitive samples responded similarly to a reference reduced sensitive isolate obtained from a German orchard (Stanis & Jones, 1985) where the control achieved with triazole fungicides was reported to be inadequate. All orchard samples were from trees in orchards where triazole fungicides comprised part of a routine spray schedule. The assay results for the orchard samples indicated that those from three of the four could be classified as sensitive or intermediate but those from one orchard (site 4, Table 2) tended towards reduced sensitivity. Reduced sensitivity evident *in vitro* was reflected in an *in vivo* seedling assay (Table 3).

The significance of these observations with respect to field efficacy of both dodine and triazole fungicides in UK orchards remains to be determined.

TABLE 1. Range of response to dodine of mass spore samples and single spore isolates of *Venturia inaequalis* from the UK.

	LD50 (mg/l)	
	Range	Mean±SE
Single spore isolates	1.12 - 1.85	1.35±0.12
Mass spore samples		
a) Commercial orchards	1.09 - 2.10	1.62±0.10
b) Untreated plots	0.85 - 1.59	1.23±0.08

TABLE 2. Range of responses to myclobutanil of mass spore samples and single spore isolates of *Venturia inaequalis* from the UK.

	% reduction in hyphal length at 0.5 mg/l	
	Range	Mean±SE
Single spore isolates		
Reference isolates		
E1 (sensitive)	60.3 - 80.0	73.7±4.4
TR77 (reduced sensitive)	-18.9 - 48.8	19.9±3.3
Other UK isolates (13)	24.3 - 78.9	61.1±4.0
Mass spore samples		
Commercial orchard 1.	43.5 - 71.7	59.3±4.3
2.	59.1 - 85.2	75.4±5.3
3.	50.0 - 78.8	70.5±5.9
4.	7.5 - 49.6	30.9±5.0

TABLE 3. Response of two isolates of *V. inaequalis* to myclobutanil, in an *in vivo* assay, on fungicide treated apple seedlings.

	Myclobutanil (mg/l)							
	0		10		20		40	
	A*	B*	A	B	A	B	A	B
	% infected seedlings							
E1 (sensitive)	100	100	8	43	0	0	0	0
TR77 (reduced sensitive)	86	97	50	57	22	42	3	0

* A & B refer to two separate experiments

ACKNOWLEDGEMENTS

This work is funded by the Ministry of Agriculture, Fisheries and Food (England and Wales).

REFERENCES

Fiaccadori, R.; Gielink, A.J.; Dekker, J. (1987) Sensitivity to inhibitors of sterol biosynthesis in isolates of *Venturia inaequalis* from Italian and Dutch orchards. *Netherlands Journal of Plant Pathology* **93**, 285-287.

McKay, M.C.R.; MacNeill, B.H. (1979) Spectrum of sensitivity to dodine in field populations of *Venturia inaequalis*. *Canadian Journal of Plant Pathology* **1**, 76-78.

Sholberg, P.L.; Yorston, J.M.; Warnock, D. (1989) Resistance of *Venturia inaequalis* to benomyl and dodine in British Colombia, Canada. *Plant Disease* **73**, 667-669.

Stanis, V.F.; Jones, A.L. (1985) Reduced sensitivity to sterol-inhibiting fungicides in field isolates of *Venturia inaequalis*. *Phytopathology* **75**, 1098-1101.

Szkolnik, M.; Gilpatrick, J.D. (1973) Tolerance of *Venturia inaequalis* to dodine in relation to the history of dodine usage in apple orchards. *Plant Disease Reporter* **57**, 817-821.

Thind, T.; Olivier, J.M.; Clerjeau, M. (1986) Tavelure du pommier: mise en evidence d'une resistance aux fongicides inhibiteurs de la biosynthese de l'ergosterol. *Phytoma Defense des Cultures* **381**, 13-16.

PARTIAL RESISTANCE OF WHEAT VARIETIES TO POWDERY MILDEW - A FACTOR PREVENTING RESISTANCE TO FUNGICIDES

ILIYA ILIEV

Institute of Wheat and Sunflower "Dobroudja" near General Toshevo 9521, Bulgaria

ABSTRACT

Systemic fungicides help farmers in their fight against plant disease. Their ease of application and rapid effect resulted in their wide-spread use, which brought about the development of resistance of pathogens to them. The efficacy of propiconazole (Tilt 250 EC) fungicide against the powdery mildew pathogen lasts about 15-18 days. It is possible to make this period longer if the chosen varieties possess partial (incomplete) resistance to the pathogen. In the case of the highly susceptible variety "Sadovska ranozreika-4", one generation of the pathogen is accomplished in 8 days. With a combination of the variety Charodeika and the fungicide, the generation time of the pathogen is doubled, while combination of the variety "Dobroudja" and the fungicide increased the generation time by a factor of approximately six. The probable number of generations is inversely proportional to the varietal resistance, but it does not vary with the fungicide applied.

INTRODUCTION

The application of the new and efficient systemic fungicides seemed to offer a solution to the problem of powdery mildew, but their wide-spread use resulted in the development of resistance by the pathogen. Thus, it became necessary to develop schemes for applying the fungicides and a basic principle proved to the rotation of the latter, depending on the mode of action of the active ingredient. Wheat growers found this problematical.

The present study aims to illustrate the role and the importance of partial resistance of some common winter wheat varieties against the pathogen of powdery mildew *Erysiphe graminis* f.sp. *tritici* in prolonging the efficacy of systemic fungicides and preventing the pathogen from developing resistance to them.

MATERIALS AND METHODS

The investigation was carried out at the "Dobroudja" Institute of Wheat and Sunflower near General Toshevo. The materials tested were common wheat varieties, bred at the Institute and possessing partial (incomplete) resistance to the powdery mildew pathogen (i.e. varieties Charodeika and Dobroudja-1). Sadovska ranozreika-4, a highly susceptible wheat variety, served as a control variety (C).

The trial was carried out under laboratory conditions with plants at the 2nd-leaf stage of development. Seeds were sown in pots of size 28x28x6cm and grown under controlled temperature at 18-20°C with natural light. A genetically homogeneous powdery mildew culture, race 7430, was used in the investigation (Iliev, 1992). The two varieties used in the trial are susceptible to that isolate. Prior to the inoculation with the pathogen, wheat plants were treated with the systemic fungicide, propiconazole (50ml/ha), by spraying the leaves. Several hours later, when the vapour effect had faded away, plants were inoculated by shaking inoculated wheat material sharply over them. Up to the 10th day, inoculation was carried out every second day, then the same procedure was followed every 8th day. The type of infection, serving as a qualitative expression of pathogen development, was duly recorded, using Mains and Dietz scale (by Kounovski, 1973) for types of infection from 0 to 4. (0 - no infection, 1 - formation of mycelium without formation of spores, 2 - formation of mycelium with evident sporulation [small tufts], 3 - formation of medium large tufts with normal sporulation, 4 - formation of large tufts with abundance of sporulation).

Trials were carried out with 3 replicates, using untreated wheat plants of the three varieties as controls. Recording began for the first time on the second day after the application of the fungicide and continued every second day.

RESULTS AND DISCUSSION

Results showing powdery mildew development on untreated wheat plants of varieties Sadovska ranozreika-4, Charodeika and Dobroudja-1 are presented in the columns on the left for the 1st, 2nd, 3rd and 4th type of infection respectively. Results obtained with plants treated with propiconazole have been put in the columns on the right (Fig. 1). The development of the pathogen to type-1 infection can be achieved in 3 days with varieties Sadovska ranozreika-4 and Charodeika. Variety Dobroudja-1 needed 1 more day. When lacking genetically controlled resistance to the pathogen, as in the case of variety Sadovska ranozreika-4, the activity of the fungicide remains good. The maximum period of time during which the beneficial effect of the fungicide is in force is 15 days, i.e. 1.9 "standard generations" of the pathogen. With Charodeika, a variety of moderate susceptibility to the pathogen, the effect of the propiconazole in combination with the presence of genetic resistance to the pathogen can increase the generation time by a factor of 3.1 to 25 days. With variety Dobroudja-1, possessing greater genetically controlled resistance to the pathogen, the generation time increases by a factor of 3.9, to 31 days.

The development of the pathogen up to type-2 infection occurs on the fourth day after inoculating varieties Sadovska ranozreika-4 and Charodeika with conidiospores (in the absence of fungicide treatment).

With variety Dobroudja-1, the period needed is up to 6 days. The presence of that type of infection is characterised by the presence of developed conidia-carriers with conidiospores. In plants treated with propiconazole, type-2 infection appears on the 18th day with variety Sadovska ranozreika-4; on the 28th day with variety Charodeika and on the 52nd day with variety Dobroudja-1.

Type of infection

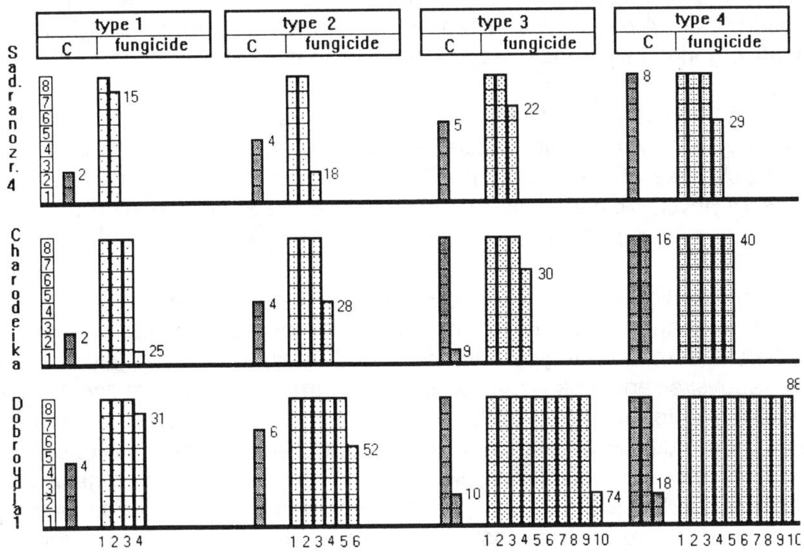

Number of generations of the pathogen (one generation is completed in 8 days with the standard variety S.r.-4, in the absence of fungicide).

Fig. 1. Powdery mildew development expressed as the type of infection at the second leaf stage, under laboratory conditions.

The squares represent the number of days required for development of the pathogen to types 1, 2, 3 and 4 infection respectively. Eight days are normally necessary for the completion of one generation (type-4 infection) of the pathogen on the variety "Sadovska ranozreika-4", in the absence of fungicide treatment. This interval is termed "standard generation time" for the purposes of this discussion. Type 2 infection occurs after 4 days.

When fungicide (propiconazole) is applied to this variety, the generation time is extended, so that type 2 infection develops after 18 days.

When the fungicide is applied to the variety "Charodeika", the time taken for "type 2" infection to develop is increased to 28 days. This represents 3.5 "standard generations". In the case of Dobroudja-1, the generation time is extended to 52 days (6.5 times greater than the standard generation time).

Thus, variety Charodeika is able to increase the generation time of the pathogen by a factor of 1.3 (10 days).

In the presence of propiconazole, the time taken to reach type 2 infection is 34 days longer with variety Dobroudja-1 than Sadovska ranozreika-4. The difference between the time taken for the varieties Dobroudja-1 and Charodeika to develop type 2 infection when treated with propiconazole is 3 "standard generations", or 24 days. This difference is due to the different qualitative and quantitative genetically conditioned resistance of the varieties studied.

Type 3 infection on the highly susceptible variety Sadovska ranozreika-4, with no treatment with propiconazole, is achieved on the 5th day after inoculation with conidiospores of the pathogen. Type-3 infection is characterised by the presence of physiologically mature conidiospores on a well-developed pustule of the pathogen. At this stage, secondary infections by the pathogen are possible.

With varieties Charodeika and Dobroudja-1, type-3 infection develops on the 9th or 10th day after inoculation. The application of the fungicide prolongs the period before formation of type-3 infection to 22 days with the highly susceptible variety Sadovska ranozreika-4, to 30 days with the moderately susceptible variety Charodeika and to 74 days with the resistant variety Dobroudja-1. When calculating the number of "standard generations" likely to occur in variety Dobroudja-1, the ratio established was 1:9.3. One fully-developed generation of the pathogen on that variety, when treated with propiconazole, protects the plants against 9.3 "standard generations" of the pathogen. This is 6.6 "standard generations" longer than with propiconazole-treated Sadovska ranozreika-4, and 6 "standard generations" longer than with propiconazole-treated Charodeika.

One of the differences between type-4 and type-3 infection is the size of the pustule. This shows that the vegetative (mycelial) part of the pathogen is larger with type-4 infection, which means that the reproductive abilities of a type-4 pustule are correspondingly greater than type-3 pustules. When considering the results concerning type-4 infection, possible generations are measured in 1 unit quantities, without considering the multiplication effect, which is considerably higher and is determined by the number of conidiospores formed from a single pustule per unit time. It proved to be physically impossible to include this index in the present investigation. Infection type-4 was recorded on the 8th day after inoculating the highly susceptible variety Sadovska ranozreika-4 with conidiospores of the pathogen, when no fungicide was applied. When fungicide was applied, type-4 infection was recorded on the 29th day (Fig. 1). The generation time on treated plants (cv. Sadovska ranozreika-4) was increased by a factor of 3.6 compared with untreated plants. With the moderately susceptible variety Charodeika, type-4 infection occurred on the 16th day after inoculating untreated plants. When propiconazole was applied, no formation of type-4 infection was established. Throughout the experiment, the growth of the mycelium was limited to type-3, hence the quantity of developed conidiospores, and hence the number of possible pathogen generations were limited too. The ratio here is similar to that of variety Charodeika, i.e. one generation to great numbers of generations for the untreated variant. In this case, type-4 infection cannot be taken as a criterion, as secondary infections are quite possible at infection type-3 or type-2. Here, the presence of type-4 infection is an index for the significance of the resistance possessed by the variety, i.e. for the increase of the total effect giving protection against the pathogen.

When summing up the results of the present investigation, it can be concluded that the effectiveness of the fungicide applied is constant, varying little in its value. The genetically conditioned varietal resistance affects the duration of activity of the fungicide. The effect is most vigorously expressed in variety Dobroudja-1, followed by variety Charodeika and is least expressed in variety Sadovska ranozreika-4. The effect of the genetically controlled resistance is greater than the independent effect of the fungicide. On the other hand, the greater duration of the fungicidal effect prevents the reproduction of the pathogen. Thus, conditions can be manipulated to prevent the development of resistance to the fungicide.

CONCLUSIONS

The efficacy of the applied fungicide is a constant, varying from 15 to 18 days.

The combination of fungicide and a variety with genetically controlled resistance to powdery mildew lengthens the period of protection against the pathogen. This period is directly proportional to the degree of varietal resistance.

Growing varieties with partial (incomplete) resistance to powdery mildew limits the number of reproductive generations of the pathogen, and helps prevent development of resistance to the fungicide.

REFERENCES

Iliev, I. (1989) Racial variety of the *Blumeria graminis tritici* populations. Plant Science vol. XXVI N9, 85-91.

Iliev, I. (1992) New Designation of *Blumeria graminis* (DC) *Speer* f.sp. *tritici* (*Erysiphe graminis* f.sp. *tritici*) Races established. Plant Science, vol. XXIX, N1-2, 94-105.

Kounovski, Z. (1973) Study of the physiological specialization and some biological and physiological characters of powdery mildew (*Erysiphe graminis* f.sp. *tritici*) Dissertation.

variant. In this case two of the [illegible] type 3 and [illegible] Type 1 [illegible] the significance of [illegible] possessed by the variety [illegible] the [illegible] component.

[illegible]

ACKNOWLEDGEMENTS

[illegible]

[illegible]

[illegible]

REFERENCES

[illegible]

[illegible]

[illegible]

RESISTANCE TO INHIBITORS OF STEROL C-14 DEMETHYLATION IN THE CEREAL EYESPOT FUNGUS, PSEUDOCERCOSPORELLA HERPOTRICHOIDES

P. LEROUX

INRA, Unité de Phytopharmacie, 78026 Versailles Cédex (France)

P. MARCHEGAY

ACTA, 149 rue de Bercy, 75595 Paris Cédex 12 (France)

J.L. MIGEON

SRPV, BP 47, 62750 Loos-en-Gohelle (France)

C. MAUMENE

ITCF, Station Expérimentale, 91720 Boigneville (France)

ABSTRACT

According to their mycelial growth rate and their in-vitro sensitivity towards prochloraz and other inhibitors of sterol C-14 demethylation, five types of P. herpotrichoides were characterized. Among them, the slow-growing ones could result in inadequate control of eyespot in winter wheat treated by these fungicides.

INTRODUCTION

In France, eyespot caused by P. herpotrichoides, is the major stem-base disease in winter wheat. In the last ten years, inhibitors of sterol C-14 demethylation (DMIs) have replaced the benzimidazole fungicides (e.g. carbendazim) because of the development of benzimidazole-resistant strains. The DMIs registered in France are the imidazole prochloraz and several triazoles (e.g. bromuconazole, flusilazole). From the survey assessing the susceptibility of P. herpotrichoides to these DMIs, several types of resistant strains were characterized.

IN-VITRO EFFECTS OF FUNGICIDES

Field isolates of P. herpotrichoides produced either fast-growing (type I) or slow-growing (type II) mycelial colonies when cultivated on agar media (Leroux and Gredt, 1988). According to their in-vitro response towards DMIs, five types of strains were identified (Leroux and Migeon, 1993 ; Table 1).

- Type Ia : fast-growing strains sensitive to both prochloraz and triazoles,
- Type Ib : fast-growing strains sensitive to prochloraz but resistant to triazoles to some extent (Leroux and Gredt, 1988). Their frequencies have increased recently since they were first detected several years ago. (Migeon et al., 1994). However, it has never been established that such strains could reduce the efficacy of triazoles in the field.

TABLE 1. Effects of some DMIs and cyprodinil on the germ-tube elongation of various strains of *P. herpotrichoides*[a]

Fungicides	Mean	EC 50	(mg/l) of	types	
	Ia	Ib	Ic	IIs	IIp
prochloraz	0.005	0.006	0.75	0.007	0.25
bromuconazole	0.03	0.50	2.0	0.50	0.50
epoxyconazole	0.02	0.30	1.2	0.25	0.40
flusilazole	0.008	0.10	1.5	0.10	0.40
hexaconazole	0.01	0.50	1.0	0.30	0.15
cyprodinil	0.006	0.005	0.005	0.007	0.006

[a] : the tests were conducted according to the method of Leroux and Gredt (1988), with 2 to 8 strains of each type. The strains Ib were chosen among the most triazole-resistant ones.

- Type Ic : fast-growing strains resistant to both prochloraz and triazoles. They were found in 1992 and 1993 but remained rare in France.

- Type IIs : slow-growing strains naturally resistant to triazoles but sensitive to prochloraz. When such strains were prevalent in the field, triazoles were less effective than prochloraz (Leroux *et al*., 1990).

- Type IIp : slow-growing strains resistant to prochloraz (resistance levels between 30 and 40) and to triazoles. They were detected in France for the first time in 1990 (Leroux and Marchegay, 1991).

All these types were equally sensitive to cyprodinil, an anilinopyrimidine recently introduced in France (Bocquet *et al*., 1994).

FIELD RESISTANCE TO PROCHLORAZ

Twenty one field trials were conducted in 1992 and 1993, on winter wheat, by the Plant Protection Service. According to the efficacy of prochloraz (450 g/ha), these trials could be subdivided into two categories (Table 2). The first one corresponded to locations where this fungicide was effective and where the percentages of prochloraz-resistant strains in the control plots were low. In the other trials, located in Northern France, the presence of high frequencies of prochloraz-resistant strains (mainly IIp) resulted in inadequate control of eyespot.

LONG-TERM FIELD EXPERIMENTS

In 1988, a long-term field trial was started by ACTA, ITCF and INRA at Villiers-le-Bâcle near Versailles. (rotation : winter wheat/rape).

Prochloraz one or two times per year and flusilazole (twice per year) were applied every year to the same plots until 1992 ; in 1993 no fungicide was used in the plots previously treated twice a year (Table 3).

TABLE 2. Efficacy of prochloraz and population composition of P. herpotrichoides in trials conducted by the Plant Protection Service in 1992 and 1993

Number of trials	% strains in control plots[a] type II	type IIp	% efficacy of prochloraz[b]
11	41 (16 - 87)	3 (0 - 15)	62 (48 - 79)
10	89 (65 - 100)	63 (43 - 95)	9 (0 - 21)

[a] : extreme percentages in parentheses.
[b] : prochloraz (450 g/ha) was applied at GS 31, whereas the estimates of necrosed sections and the analyses of strains were carried out at GS 71-75.

TABLE 3. Evolution of fungicide efficacies and populations of P. herpotrichoides in a long-term field trial in winter wheat

years	control		prochloraz (x 1)[a]		prochloraz (x 2)		flusilazole (x 2)	
	% IIs	% IIp	% Ef.[b]	% IIp	% Ef.	% IIp	% Ef.	% IIp
1988	68	0	67	0	82	0	23	0
1989	93	0	60	0	62	0	28	0
1990	83	0	56	24	95	50	16	0
1991	80	0	18	50	27	86	7	13
1992	63	23	19	88	7	100	24	94
1993	42	42	22	100	-[c]	72	-[c]	52

[a] : prochloraz (450 g/ha) or flusilazole (200 g/ha) were applied between GS 29 and GS 32 whereas the estimates of necrosed sections and the analyses of strains were carried out at GS 75.

[b] : percentages of efficacy in comparison to the control plots for necrosed sections.

[c] : no treatment in 1993.

In this location the slow-growing strains (IIs) were prevalent even in the control plots from 1988. This probably explains why, between 1988 and 1990, flusilazole was less effective than prochloraz. Since 1991, prochloraz has failed to control eyespot because of the development of type IIp strains. The selection pressure was greater with two applications of prochloraz per year than with one application as judged from the higher frequency of type IIp isolates in plots receiving two applications of prochloraz (Table 3). Application of flusilazole also selected for these prochloraz-resistant strains. The relative increase of strains of type II p in the control plot in 1992 and 1993 suggests that they are more fit than those of type IIs. Conversely the slight decrease in the percentage of type IIp strains, observed in 1993 when prochloraz or flusilazole were stopped, suggests that strains IIp are less competitive than the type IIs strains (Table 3).

CONCLUSION

The reduced efficacy of triazoles is a well known phenomenon in locations where the slow-growing strains of P. herpotrichoides (type II) are prevalent (Leroux et al., 1990). More recently, in several departments of Northern France, the development of a sub-population of slow-growing strains resistant to prochloraz (type IIp) resulted in field resistance to this fungicide. Prochloraz and also the triazole flusilazole can select for these II p type strains. However this phenomenon does not seem to occur with all triazoles (e.g. hexaconazole) (Leroux, unpublished data).

REFERENCES

Bocquet, G. ; Sylvestre, M. ; Speich, J. (1994). Le cyprodinil, fongicide céréales. Phytoma, **458**, 53-55.

Leroux, P. ; Gredt, M. (1988). Caractérisation des souches de Pseudocercosporella herpotrichoides, agent du piétin-verse des céréales, résistantes à substances antimitotiques et à des inhibiteurs de la biosynthèse des stérols. Agronomie, **8**, 719-729.

Leroux, P. ; Cavelier, N. ; Migeon, J.L. (1990). Resistance to inhibitors of sterol biosynthesis in Pseudocercosporella herpotrichoides, the causal agent of cereal eyespot. Tagungsbericht Akademie der Landwirtschaftswissenschaften, Berlin (DDR), **291**, 137-144.

Leroux, P. ; Marchegay, P. (1991). Caractérisation des souches de Pseudocercosporella herpotrichoides, agent du piétin-verse des cérales, résistantes au prochloraze, isolées en France sur blé tendre d'hiver. Agronomie, **11**, 767-776.

Leroux, P. ; Migeon, J.L. (1993). La résistance en pratique du piétin-verse des céréales vis-à-vis du prochloraze, une certitude en 1992. Phytoma, **448**, 20-24.

Migeon, J.L. ; Mathon, M.P. ; Chudzicki, A.M. ; Leroy, J.P. (1994). Piétin-verse des céréales : cartographie 1993, définition d'un risque régional en matière de type souches. Phytoma, **458**, 24-26.

SENSITIVITY DISTRIBUTION TO VARIOUS FUNGICIDES IN EYESPOT (*PSEUDOCERCOSPORELLA HERPOTRICHOIDES*) IN THE RHINELAND/GERMANY

H. HINDORF, C. POHL

Institut für Pflanzenkrankheiten der Universität, Nuβallee 9, D-53115 Bonn

ABSTRACT

Fungal isolations were made in 1992 from leaf sheaths of winter wheat at GS 30-32 growing at different locations in the Rhineland (Cologne-Aachen). In vitro sensitivity tests were carried out with carbendazim fungicides (MBC), either alone (Derosal), or in combination with prochloraz (Sportak alpha), or flusilazole (Harvesan). Carbendazim alone had no fungicidal effect on any of the *Pseudocercosporella* isolates from 7 localities, even at the highest concentration tested (20 mg/l). The combination of carbendazim and prochloraz inhibited mycelium growth and conidium production in isolates from 5 localities at a concentration of 0.2 mg/l (the lowest concentration tested), and all isolates at a concentration of 20 mg/l. Isolates of *P. herpotrichoides* from 7 localities reacted differently to the combination of carbendazim and flusilazole than to carbendazim alone.

INTRODUCTION

The first fungicides used for the control of eyespot (*Pseudocercosporella herpotrichoides*) in winter wheat were multi-site inhibitors with an unspecific mode of action (Buchenauer, 1984b). The control of eyespot became more effective after the introduction into cereal production of systemic fungicides, which were mostly single-site inhibitors (Obst, 1989). Carbendazim fungicides were first approved for use in Germany for the control of eyespot in winter wheat in 1973 (Fehrmann, 1976). At that time it was well known that after continuous applications of MBC fungicides their controlling effect was lost against some pathogens and resistant pathotypes could be identified in several groups of fungi (Buchenauer, 1984a).

In 1974 Rashid and Schlösser (1975) detected MBC resistance in eyespot for the first time in Germany, in a pathogen population at Giessen; resistance was present in 2 to 4% of their isolates. Owing to the fact that only a single application was made to control eyespot each season, Fehrmann and Horsten (1980) and Fehrmann (1984) considered the threat of the development of MBC resistance in eyespot to be unimportant. However, in a long term monitoring experiment which started in 1983 in Germany, Fehrmann *et al.* (1989) showed a reduction in sensitivity to MBC fungicides in *P. herpotrichoides*. Since 1982 a rapid increase in the frequency of MBC resistant isolates has been observed in northern Germany (Fehrmann, 1985). More recently MBC resistance in *Pseudocercosporella* was found in Mecklenburg-Vorpommern (Seidel *et al.*, 1988) and Thuringia (Schinke and Breitenstein, 1992).

MATERIALS AND METHODS

In 1992, fungal isolations were made in winter wheat at 8 localities (Büsdorf, Kapellen, Kuchenheim, Erkelenz, Hürth, Wiedenfelder Höhe, Weilerswist and Beckrath) in the Rhineland where an intensive cereal crop was included in the rotation. One hundred plants at GS 30-32 were collected, and fungal isolates were obtained from the outer and inner leaf sheaths. The fungal population was differentiated into species and varieties within species known to be involved in the foot rot disease complex. Thereafter, single conidial isolates were made for each species and 4 replicates of each were used in all further investigations.

To test the fungicide sensitivity of *Pseudocercosporella* isolates mycelial growth rate (colony area) was measured on potato dextrose agar (PDA) supplemented with 0.2, 2.0, and 200 mg/l of carbendazim (360 g/l), or carbendazim (80 g/l) + prochloraz (300 g/l) and of carbendazim (125 g/l) + flusilazole (250 g/l). Unsupplemented PDA was used as a control in each case. The cultures were incubated at 20°C in darkness until 21 days after inoculation, when the radial growth of the fungal colonies was measured by an electronic planimeter.

RESULTS

P. herpotrichoides var. *acuformis* was isolated at 6 of the 8 localities (Büsdorf, Kapellen, Kuchenheim, Erkelenz, Hürth, and Weilerswist), whereas *P. herpotrichoides* var. *herpotrichoides* was isolated at only 2 (Wiedenfelder Höhe and Beckrath). All isolates collected at 7 of the 8 localities proved to be resistant to carbendazim (Table 1). In a number of cases, there was a stimulation of fungal growth when carbendazim alone was added to the PDA, with the largest fungal colonies recorded at the highest concentration of the fungicide. The mixture of carbendazim and prochloraz showed a strong inhibition of growth of all *P. pseudocercosporella* isolates. The mixture of carbendazim and flusilazole did not suppress growth significantly at low concentrations. These results are constistent with those obtained by Bateman (1990).

TABLE 1: Mycelial growth (mm^2) of *P. herpotrichoides* after fungicide treatment

		Location							
Fungicide	mg/l	Büsd.	Kap.	Kuch.	Erk.	Hürth	Weil.	Wied.	Beck.
	0	319.0	391.0	145.5	342.1	160.9	371.3	1842.2	1542.0
carben-	0.2	298.3	420.8	221.5	340.3	261.8	477.9	1777.4	0.0
dazim	2.0	406.0	471.1	224.6	449.4	197.9	547.8	1788.1	0.0
	20.0	424.4	556.7	244.0	578.5	107.6	519.2	1625.2	0.0
	0	482.0	814.1	531.1	509.6	212.3	426.2	2401.2	1682.4
prochloraz	0.2	0.0	314.8	0.0	0.0	0.0	221.8	0.0	0.0
	2.0	0.0	0.0	0.0	0.0	0.0	22.4	0.0	0.0
	20.0	0.0	0.0	0.0	0.0	0.0	0.0	0.0	0.0
prochloraz	0	432.0	814.0	525.4	502.3	233.2	426.2	2420.0	1079.4
+	0.2	0.0	301.8	0.0	0.0	28.5	325.8	0.0	0.0
carben-	2.0	0.0	0.0	0.0	0.0	0.0	37.3	0.0	0.0
dazim	20.0	0.0	0.0	0.0	0.0	0.0	0.0	0.0	0.0
flusilazole	0	431.2	814.0	576.5	538.6	233.2	426.2	2406.0	1709.4
+	0.2	497.0	313.8	358.1	349.8	326.6	276.6	373.3	0.0
carben-	2.0	29.2	120.4	52.2	44.5	60.8	89.7	0.0	0.0
dazim	20.0	0.0	0.0	0.0	0.0	0.0	0.0	0.0	0.0

REFERENCES

Bateman, G.L. (1990) Comparison of the effect of prochloraz and flusilazole on foot rot disease and on populations of the eyespot fungus *Pseudocercosporella herpotrichoides*, in winter wheat. *Zeitschrift für Pflanzenkrankheiten und Pflanzenschutz*, **97**, 508-516.

Buchenauer, H. (1984a) Resistenzentwicklung von Pilzen gegenüber Fungiziden und Strategien zur Vermeidung von Fungizidresistenz. *Gesunde Pflanzen*, **36**, 132-142.

Buchenauer, H. (1984b) Stand der Fungizidresistenz bei Getreidekrankheiten am Beispiel der Halmbruchkrankheit und des Echten Mehltaus. *Gesunde Pflanzen*, **36**, 161-169.

Fehrmann, H. (1976) Systemische Fungizide - ein Überblick. I. Zur Wirkungsweise praxisüblicher Substanzen. *Phytopathologische Zeitschrift*, **86**, 67-89.

Fehrmann, H. (1984) MBC-resistente Freilandpopulationen von *Pseudocercosporella herpotrichoides* im langjährigen Überwachungsversuch. *Phytopathologische Zeitschrift*, **110**, 82-86.

Fehrmann, H. (1985) Resistance to benzimidazoles in *Pseudocercosporella herpotrichoides*. *EPPO Bulletin*, **15**, 477-483.

Fehrmann, H.; Horsten, J. (1980) Gibt es bei *Pseudocercosporella herpotrichoides* eine Gefahr der Fungizidresistenz in der Praxis? *Gesunde Pflanzen*, **32**, 19-23.

Fehrmann, H.; Berndt, H.; Manns, G. (1989) Langfristiger Monitoring-Versuch in Weizen zur Sensitivität von *Pseudocercosporella herpotrichoides* und *Septoria nodorum* für DMI-Fungizide: erste Ergebnisse. *Gesunde Pflanzen*, **41**, 38-43.

Obst, A. (1989) Zur Fungizidempfehlung im Getreidebau unter Berücksichtigung des Resistenzrisikos. *Gesunde Pflanzen*, **41**, 159-169.

Rashid, T., Schlösser, E. (1975) Resistenz von *Cercosporella herpotrichoides* gegenüber Benomyl. *Zeitschrift für Pflanzenkrankheiten und Pflanzenschutz* **82**, 765-766.

Schinke, C.; Breitenstein, G. (1992) Carbendazimresistenz erfaβt jetzt auch Südthüringen. *Nachrichtenblatt des deutschen Pflanzenschutzdienstes*, **44**, 220.

Seidel, D.; Amelung, D.; Rothacker, D. (1988) Untersuchungen zur MBC-Resistenz von *Pseudocercosporella herpotrichoides*-Arten und -Varietäten im Norden der DDR. *Tagungsbericht der Akademie der Landbauwissenschaften der DDR*, **271**, 139-14

SENSITIVITY OF BROWN AND YELLOW RUST POPULATIONS ON WHEAT TO CYPROCONAZOLE

L. OHL and U. GISI

Agrobiological Research Station, SANDOZ AGRO LTD., CH-4108 Witterswil, Switzerland

ABSTRACT

In order to evaluate whether cyproconazole treatments cause changes in sensitivity of *Puccinia recondita* f. sp. *tritici* and *Puccinia striiformis* populations, a monitoring program was carried out with bulk samples collected from wheat fields in several countries for several years. The mean EC 50 values of Swiss brown rust populations varied from 0.3 mg/l (1990) to 1.2 mg/l (1991) to 1.7 mg/l (1992) to 0.4 mg/l (1993). In 1993, the French and the German brown rust populations had average EC 50 values of 0.4 and 2.0 mg/l, respectively. No influence of cyproconazole on the sensitivity distribution was detected when DMI-treated and -untreated samples were compared. The sensitivities of yellow rust populations from France and Germany were not different from each other, but was unambiguously higher than those of brown rust.

INTRODUCTION

The aim of this survey was to evaluate the variability in sensitivity to cyproconazole among strains of *Puccinia recondita* f. sp. *tritici* and *Puccinia striiformis* from Switzerland, France and Germany, as well as to compare the sensitivity of DMI-treated and-untreated samples over several years. The sensitivity of Swiss brown rust populations has been analysed since 1990. In 1993, brown rust populations were also monitored for France and Germany. For yellow rust, the survey started in 1993 with populations from France and Germany.

MATERIALS AND METHODS

Infected leaf samples were taken from fields treated with DMIs and from fields not treated with DMIs. In a modified leaf piece test (brown rust) and with whole plants (yellow rust), the sensitivity of bulk samples was determined with a range of fungicide concentrations; dose-response correlations were established to calculate EC 50 values (effective concentrations resulting in 50 % disease suppression). For the leaf piece test 7-day-old wheat plants were treated, the leaves cut into 3 cm pieces after 24 hours, placed on water agar, inoculated with an urediospore suspension of the field sample of brown rust and incubated for 10 days at 18°C. Since yellow rust has a long incubation period, this test was carried out on whole plants in a similar manner as described previously for brown rust (Stähle-Csech and Gisi, 1991).

RESULTS AND DISCUSSION

The sensitivity of Swiss brown rust populations has been analysed since 1990 (Table 1). Until 1992, the lowest EC 50 values were around 0.01 mg/l and the highest EC 50 values were 6.8 mg/l (1991) and 5.0 mg/l (1992), respectively. The sensitivity of the 1993 populations were unchanged, since both the mean and the highest values were in the same range as in previous years. Nevertheless, the most sensitive strains were no longer detected, which might be a result of the small number of samples in 1993.

In 1993, the sensitivity of the brown rust populations from Switzerland and France showed a mean EC 50 value of 0.4 mg/l, whereas in Germany the value was 2.0 mg/l (Table 2). The highest EC 50 values for single samples were found in France and Germany. The sensitivity of the German isolates was generally lower compared to Switzerland and France, a phenomenon which is also known to occur for powdery mildew on wheat and barley but in a more pronounced manner (Felsenstein, this volume).

The samples originating from DMI-treated fields in France showed sensitivities between 0.05 and 10 mg/l as compared to 0.07 and 2 mg/l for samples from fields not treated with DMIs (Fig. 1). Although the mean and upper end of the treated distribution are numerically slightly higher than those of the untreated distribution, no difference in sensitivity can be claimed between the two distributions.

TABLE 1. Sensitivity of *P. recondita* f. sp. *tritici* samples to cyproconazole in Switzerland, 1990-1993

year	no. of samples	$\bar{x}$ EC50	lowest EC50	highest EC50
1990	35	0.3	0.01	1.7
1991	42	1.2	0.01	6.8
1992	22	1.7	0.01	5.0
1993	13	0.4	0.07	1.8

TABLE 2. Sensitivity of *P. recondita* f. sp. *tritici* samples to cyproconazole in Switzerland, France and Germany in 1993

country	no. of samples	$\bar{x}$ EC50	lowest EC50	highest EC50
Switzerland	13	0.4	0.07	1.8
France	87	0.4	0.05	9.8
Germany	46	2.0	0.10	10.0
standard isolate[a)]	1	0.3	-	-

[a)]standard isolate was collected before 1967 in Switzerland

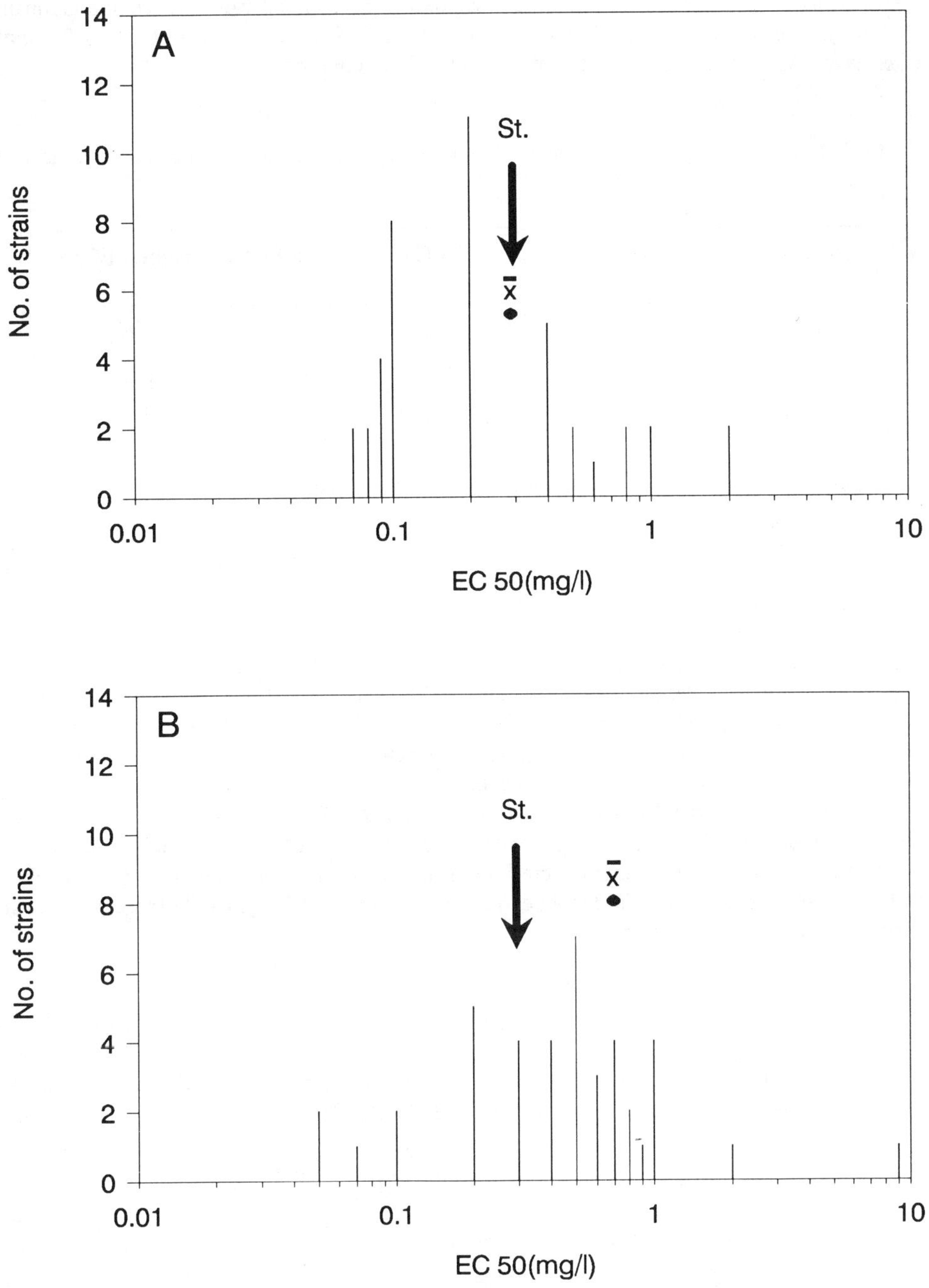

FIGURE 1: Sensitivity distribution of French brown rust populations in 1993 from DMI-untreated fields (A) and fields treated with DMIs (B). St.= standard isolate, $\overline{x}$ = mean

Only a small number of yellow rust samples were analysed from France and Germany (Table 3). The sensitivity ranged from 0.01 to 0.6 mg/l (France) and from 0.01 to 0.3 mg/l (Germany), with a mean EC 50 value of 0.1 mg/l in both countries.

TABLE 3. Sensitivity of *P. striiformis* samples to cyproconazole in France and Germany in 1993

country	no. of samples	$\bar{x}$ EC50	lowest EC50	highest EC50
France	12	0.10	0.01	0.6
Germany	7	0.10	0.01	0.3
standard isolate [a)]	1	0.15	-	-

[a)]standard isolate was collected before 1967 in Switzerland

CONCLUSIONS

No differences in sensitivity to cyproconazole were found for the Swiss brown rust populations between 1990 and 1993. The mean EC 50 value varied over the years between 0.3 and 1.7 mg/l and the width of the distribution (between the lowest and highest EC 50 values) was a factor of 25 (1993) to 680 (1991). This may represent rather broad variation within the populations. The populations from France and Switzerland had the same mean EC 50 value, whereas the value of the German population was higher. The range of EC 50 values from Germany was higher than those of Switzerland and France. The sensitivities of yellow rust samples from France and Germany were both in the same range and the mean EC 50 values were close to that of the standard isolate. Yellow rust seems to be generally more sensitive to cyproconazole than brown rust.

REFERENCES

Stähle-Csech, U.; Gisi, U. (1991) Determination of the sensitivity to DMI fungicides of *Puccinia recondita* on wheat. *EPPO Bulletin* **21,** 315-317.

Session 2
Phenylamides

Chairman &
Session Organiser M SMITH

EARLY EXPERIENCES WITH PHENYLAMIDE RESISTANCE AND LESSONS FOR CONTINUED SUCCESSFUL USE

T. STAUB

Plant Protection Division, R & D Disease Control, Ciba-Geigy Ltd., CH 4002 Basle, Switzerland

ABSTRACT

The first and most dramatic cases of phenylamide resistance occurred in cucumber downy mildew in plastic houses in Israel and in potato late blight in Europe in 1980. They were associated with exclusive use of solo products under high disease pressure. This led to the withdrawal of the solo product in favor of prepack mixtures with residual compounds. Where phenylamides were used in mixtures from the start, as against grape downy mildew and in the UK against potato late blight, resistance was slower to emerge and spread. In addition, where mixtures were used, performance problems remained rare when resistant strains began to appear in the target pathogens. The successful use strategies throughout the 80s were based on the use of prepack mixtures, the avoidance of curative use and a limitation to 2-4 treatments early in the season. Where these strategies were implemented performance of phenylamides mixtures remained good even where resistance could readily be detected. Resistance tends to increase during seasons and recede again between seasons. This indicates a fitness deficit of resistant strains compared to the wild-type populations from which they emerged. Special studies confirmed that resistant strains are, as a rule, less fit for survival from season to season. In *P. infestans* a special situation may exist when resistant strains are imported with seed or plant material from areas with more virulent populations of late blight. Studies with DNA fingerprinting and other genetic markers show that for *P. infestans* migration with infected tubers may play a much bigger role in the initial establishment of phenylamide resistance in a region than previously thought. Early experiences showed that the phenylamides can be preserved as valuable tools for Oomycete control, if the anti-resistance strategies are implemented.

INTRODUCTION

Metalaxyl and furalaxyl were the first phenylamide fungicides to be introduced in 1977. Other companies presented four more representatives of this group over the next five years (Table 1) The six phenylamides differ considerably in their level of activity and in the versatility of their use (Schwinn and Staub, 1987). The high level of specific and systemic activity of the first phenylamides represented an exciting new technology for the control of most Oomycete pathogens; novel features were their high inherent level of activity, their rapid uptake and acropetal transport which led to extended spray intervals and to the protection of new growth. For systemic soil- and seedborne pathogens they offered effective chemical control for the first time. The biological properties of the phenylamides were so attractive that they were often used exclusively, especially in situations of high disease pressure which were difficult to handle with the old residual fungicides. It was also tempting for farmers to exploit their systemic and curative properties to the maximum, e.g. delay treatments until the epidemics were well established and hard to stop.

TABLE 1: Phenylamide introductions

Common name	Code	Year of presentation	Company
metalaxyl	CGA 48988	1977	Ciba-Geigy
furalaxyl	CGA 38140	1977	Ciba-Geigy
ofurace	RE 20615	1978	Chevron
benalaxyl	M 9834	1981	Montedison
cyprofuram	SN 78314	1982	Schering
oxadixyl	SAN 371F	1983	Sandoz

EARLY CASES OF PHENYLAMIDE RESISTANCE

The first cases of phenylamide resistance in practice occurred in the winter season 1980 in *Pseudoperonospora cubensis* on plastic-house grown cucumbers in Israel (Table 2). Up to that time metalaxyl had given excellent control under these conditions of heavy and continuous disease pressure both in trials and in the first seasons of use. Compared to the previously used residual fungicides, metalaxyl was so much better that it was used exclusively in many plastic houses. Therefore, the control failures in plastic houses where resistance appeared were complete and the solo product had to be withdrawn from this use.

TABLE 2: First major cases of phenylamide resistance

Year	Countries	metalaxyl use	disease control	special observations
Pseudoperonospora cubensis				
1980	ISL, CR	solo	lost	intensive use in plastic houses
Phytophthora infestans				
1980	CH, NL, IRL	solo	lost	heavy attacks, curative use
1981	UK, F , D	mixtures	good	few R samples in monitoring
Plasmopara viticola				
1981	SA, F	mixtures	good	few R samples in monitoring
Peronospora tabacina				
1981	Centr. Am.	solo	lost	problems restricted to shaded tobacco

The first occurrence of phenylamide resistance in a field crop was in *Phytophthora infestans* on potatoes during the 1980 season in Europe where metalaxyl had gained a high market share in the first two years of use (Table 2). Resistance appeared in an explosive fashion almost simultaneously in Switzerland, the Netherlands and in Ireland. In all three countries metalaxyl had been used season long as a solo product. The disease pressure was very high that summer and many farmers had problems spraying in time due to the bad weather. Therefore many applications were made in curative or eradicative way in potato fields with substantial levels of late blight attack. As soon as the surprisingly fast occurrence of phenylamide resistance was realized, the solo product was withdrawn from the market and later replaced by mixtures with residual fungicides. No performance problems occurred in the UK, where metalaxyl was available as a prepack mixture with mancozeb from the start. The rationale for the different use in the UK had been the improved late blight control by mixtures on older foliage late in the season where phenylamides alone were found to be less active than on younger, vigorously growing foliage.

The sudden appearance of phenylamide resistance so soon after the product introduction was most surprising since several risk analysis studies on the propensity of resistance development indicated a lower risk level than that known from experience for the benzimidazoles (see below). Equally surprising was the explosive nature of the first occurrences of resistance in the cases described above. In potato fields near Berne it took merely 10 days from the first observations of a possible problem until the foliage was completely destroyed in the foci where resistance appeared first (Table 3). This illustrates the futility of the "wait and see" attitude with such explosive diseases like late blight. A first lesson was that strategies have to be implemented early and that sensitivity methods have to be available for quick testing of samples from critical fields.

TABLE 3. Time course of events in 1980 related to the first occurrences of phenylamide resistance in *Phytophthora infestans* (Staub and Sozzi, 1983)

Day	1	- First indications of problems (active sporulation on treated plants)
Day	5	- 50% attack in actively sporulating foci (samples collected)
Day	10	- 100% attack in first foci
Day	20	- Resistance confirmed in leaf disc test in laboratory

In the following year phenylamide resistance appeared also in *Plasmopara viticola* on grapes in South Africa and southwestern France and in *Peronospora tabacina* on tobacco in Central America (Table 2). On tobacco, the phenylamides were so outstanding compared to the previously known residual fungicides, that the season-long exclusive use of the solo products led to resistance selection under the high disease pressure in that area. On grapes the first foci might have appeared in nurseries and spread from there to the vineyard, where mixtures with copper or folpet were used (Staub and Sozzi, 1981). The use of mixtures on grapes did not prevent the appearance and spread of resistant strains, but it did prevent wide spread performance problems.

WHERE HAD RESISTANCE-RISK STUDIES FAILED ?

The unexpected and dramatic occurrence of phenylamide resistance raised the question of the usefulness and the nature of the resistance risk studies done. Several studies done before the introduction of the first phenylamides showed that for *P. infestans* strains with decreased sensitivity could be selected in vitro, but such strains were either not pathogenic or they could not infect plants treated with phenylamides (Bruck et al, 1980; Staub et al, 1979). Furthermore, selection experiments on treated potato plants over 14 generations did not yield any resistant strains. Other labs obtained similar results with other pathogens (Bruin, 1980; Lukens et al, 1978). Mutagens were used only in some of these studies for fear of either creating a resistance problem or ending up with "unnatural" resistant mutants that would not occur in nature as is well known in the case of dicarboximides. Mutagens were included systematically in a detailed study on the possible development of phenylamide resistance in the soilborne pathogen *Phytophthora megasperma* (Davidse, 1981). In this study phenylamide resistant strains were produced that were fully pathogenic and that were not controlled on treated soybean plants. The lesson from these experiences is that resistance risk studies should include the use of mutagens, when other studies don't give sufficient information. However, as was the case with *P. megasperma*, such studies have to be done with pathogens that don't represent a risk for the farmers in the region. Resistance risk studies should assess the risk without increasing it.

RECOVERY THROUGH ADAPTED USE STRATEGIES

The dramatic developments in the second year after the first introduction of phenylamides led to a drastic revision of the use strategies against foliar pathogens. It was evident that all phenylamides belonged to the same cross-resistance group (Diriwächter et al, 1987), which meant that common anti resistance strategies had to be defined. In 1981 the Fungicide Resistance Action Committee (FRAC) was established to coordinate the anti-resistance strategies of cross-resistant fungicides. This met the urgent need of the phenylamide producers, who established a working group to deal with the rather dramatic situation facing the further use of these novel fungicides. The working group established the following general guidelines for use of phenylamides against foliar pathogens (Urech and Staub, 1985):

- sell only prepack mixtures with residual partner fungicides
- include high rates of residual partner: ¾ to full rate
- intervals should not exceed 14 days
- limitation to 2-4 sprays early in the season
- no curative or eradicative applications
- no soil applications against foliar pathogens
- and no use on seed potatoes and nurseries

The strategies established in the early 80s proved to be very successful. A crucial factor for the success was the readiness of all members of the phenylamide working group to cooperate in designing reasonable ant-resistance strategies and to fight for their implementation within their companies and with officials and farmers through effective communication. Please refer to the next chapter for further details on the role of FRAC in dealing with phenylamide resistance.

Against late blight on potatoes the performance of the phenylamide mixtures remained stable in spite of the presence of resistant strains; this was also true for the Netherlands and Ireland, where phenylamides were reintroduced in mixture with residual fungicides in 1985 after the residual compounds failed to control the heavy epidemics of the previous year. L. Dowley describes the Irish experience with phenylamides against late blight since 1985 in more detail in a separate chapter in these proceedings. In the rest of Europe phenylamide mixtures continued to provide late blight control at a level clearly superior to that of residual compounds alone. Fig. 1 illustrates this for an area in Switzerland where phenylamide resistance in *P. infestans* could commonly be detected. In this trial area the resistance level was estimated at 30 % at the beginning of the epidemic. For this estimate, samples were taken from the first infections in untreated border rows and analyzed with a semi-quantitative sensitivity assay (Nuninger et al, 1992). The control in the RIDOMIL MZ plots was clearly superior to that in the mancozeb plots. In other similar trials comparable results were obtained as long as the first treatments with the metalaxyl mixture was applied protectively before the onset of the epidemic. Where curative or eradicative situations occurred because treatments could not be applied in time, neither the mixtures nor the residuals alone provided sufficient control under continuous high disease pressure. This example illustrates that phenylamides used strictly according to the FRAC guidelines continue to contribute significantly to disease control even when detectable levels of resistance are present.

FIGURE 1: Late blight control by Ridomil MZ and Mancozeb in large plot trials
(FR 125-90, 250m2/plot; 30 % resistance in initial inoculum on border rows)

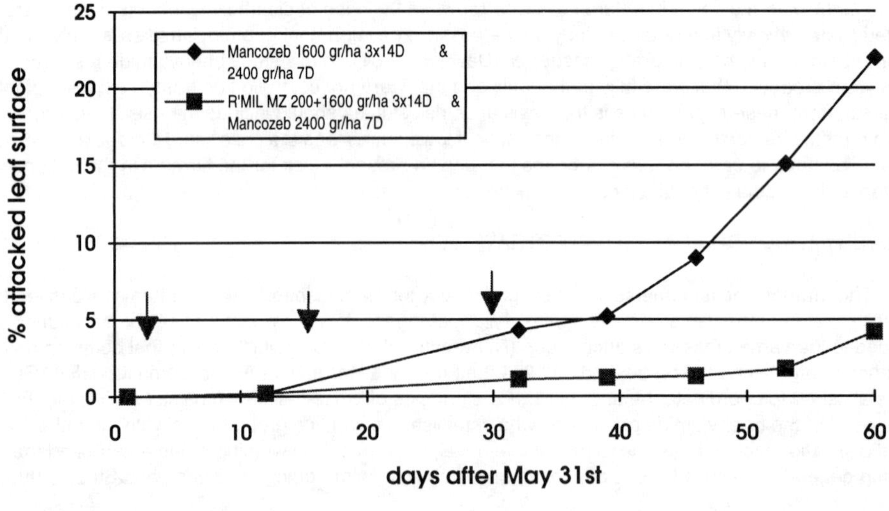

It was frequently surprising how well phenylamide mixtures still worked, even where monitoring revealed high levels of resistance. This apparent discrepancy was often a consequence of the late monitoring and an overestimation of the resistance levels with the simple leaf disc test. Therefore, a semi-quantitative test was developed (Nuninger et al, 1992) and monitoring has been increasingly focused on the first infections of epidemics, before treatments are made.

In grape downy mildew phenylamide resistance appeared first in South Africa and south western France. The cases were associated with high disease pressure and it appeared that nurseries might have been the initial foci (Staub and Sozzi, 1981). In France, phenylamide resistance fluctuated somewhat from 1983-87 depending on the disease pressure and the extent of phenylamide use. Resistance was confined for several years to the western part, where downy mildew pressure is heavier, and has only more recently reached the eastern grape growing regions in France. Time of appearance of phenylamide resistance in the vine growing regions of France was clearly correlated with the general downy mildew pressure of an area. With moderate use of phenylamide mixtures, they continue to provide reliable downy mildew control. In grapes the best use of the special strength of the phenylamides is around flowering when downy mildew attacks on the young bunches can cause big yield losses.

For tobacco blue mold, special strategies were necessary for different regions because of extreme disease pressure and the relatively poor performance of the residual mixture partner available. In the critical regions of central America, it was recommended to use phenylamide mixtures with the full dose of the residual partner in alternation with applications of the residual fungicides alone. This strategy has been successful where both alternations and residual spray intervals were implemented.

WHY DID THE ANTI-RESISTANCE STRATEGIES WORK ? -- FITNESS OF R STRAINS

In addition to the proper design and the vigorous implementation of anti-resistance strategies for phenylamides, some basic biological properties of the resistant strains seem to have contributed to the success of these strategies; they are related to the fitness and the population dynamics of the resistant strains. The examples described above suggest that phenylamide resistance tends to decrease in the absence of selection pressure. It is this phenomenon that makes it possible to design use strategies that lead to relatively stable resistance situations and to continued good contributions by the phenylamide partners in mixtures in spite of detectable levels of resistance in the target pathogen populations. This is a more favorable situation than that of benzimidazole resistance which tends to persist at high levels and where interruption of selection pressure does not lead to a decrease in the resistance level in the population.

The basis for the reduction of phenylamide resistance in the absence of selection pressure is not well understood, but it may be due to decreased fitness of the resistant strains compared to the wild type populations from which they emerged through mutation and selection. Fitness parameters that were identified as playing a role in this context are increased sensitivity to high temperatures for *P. viticola* (Piganeau and Clerjeau, 1985) and decreased survival in potato tubers for *P. infestans* (Walker and Cooke, 1990). Our own studies on the survival of *P. infestans* in potato tubers at low temperatures gave similar indications (Table 4). Twenty-two populations from the 1989 monitoring program in Switzerland and a few populations from other countries with resistance levels between 0.04 and 78% were inoculated into potato tubers and stored at 6 C for 8 months. At the beginning, the middle and the end of the storage period, the populations were analyzed for the resistance levels with the semi-quantitative monitoring assay (Nuninger et al, 1992).

TABLE 4: Development of phenylamide resistance in field samples[1] of *Phytophthora infestans* collected in 1989 and stored for 8 months at 6°C in potato tubers.

starting populations		development of resistance during storage				
% resistance	samples found	decrease		stable	increase	
		below DL[2]	2-5x	< 2x	2-5x	to 100%
0.04-0.39	4	4	.	.	.	.
0.4-3.9	3	3	.	.	.	.
4.0-39	11	6	.	1	3	1[3]
40-59	6	1	1	2	.	2[4]
>=60	3	3	.	.	.	.
Totals	27	17	1	3	3	3

[1] bulk samples collected 1989 in Switzerland (22), NL (2), IRL (2) and UK (1)
[2] DL = detection limit of semi-quantitative assay is 0.02%
[3,4] initial R frequencies were 35[3], 50 and 51%

At the end of the storage period 17 of the 27 populations had lost the resistant part of the population including several that had initial R frequencies of >40% (Table 3). In most of these cases resistance was no longer detectable after only 4 months of storage. This indicates a rapid decrease of the resistant portion in these populations by a factor of at least 2000x. In seven populations with relatively high initial R frequencies, the resistance level changed only slightly (less than 5x). In the three samples that reached 100% resistance at the end of the storage period, the initial R frequencies ranged from 35 to 51%. The cases where resistance was lost clearly dominate, so in most cases fitness for survival in potato tubers at 6 C seems to be reduced in resistant strains. The relatively small increases observed in some populations are most likely variation that occurred by chance or by the sampling procedures and they don't necessarily indicate fitness advantages of the resistant strains in these populations.

Reduced fitness for overwintering had also been described for strains of *P. infestans* collected in Israel (Kadish and Cohen, 1992) and Ireland (Walker and Cooke, 1990). The study from Israel showed a difference in the behavior of resistant strains in tubers and on the foliage. While survival of the resistant strains in tubers was clearly reduced they tended to be more aggressive than the sensitive ones on the foliage. This behavior was also confirmed by monitoring field populations between and during growing seasons.

THE ROLE OF MIGRATION VS INDEPENDENT MUTATION AND SELECTION EVENTS.

Already after the first cases of phenylamide resistance in 1980, the question was addressed whether resistance had appeared in many places independently or whether it appeared in one or a few places and spread from there throughout larger regions where it eventually led to the problems described above (Davidse et al, 1983). At the time race-typing with differential cultivars was used to look at some resistant Dutch *P. infestans* isolates collected in 1981. Phenylamide resistance was found in 10 of 23 physiological races detected in a survey. This indicated that several independent mutation and selection events had occurred in different places and that phenylamide resistance was genetically independent of any virulence traits in the different races. Only two of eight virulence genes detected in 1981 were not associated with phenylamide resistance at that time.

Improved methods of distinguishing between local developments and migrations were used to characterize the *P. infestans* populations on a world-wide basis (Fry et al, 1993). Interest in such studies was stimulated by the discovery of the A2 mating type in various parts of the world (Hohl and Iselin, 1984); previously, the A2 mating type was thought to be confined to regions of Mexico. The improved methods rely on allozymes patterns and on DNA fingerprinting to define clonal lineages. The main conclusion from the studies by Fry's group was that a large migration occurred during the 70s from Mexico into and through Europe and in the 80s from Europe to many other parts of the world (Fry et al, 1993). In the US and Canada a similar migration seems to have started in the last three years. The "new" population seems to be more aggressive than the "old" one. The long-distance spread is assumed to occur mainly via shipments of potato tubers and tomato seedlings infected by phenylamide resistant late blight strains.

For phenylamide resistance it is significant that it was mainly found in the "new" populations in Europe and that the "new" populations in the US seems to be largely resistant (Goodwin et al, 1994). Therefore, future anti-resistance strategies should include as a new element the prevention of the spread of resistant strains via seed and plant material. The presence of the A2 mating type is not linked genetically to phenylamide resistance, but it can be associated with it if clones with both A2 and phenylamide resistance happen to dominate an epidemic. The studies also showed that epidemics are often caused by only a few clones. Sexual recombination is still rare in Europe and the US in spite of the presence, side by side, of the two mating types. In Europe, the first indications have come from a study in Poland that sexual recombination may start contributing to the diversity in the populations of *P. infestans* (Sujkowsi et al, 1994). Increased sexual recombination will allow *P. infestans* to adapt more quickly to new sets of environmental factors. A new fungicide would be one such factor, fitness traits would be others that could be recombined through sexual recombination to the advantage of the pathogen and to the detriment of fungicidal efficacy.

CONCLUSIONS

From the early experiences with phenylamides, the following conclusion can be drawn that will help preserve the effectiveness of this valuable group of fungicides. Some of the lessons may be of a more general nature and apply also to other groups of fungicides with an inherent resistance risk.

<u>Phenylamide resistance is manageable</u>
Experience has shown that phenylamide resistance is manageable, if the proper anti-resistance strategies are employed. Depending on the pathogen, they emphasize the use of effective mixtures with residuals, a limitation of the application to the most crucial part of the epidemics, and the avoidance of curative use.

<u>Adherence to anti-resistance strategies is essential</u>
The strict implementation of the established anti-resistance strategies is absolutely essential. Non-compliance leads inevitably to performance problems and to a higher risk of a rapid build-up of resistant populations. This is especially crucial where detectable levels of resistance are already present.

<u>Fitness of phenylamide resistant strains is usually reduced</u>
Both circumstantial evidence and experimental results indicate that resistant strains are, as a rule, less fit than the wild type populations from which they emerged; resistant population diminish again when selection pressure is absent. The above strategies help prevent the selection of strains with both phenylamide resistance and normal fitness.

<u>For *P. infestans* migration is an important factor for the development of resistance</u>
This may lead to situations where the "new" imported resistant strains appear more fit and aggressive than the "old" sensitive ones. More care is required to minimize the spread of infected plant material that can carry with it resistant strains to areas previously free of resistance.

Phenylamides remain valuable part of arsenal against Oomycetes
Even where resistance is readily detectable, phenylamides used properly continue to contribute significantly to disease control over and above the one that can be achieved with residuals alone.

REFERENCES

Bruck, R.E.; Fry , W.E.; Apple, A.E. (1980) Effect of metalaxyl, an acylalanine fungicide, on developmental stages of *Phytophtora infestans*. Phytopathology, 70, 597-601.

Bruin, G. (1980) Resistance in Peronosporales to acylalanine type fungicides. Ph.D. Thesis, University of Guelph, Ontario, Canada, 110 p.

Davidse, L.C. (1981) Resistance to acylalanine fungicides in *Phytophthora megasperma f.sp. medicaginis*. Neth.J.Pl.Path., **87**, 11-24.

Davidse, L.C.;Danial, D.L.;van Westen, C.J. (1983) Resistance to metalaxyl in *Phytophthora infestans* in the Netherlands. Neth.J.Pl.Path., **89**, 1-20.

Diriwächter, G.; Sozzi, D.; Ney, C.; Staub, T. (1987) Cross-resistance in *Phytophthora infestans* and *Plasmopara viticola* against different phenylamides and unrelated fungicides. Crop Protection, **6**, 250-255.

Fry, W.E.; Goodwin, S.B.; Dyer, A.T.; Matuszak, J.M.; Drenth, A.; Tooley, P.W.; Sujkowski, L.S.; Koh, Y.J.; Cohen, B.A.; Spielamn, L.J.; Deahl, K.L.; Inglis, D.A.; Sanlan, K.P. (1993) Historical and recent migrations of *Phytophthora infestans*: Chronology, pathways and implications. Plant Disease, **77**, 653-6661.

Goodwin, S.B.; Sujkowski, L.S.; Dyer, A.T.; Fry, W.E. (1994) Recent migrations of the Irish famine fungus and its previous genetic isolation in the USA and Canada. Science (in Press)

Kadish, D.; Cohen, Y. (1992) Overseasening of metalaxyl-sensitive and metalaxyl-resistant isolates of *Phytophthora infestans* in potato tubers. Phytopathology, **82**, 887.889.

Lukens, R.J.; Cham, D.C.K.; Etter, G. (1978) Ortho 20615, a new systemic for the control of plant diseases caused by Oomycetes. Phytopath. News, 12, 142 (Abstract)

Nuninger, C.; Bolton, N.; de Wijs, J.J.; Staub, T. (1992) Determination of the sensitivity of *Phytophthora infestans* to phenylamides: a semi-quantitative test method. EPPO Bulletin, **22**, 309-311.

Piganeau, B.; Clerjeau, M. (1985) Influence différentielle de la température sur la sporulation et la germination des sporocystes de souches de *Plasmopara viticola* sensibles et résistantes aux phenylamides. In: Fungicides for Crop Protection, 100 Years of Progress. Vol. 2. British Crop Protection Council, Croydon, UK, pp. 327-330.

Schwinn, F.J.; Staub, T. (1987) Phenylamides and other fungicides against Oomycetes. In: Modern Selective Fungicides, H.Lyr (Ed.), Jena: VEB Gustav Fischer, pp. 259-273.

Staub, T.;Dahmen, H.; Urech P.A.; Schwinn, F. (1979) Failure to select for in vivo resistance in *Phytophthora infestans* to acylalanine fungicides. Plant Dis. Rep., **64**, 385-389.

Staub, T.; Sozzi, D. (1981) Résistance au métalaxyl en pratique et les consequences pour son utilisation. Phytiatrie-Phytopharmacie, **30**, 283-291.

Staub, T.; Sozzi, D. (1983) Recent practical experiences with fungicide resistance. Proc. 10th Internat. Congr. Pl. Prot., Brighton, UK, pp. 591-598.

Staub, T.; Sozzi, D. (1984) Fungicide resistance: A continuing challenge. Plant Disease, **68**, 1026-1031.

Sujkowski, L.S.; Goodwin, S.B.; Dyer, A.T.; Fry, W.E. (1994) Increased genetic diversity via migration and possible occurrence of sexual reproduction of *Phytophthora infestans* in Poland. Phytopathology, **84** (in press).

Urech, P.A.; Staub, T. (1985) The resistance strategy for acylalanine fungicides. EPPO Bulletin, **15**, 539-543.

Walker, A.S.L.; Cook, L. (1990) The survival of *Phytophthora infestans* in potato tubers - the influence of phenylamide resistance. Proc. Brighton Crop Prot. Conf. - Pests and Diseases, pp. 1109-1114.

ACKNOWLEDGEMENTS:

The author thanks B. Forster, C. Nuninger and C. Steden for critically reading the manuscript and for their useful suggestions.

DYNAMICS OF PATHOGEN RESISTANCE AND SELECTION THROUGH PHENYLAMIDE FUNGICIDES

U. GISI and L. OHL

Agrobiological Research Station, SANDOZ AGRO Ltd., CH-4108 Witterswil, Switzerland

ABSTRACT

Phenylamide-based products continue to successfully control diseases caused by fungi of the Peronosporales. Pathogen resistance to phenylamides however, has developed since the early 1980's in *Phytophthora* and *Plasmopara* species and may reduce efficacy if products are not used properly. Various antiresistance strategies contribute to lower the risk of resistant pathogen subpopulations becoming predominant. Mixtures of phenylamides with fungicides possessing a different mode of action delay resistance build-up significantly. The delaying effect is more pronounced if synergistic interactions between the components are at least as high for resistant as for sensitive strains. Sensitivity monitoring can assist resistance management and help explain problems of product performance. However, the sampling and sensitivity test methods used greatly influence the relevance of the data. Sensitivity monitoring by itself cannot be used to predict resistance development. It provides information on geographical and seasonal distribution, as well as on changes in pathogen resistance levels from year to year. Nevertheless, knowledge of the initial proportion of the resistant subpopulation at the beginning of the epidemics will help to better understand the subsequent process of selection. In order to interpret the sensitivity results, detailed studies should be made on the fitness and mating type distribution of field strains as well as their migration and appearance of novel pathogen genotypes.

INTRODUCTION

Phenylamide (PA-) fungicides, e.g. oxadixyl, metalaxyl, benalaxyl and ofurace, are single-site inhibitors with a high specific efficacy against fungi of the order Peronosporales. Their high specificity and consequent widespread use in agriculture selected phenylamide-resistant individuals within pathogen populations soon after introduction. Phenylamide resistance is a major-gene resistance and the selection process is disruptive. Phenylamide-resistant isolates may be as competitive as sensitive ones. Use strategies have been developed for phenylamide-based products to delay and reduce the development of resistance. As a result, they continue to contribute significantly to the control of diseases such as potato late blight and downy mildew of grape. Conditions under which such mixtures can be used successfully and their effect in delaying resistance build-up will be discussed in detail. Also, results on sensitivity monitoring from different countries generated according to the official PA-FRAC method descriptions (Gisi 1992) will be shown.

DISTRIBUTION OF PHENYLAMIDE-RESISTANT ISOLATES

The average amount of phenylamide resistant isolates of *P. infestans* collected in potato fields over the last years yielded the following approximate values: Switzerland 40%, The Netherlands 10-50%, England 10-30% and 50-80% in the early and late 80`s, respectively, Ireland 40-80% and France 50-90% (Table 1). The amount of resistant isolates significantly increased during the season (to 60-80%) according to the selection process, but started again at a low level (20-30%) at the beginning of the next season (Table 2). This observation was made in both treated and untreated fields in Switzerland. Samples of fields treated once, twice, three, and more than three times with PA-based products contained finally about 30%, 60%, 50%, and 80% resistant isolates, respectively (Majoros et al.,1993). Resistant isolates seem to be less fit than sensitive isolates in regard to overwintering capability (Walker and Cooke, 1990) but more fit in regard to colonization capability of leaves during epidemics (Cohen and Samoucha, 1990). According to O`Sullivan and Dowley (1993), resistant isolates became less frequent at the very end of the season (September) long after the last spray application of PA-based products.

In vineyards, phenylamide-resistant isolates of *P. viticola* are known to have occurred for at least 10 years. In vineyards not treated with PA-based products, sites containing resistant isolates were identified between 1983 and 1988 allover France, e.g. in the Bordeaux area

---

TABLE 1. Average amount (%) of phenylamide-resistant isolates of *P. infestans* from potato fields (annual means) in different european countries[a)]

year	% phenylamide resistant isolates					
	N.IRL[b)]	IRL[c)]	UK[d)]	NL[e)]	CH[f)]	F[g)]
1981	15	75	25	11	-	100
1982	25	21	15	5	-	90
1983	10	6	6	ld	-	97
1984	50	12	28	8	-	77
1985	15	8	20	12	-	46
1986	30	31	32	0	-	ld
1987	85	68	69	15	38	48
1988	90	84	81	43	45	67
1989	80	84	81	46	42	88
1990	67	57	78	-	41	71
1991	46	40	64	ld	ld	49
1992	65	38	46	-	73	93
1993	53	55	66	-	98	83

[a)] resistant isolates were also found in D, J, USA, CAN, Russia, South Africa, Israel
[b)]L. Cooke; [c)]L. Dowley; [d)]Ciba (80-85), ADAS (86-92), FRAC (93); [e)]L. Davidse; [f)]U. Gisi; [g)]S. Duvauchelle
- = no values available ld= low disease

17, 7, 6, 3, 30, and 60% of sites contained resistant isolates in the respective year, in the Bourgogne area 0, 20, 29, 17, 53, and 47% and in the Charentes area 93, 97, 74, 70, 48, and 67% of sites were found (Clerjeau, GRISP, 1988, pers. communication). In treated fields in France we have detected in 1988 up to 70% resistant isolates and in 1993, after several years of low downy mildew pressure, again a similar amount of isolates were resistant. Also in other countries (Italy, Switzerland, Spain, Portugal), resistant isolates were detected. Resistant isolates are known to occur also in *Pseudoperonospora cubensis, Bremia lactucae* and some other downy mildews (PA-FRAC information).

ESTIMATION OF RESISTANCE IN A POPULATION

Experience has shown that rather high numbers of resistant isolates can be found in the field but product performance remains good in most cases. An important question is then, what percentage of resistant sporangia in a field population may lead to performance problems.

TABLE 2. Seasonal variation of the amount (%) of phenylamide resistant isolates of *P. infestans* from potato fields treated with (PA+) or not treated with (PA-) phenylamide-based products [a] over seven years in Switzerland

year		% phenylamide resistant isolates			No of
		PA-	PA+	m	isolates
	Jn	17	23		
1987	Jl	-	50	37	35
	Ag	-	(50)		
	Jn	(27)	33		
1988	Jl	-	48	41	27
	Ag	-	-		
	In	(40)	20		
1989	Jl	25	59	42	69
	Ag	69	61		
	Jn	15	15		
1990	Jl	33	61	41	106
	Ag	87	66		
	Jn	-	-		
1991	Jl	-	-	-	-
	Ag	-	-		
	Jn	-	-		
1992	Jl	(75)	(100)	-	21
	Ag	-	-		
	Jn	100	99		
1993	Jl	99	99	99	148
	Ag	100	100		

[a] Phenylamide-based products (1-3 sprays per season) are used mainly during June and July
Jn = June, Jl = July, Ag = August, m = annual mean, - = no samples available
( ) = data base questionable, e.g. few isolates

So far, all routine sensitivity tests with *P. infestans* yield a resistant result for bulk samples if they contain at least 1-10% resistant sporangia. This figure was obtained by testing different mixtures of a sensitive and a resistant isolate of *P. infestans* against a range of fungicide concentrations (Table 3). The EC50 values increased more than 10 fold when amounts of r increased from 1 to 10%. For oxadixyl, a sharp increase of the RC50 values occurred between 10 and 100 mg/l (Table 3). Therefore, we should concentrate on the determination of amounts of resistant sporangia in a field isolate between about 1 and 10% rather than try to find levels down to 0.01%. More experimental work has to be done to determine the percentage of resistant sporangia present in field populations prior to the use of fungicide. In most cases, field isolates of *P. infestans* provide either a sensitive or resistant response, whereas in *P. viticola*, also intermediately resistant isolates can be found.

TABLE 3. Fungicidal activity of oxadixyl against a phenylamide-sensitive (0% r) and a phenylamide-resistant (100% r) isolate as well as against three different mixtures of both isolates (0.1, 1, 10%) of *P. infestans* in the potato leaf disc test

amount of r (%) in sporangium suspension	% inhibition of sporulation on leaf disc at .. mg/l oxadixyl					
	1000	100	10	1	0.1	EC50[a]
0	100	100	60	20	0	3
0.1	100	95	40	10	0	9
1	95	80	30	5	0	26
10	60	50	10	0	0	270
100	10	0	0	0	0	>500
RC50[b]	13	9	0.04	<0.01	<0.01	

[a]EC50 = calculated concentration of oxadixyl resulting in 50% inhibition of sporulation
[b]RC50 = calculated amount of r (%) allowing 50% inhibition of sporulation at a given fungicide concentation

TABLE 4. Percent phenylamide resistant sporangia of *P. infestans* in a mixed population after repeated fungicide treatment [a] with oxadixyl + mancozeb (o+ma = 1 + 7) and oxadixyl + mancozeb + cymoxanil (o+ma+c = 1 + 7 + 0.4) (Samoucha and Gisi, 1987)

sporangium generations	percent resistant sporangia collected from plants treated with oxadixyl + mancozeb					o+ma+c
0	0.01	0.1	1	10	50	50
1	<1	<1	8	12	63	45
2	1	6	12	20	72	46
3	8	12	22	40	90	40
4	10	32	45	60	100	45
5	22	70	90	100	100	43

[a] each fungicide treatment (100+700±40 mg/l) was made on a new set of plants inoculated with sporangia of the previous generation (cycling every 7 days)

SELECTION PROCESS THROUGH FUNGICIDE APPLICATION

A sensitive and a phenylamide-resistant strain of *P. infestans* were mixed in proportions of 0.01, 0.1, 1, 10, 20, and 50% r in the sporangium suspension (Table 4). Plants were inoculated and produced symptoms with a new sporangium generation after one week in a growth chamber. The first sporangium generation was harvested and used for inoculation of the second plant set. The applications of a PA-based mixture was made for each inoculation, providing a repeated fungicide selection pressure. The resistant subpopulation did not increase to more than 12% after four, three, and two sporangium generations, when initial r populations were 0.01, 0.1, and 1%, respectively (Table 4). Therefore, when there is not more than 1% resistance in a population, two to four applications of fungicide mixtures per season are justifiable without getting into serious resistance problems. Thus, the FRAC recommendations (Urech and Staub, 1985) to restrict the number of applications are fully supported by experimental data. The population treated with the two-way mixture was fully resistant after four cycles, whereas no increase of the resistant subpopulation was detected with the three-way mixture, even after eight sporangium generations (Table 4). Results similar to those with *P. infestans* were also found with *P. viticola* on grapes (Samoucha and Gisi, 1987). Thus, the addition of cymoxanil in mixtures containing phenylamides and mancozeb strongly delay the build-up of resistance.

Field-grown potato plants in plastic houses were inoculated with a population containing 10% resistant sporangia (Fig. 1). The developing disease was treated four times with two different mixtures, either oxadixyl+mancozeb or oxadixyl+mancozeb+cymoxanil. Percentage of resistant sporangia in the population were recorded over a period of 60 days. As seen in the growth chamber results (Table 4), the three-way mixture imposed only a very low selection pressure, whereas treatments with the two-way mixture produced complete resistance after about 35 days (Fig. 1). The fungicide mixture oxadixyl+mancozeb still provided about 90% disease control after two applications, despite a proportion of about 60% resistance sporangia in the population (Cohen and Samoucha,1990).

SYNERGISTIC INTERACTIONS

Synergistic interactions occur when mixtures of a phenylamide fungicide and one or several other fungicides (contact or systemic) active against *Phytophthora* or *Plasmopara* are applied (Gisi *et al.*, 1985, Gisi 1991). If interactions between fungicides are to be investigated under field conditions, it is essential to apply the components alone and in the mixture at identical rates and intervals. As an example of many field trials, Table 5 illustrates interactions between oxadixyl, mancozeb and cymoxanil which gave synergy ratios between 1.7 and 5.3. Cymoxanil-containing mixtures produced higher synergism for resistant than for sensitive strains. Dosages yielding 90% disease control were reduced significantly in the mixtures. The three-way mixture o + ma + c was not much affected by different sensitivities of the strains, whereas the two-way mixture o + ma was clearly less effective against resistant than against sensitive strains. In another field experiment, the potato crop was inoculated with a *P. infestans* sporangium suspension containing initially 10% resistant sporangia. Half rates of phenylamide-based products applied weekly reduced resistance build-up more effectively than the full rates (n-rate) of the mixtures applied at biweekly intervals and it provided also much

Fig 1.: Percentage resistance frequency of a *P. infestans* population containing initially 10% resistant sporangia developing on field grown potatoes treated four times (Tr, arrows) with oxadixyl (o), mancozeb (ma), oxadixyl+mancozeb (o+ma) or oxadixyl+mancozeb+cymoxanil (o+ma+c). In = Inoculation; S_{un} = first symptoms on untreated plants: S_{tr} = first symptoms on treated plants (Cohen and Samoucha, 1990)

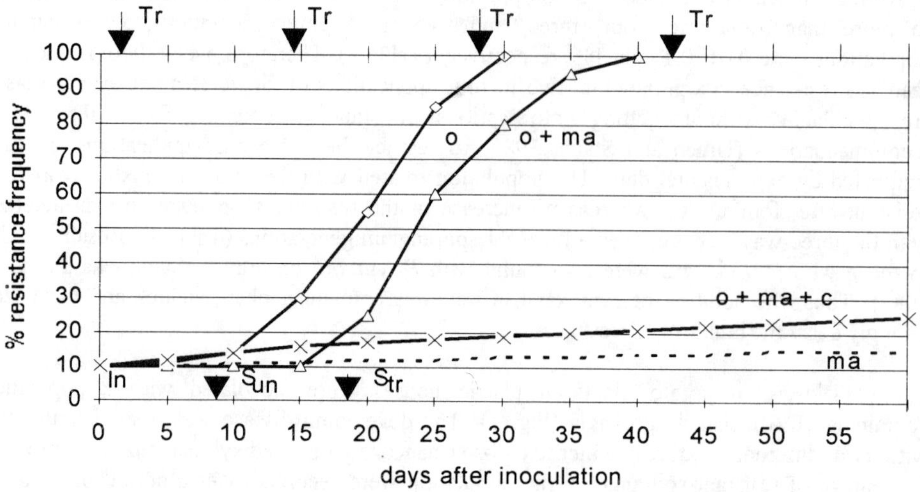

TABLE 5. Fungicidal activities (EC_{90}, g ha^{-1}) of oxadixyl (o), mancozeb (ma), cymoxanil (c) alone and in mixture and synergy ratio of mixtures against a phenylamide-sensitive and phenylamide-resistant isolate of *P. infestans* on potato under field conditions (Samoucha & Cohen, 1989)

fungicides	EC_{90} (g ha^{-1}) sensitive	resistant	Synergy ratio sensitive	resistant
o	60	>2000	-	-
ma	760	1130	-	-
c	350	310	-	-
o+ma = 1+7	100 (12+88) [a)]	550 (69+481)	3.2	2.3
ma+c = 7+2	220 (177+43)	260 (209+51)	2.8	2.9
o+ma+c = 1+7+2	110 (11+77+22)	150 (15+105+30)	2.8	5.3

[a)] Figures in parentheses are dosage of the individual components in the mixture. Recommended rates for *P. infestans* control in many countries are o+ma+c = 200+1400+80 g ha^{-1}

better disease control (Samoucha *et al.* , 1993). This observation shows that a reduced selection pressure (in this case lower dosage per treatment) delays the build-up of resistance.

CONCLUSIONS

Phenylamide-based products continue to control successfully diseases caused by fungi of the Peronosporales. Resistance to the phenylamide component can develop and cause problems if the products are not carefully used. Sensitivity monitoring can assist resistance management and help explain problems of product performance, but the sampling and sensitivity test methods used can greatly influence the relevance of the data. In the end, product performance is more important than the estimation of the amount of resistant subpopulations. Different antiresistance strategies can be used depending on the local conditions and the overall resistance risk. Mixtures of phenylamides and other fungicides represent an effective way to delay resistance build-up. Synergistic mixtures allow a decrease of the amount of active ingredients without reducing the overall activity. Decreased dosage of fungicides in synergistic mixtures lower the risk of selecting for resistant strains.

A devil's advocate may ask several questions after having read this paper: 1) Why should phenylamide-based products control the diseases in the field when sensitivity tests show that the populations are resistant against phenylamides and since contact fungicides may be removed by rain in many situations (and in case of a three-way mixture, cymoxanil may be degraded by the plant and the fungus within a few days)? 2) Why should we continue any sensitivity monitoring after having detected consistantly more than 50% of isolates to be resistant in several countries and crops? 3) What does the detection of phenylamide-resistant strains mean, since PA-based products are still showing good disease control in the field? 4) Why do we find an increasing amount of phenylamide-resistant isolates in fields never treated with phenylamide-based products?

Possible answers to the four questions may be given as follows: At the beginning of an epidemic, field populations are obviously not resistant to such a degree that products would give inadequate disease control. The overall product performance is a result of synergistic interactions also under situations of resistant subpopulations rather than a separate event of single ingredients. Mixtures extend the duration of activity of cymoxanil and delay rather than control resistance build-up. Today's routine sensitivity tests probably overestimate the amount of resistant sporangia in field populations. The frequency of resistant subpopulations fluctuates during the season and from year to year. During extensive monitoring programs one may estimate the actual amount of resistant isolates in a population, but more important, we also begin to understand many aspects of population biology like distribution and migration of the fungus in a region and over time, as well as fitness parameters (epidemiological and overwintering properties) and phenotypic and genotypic behaviour. Today's monitoring methods may not be adequate to properly estimate the amount of resistant sporangia in a field population. Furthermore, we always neglect the majority of cases in which the products gave good control, because without detectable disease left in the field after application of products, we cannot collect samples for sensitivity tests. Moreover, sensitivity tests are normally done with a single active ingredient, not with products. Parallel to the appearance of resistant subpopulations, the migration of new races and the appearance of both mating types probably became more important for epidemics of *P. infestans*. More studies like those by Fry *et al.* will be needed to better understand the permanently changing scenery of late blight in potatoes and downy mildew in grapes.

REFERENCES

Cohen, Y.; Samoucha, Y. (1990) Competition between oxadixyl-sensitive and -resistant field isolates of *Phytophthora infestans* on fungicide-treated potato crops. *Crop Protection* **9,** 15-20.

Fry, W.E.; Goodwin, S.B.; Dyer, A.T.; Matuszak, J.M.; Drenth, A.; Tooley, P.W.; Sujkowski, L.S.; Koh Y.J.; Cohen, B.A.; Spielman, L.J.; Deahl, K.L.; Inglis, D.A.; Sandlan K.P. (1993) Historical and recent migrations of *Phytophthora infestans*: Chronology, pathways, and implications. *Plant Disease* **77,** 653-661.

Gisi, U. (1991) Synergism between fungicides for control of *Phytophthora*. Chap. 24 (pp. 361-372) in J.A. Lucas, R.C. Shattock, D.S. Shaw and L.R. Cooke (Eds.), Phytophthora. *Cambridge University Press, Cambridge,* 447 pp.

Gisi, U. (1992) FRAC methods for monitoring the sensitivity of fungal pathogens to phenylamide fungicides, developed by the phenylamide fungicide resistance action committee (PA-FRAC) of GIFAP. *EPPO Bulletin* **22,** 297-322.

Gisi, U.; Binder, H.; Rimbach, E. (1985) Synergistic interactions of fungicides with different modes of action. *Transactions of the British Mycological Society* **85**, 299-306.

Majoros, K.; Gees, R.; Gisi, U. (1993) Studies of mating type and phenylamide sensitivity behaviour in *Phytophthora infestans*. pp. 175-182 in H. Lyr and C. Polter (Eds), *Proceedings of the 10th International Symposium on Systemic Fungicides and Antifungal Compounds, Ulmer,* 463 pp.

O`Sullivan, E.; Dowley, L.J. (1993) Quantitative assessment of phenylamide resistance in *P. infestans* and its influence on fungicide use. pp. 24-25 in *Abstracts of the 12th Triennial Conference of the European Association for Potato Research, INRA,* 511 pp.

Samoucha, Y.; Baider, A.; Cohen, Y.; Gisi, U. (1993) Control of late blight in potato by full and reduced rates of oxadixyl mixtures. *Phytoparasitica* **21,** 69-73.

Samoucha, Y.; Cohen, Y. (1989) Field control of potato late blight by synergistic fungicidal mixtures. *Plant Disease* **73,** 751-753.

Samoucha, Y.; Gisi, U. (1987) Use of two- and three-way mixtures to prevent build-up of resistance to phenylamide fungicides in *Phytophthora* and *Plasmopara. Phytopathology* **77,** 1405-1409.

Urech, P.A.; Staub, T. (1985) The resistance strategy for acylalanine fungicides. *EPPO Bulletin* **15,** 539-543.

Walker, A.S.L.; Cooke, L.R. (1990). The survival of *Phytophthora infestans* in potato tubers: the influence of phenylamide resistance. *Brighton Crop Protection Conference - Pests and Diseases,* 1109-1114.

PRACTICAL ASPECTS OF PHENYLAMIDE RESISTANCE MANAGEMENT

L. J. DOWLEY

Plant Pathology/Entomology Department, Oak Park Research Centre, Carlow, Ireland.

ABSTRACT

Phenylamide resistant (PA) strains of *Phytophthora infestans* were first confirmed in Ireland in 1980 and PA based products were withdrawn from the market. A programme to monitor the distribution of PA resistance confirmed that in the absence of PA based products the distribution of PA resistance decreased rapidly between 1981 and 1983. This apparent lack of fitness was later associated with lower sporangia production per unit leaf-area and the reduced ability to overwinter in tubers stored at low temperatures.

Records of the date of initial outbreak of late blight in unsprayed crops of the cv. "Kerr`s Pink" confirmed that while the average date of outbreak varied between different areas of the country, it was normally after the middle of July. The monitoring programme also confirmed that the distribution of PA resistance was lowest at the beginning of the season and increased thereafter as the season advanced.

Initial experiments on disease control confirmed that there was no significant benefit in using more than 3 sprays of a phenylamide/mancozeb mixture (PAM) in a 10 spray programme based on mancozeb. It was also confirmed that the three systemic sprays were most effective when applied at the beginning of the spraying programme. No further benefit was observed in decreasing the number of systemic sprays below three.

These facts contributed towards the development of an agreed label recommendation which was based on a maximum of 3 early applications of PAMs followed by mancozeb alone to the end of season. This was in conjunction with normal spray intervals and avoiding curative use and use on early potatoes.

Commercial introduction of the PAMs was initially agreed for 1986. However, a severe blight epidemic in early 1985 resulted in the importation and widespread curative use of PAMs in July of that year. Their dramatic effect in controlling the 1985 epidemic when used curatively had a seriously detrimental effect on the official advice for only prophylactic use. A major education programme over a number of years was required before growers were finally convinced of the advantages of following the label recommendations.

INTRODUCTION

Late blight caused by *Phytophthora infestans* (Mont.) de Bary is the most serious disease affecting potatoes in Ireland. Under Irish conditions it has been shown that the disease reaches epidemic proportions in 7 out of 10 years (Cox and Large, 1947). The first systemic fungicide for the control of late blight in potatoes was commercially introduced to Ireland in 1978 in the form of Ridomil 25 WP. The fungicidal activity of this product was based only on the phenylamide (PA), metalaxyl. The early trial results confirmed that metalaxyl was very effective in controlling foliage and tuber blight even at a 21 day spraying interval. The improved control of late blight following the application of metalaxyl led to it`s rapid adoption by growers. By 1980 up to 90% of all potato growers were using the product and over half of this on the cultivar `Kerr`s Pink`. Late in the 1980 growing season control problems were observed in this cultivar. It was confirmed that this was due to a resistant strain of *Phytophthora infestans* (Dowley & O`Sullivan, 1981) and the PA based fungicides were withdrawn from the Irish market. This was the first practical lesson in resistance management showing that effective, long lasting control could not be achieved using the PA alone in a traditional blight control programme.

INITIAL DISTRIBUTION OF PA RESISTANCE

Using the leaf disc test (Staub & Sozzi, 1984), a programme for monitoring the distribution of PA resistance was initiated during the 1981 growing season. This programme confirmed that in the absence of PA based products the distribution of crops with PA resistance present decreased rapidly between 1981 and 1983 (Dowley & O`Sullivan, 1985). This suggested a lack of fitness on the part of the PA resistant strains. Subsequent experiments confirmed that these strains produced fewer sporangia per unit leaf-area than sensitive strains (Dowley, 1987). Further work in Northern Ireland (Walker & Cook, 1990) confirmed that PA resistant strains of *P. infestans* did not survive overwintering at low temperatures in infected tubers as efficiently as sensitive strains. This would suggest that the reduced fitness of resistant strains may be controlled by a number of different genes.

DEVELOPMENT OF AN ANTI-RESISTANCE STRATEGY

Records of the initial date of outbreak of late blight in unsprayed crops of the maincrop cultivar "Kerr`s Pink" confirmed that while the average date of outbreak varied between different parts of the country, it was normally after the middle of July (Frost, 1966). In an average year this would allow 3 sprays of a phenylamide/ mancozeb mixture (PAM) to be applied prior to the first outbreak of late blight. This would also help to reduce the curative use of PAMs.

The resistance monitoring programme also confirmed that there was seasonal variation in the distribution of PA resistance. In each year, the distribution of PA resistance was found to be lowest at the beginning of the season and subsequently increased as the season advanced (Table 1). This was evident for both the period following the withdrawal of PAs in 1980 and after the commercial reintroduction of the PAMs in 1986.

Initial experiments on disease control confirmed that the best results were obtained from applyinig the PAMs for the first three sprays in the programme and following with mancozeb to the end of the season (Table 2). There was no significant benefit in using more than 3 sprays of a phenylamide/mancozeb mixture (PAM) in a 10 spray programme based on mancozeb and no further benefit was observed in decreasing the number of systemic sprays below three (Table 2). It was also believed that because of increasing maturity, the translocation of metalaxyl became less efficient from the time of flowering to the end of the growing season and therefore should be avoided during this period.

These facts were important in helping the Irish Fungicide Resistance Action Committeee (F.R.A.C.) to formulate an agreed label recommendation for the use of phenylamide/mancozeb mixtures. The main constituents of this recommendation were as follows:-

1. PAMs should be used for the first 3 sprays in the programme
2. Spray interval should not exceed 10 days
3. Subsequent sprays should be with mancozeb alone
4. No PAMs should be used after the end of July
5. PAMs should not be used curatively at any time
6. PAMs should not be used on early potatoes
7. PAMs should not be used on seed potatoes.

INTRODUCTION OF PHENYLAMIDE/MANCOZEB MIXTURES

Commercial introduction of the PAMs was initially agreed for 1986. However, a severe blight epidemic in early 1985 resulted in the importation and widespread curative use of PAMs in July of that year. Their dramatic effect in controlling the 1985 epidemic when used curatively had a seriously detrimental effect on the subsequent prophylactic use PAMs as was officially advised. Following the experience of 1985 the growers were convinced that the PAMs were best used curatively or as a fall back position in the event of poor disease control from the protectant products. This was further aggravated by the much higer price of the systemic type products. Over five years of experience and education was required before growers began to use the PAMs according to the label recommendations (Table 6).

TRIAL RESULTS 1985-`90

Field trials to determine the effects of prophylactic use of PAMs confirmed that when they were used according to the label recommendations they had the effect of delaying the onset of the epidemic, reducing the level of foliage blight at the end of the season, decreasing the incidence of tuber blight and increasing the marketable yield (Table 3). PA resistance was confirmed for all trials sites used from 1985 to 1990.

Information on the levels of foliage blight was also collected from commercial crops through the PA resistance testing programme during the period 1985-`90. This confirmed that where PAMs had been used the level of foliage blight was lower than in crops where a protectant product only was used. Information on fungicide programmes was also collected from crops with no foliage blight present. This showed that the vast majority of crops with no foliage blight had received PAM applications and that these had been applied according to label recommendations.

CURRENT PERFORMANCE OF PAMs

During the period 1991-`93 disease pressure was low but the PAMs have performed well in both field trials and at farm level. In previous years our experience has been that PAMs tended to give their best results in years when the blight epidemics were most severe.

CURRENT DISTRIBUTION OF PA RESISTANCE

The use of phenylamides at farm level from 1986 to 1993 is given in Table 4. In 1986 PAMs were used on nearly all potato crops. In the following years there was a decrease in the use of phenylamides up to 1990 when their use again increased. The PA resistance monitoring programme using the leaf disc method was continued during the period 1986-`93 (Table 5). This confirmed that during the initial years after the introduction of the PAMs there was an increase in the number of crops where PA resistance could be detected (Dowley & O`Sullivan, 1991). This increase in resistance distribution was associated with a high proportion of growers using a phenylamide (Table 4) and also with lack of adherence to the label recommendations. The highest number of crops with PA resistance was detected in 1989 (83%), but the distribution of PA resistance has since fallen steadily.

PAMs are now used on about half the potato crops in the Republic of Ireland and the decrease in the distribution of PA resistance may be associated with the increase in the number of growers who are currently following the recommendations (Table 6).

CURRENT LEVEL OF PA RESISTANCE IN CROPS WITH RESISTANCE

During the three year period 1990-`93 the semi-quantitative (SQ) test has been used to determine the level of PA resistance within crops where PA resistance had been confirmed using the leaf disc test. This revealed that the average level of resistance for each year over this period was slightly in excess of 50% of the population. This fact may explain the superior performance of PAMs in crops where resistance was known to occur.

Continued resistance monitoring and field trials are necessary to confirm the continued advantages of the phenylamide/mancozeb mixtures for blight control in Ireland.

TABLE 1. Seasonal variation in the distribution of PA resistance in the Republic of Ireland.

Month	% Crops with PA resistance	
	1981-`85	1986-`93
June	0.5	16.6
July	5.0	54.8
August	20.0	63.7
September	24.7	52.9
Number of crops surveyed	281	1,167

TABLE 2: Comparison of different phenylamide based strategies on foliage and tuber blight (Oak Park, Carlow, 1985)

Treatment	% Foliage Blight (end of season)	Tuber Blight (t/ha)
Unsprayed	100	0.13
Mancozeb all season	46	0.21
PMM 2nd & 3rd spray	28	0.05
PMM 1st spray	26	0.02
PMM 2nd spray	23	0.13
PMM 3rd spray	31	0.07
PMM All Season	29	0.02
PMM 1st, 2nd & 3rd spray	19	0.00
LSD (5%)	27	0.12

Table 3: Effect of phenylamides on late blight control in Ireland. (Mean of National Blight Trial, Carlow, 1985-`90)

Treatment	Delay (days) in Disease Onset	% Foliage Blight end of Season	Yield in t/ha	Tuber Blight t/ha
Unsprayed	0.0	95.6	34.69	0.44
Mancozeb	9.8	36.7	42.50	0.24
Galben M	14.3	34.3	44.31	0.14
Ripost	21.5	18.4	46.93	0.07
Ridomil MZ 72	21.7	16.7	46.63	0.09

TABLE 4: Annual use of phenylamides in Ireland (Resistance monitoring programme).

Year	No. crops Sampled	% Crops on which used
1986	104	100
1987	97	92
1988	68	72
1989	23	43
1990	50	38
1991	40	42
1992	43	39
1993	49	51

TABLE 5: Frequency of PA resistance in the Republic of Ireland 1986-`93

Year	No crops with late blight	% Crops with resistance
1986	92	29
1987	88	76
1988	66	83
1989	23	83
1990	48	56
1991	37	46
1992	38	45
1993	45	64

TABLE 6: Date of application of first phenylamide spray

Year	Sample Size	% growers applying first spray before		
		June 16	June 25	June 30
1990	12	0	8	25
1991	13	0	8	39
1992	14	21	50	57
1993	23	26	52	74

REFERENCES

Cox, A. E.; Large, E. C. (1960) Potato blight epidemics through out the world. Agricultural Research Service, United States Department of Agriculture, *Agricultural Handbook No. 174.* 230 pp.

Dowley, L. J,; O`Sullivan, E. (1981) Metalaxyl-resistant strains of *Phytophthora infestans* (Mont.) de Bary in Ireland. *Potato Research* **24**:417-421.

Dowley, L. J.; O`Sullivan, E. (1985) Monitoring metalaxyl resistance in populations of *Phytophthora infestans. Potato Research* **28**:531-534.

Dowley, L. J. (1987) Factors affecting the survival of metalaxyl-resistant strains of *Phytophthora infestans* (Mont.) de Bary in Ireland. *Potato Research* **30**:473-475.

Dowley, L. J.; O`Sullivan, E. (1991). Changes in the distribution of phenylamide-resistant strains of Phytophthora infestans (Mont.) de Bary in Ireland 1985-89. *Potato Research* **34**:67-69.

Frost, C. (1966) Potato blight, effect of fungicides on tuber blight and yield. *Proc. Irish Crop Protection Conference,* 120-128.

Staub, T.; Sozzi, D. (1984). Fungicide resistance: A continuing challenge. *Plant Disease* **18**:1026-1031.

Walker, A. S. L.; Cook, L. (1990). The survival of *Phythophthora infestans* in potato tubers - the influence of phenylamide resistance. *Proc. Brighton Crop Prot. Conf. - Pest and Diseases*, pp. 1109-1114

[illegible]

[illegible] (1990). [illegible] Agricultural Research Service, United States Department of Agriculture. [illegible] 250pp.

[illegible]

[illegible]

[illegible]

[illegible]

[illegible]

[illegible]

[illegible]

VARIATION FOR RESPONSE TO PHENYLAMIDES IN UK POPULATIONS OF *BREMIA LACTUCAE* (LETTUCE DOWNY MILDEW) AND *PERONOSPORA PARASITICA* (BRASSICA DOWNY MILDEW)

I.R. CRUTE, P.L. GORDON, N.A. MOSS[a]

Plant Pathology and Weed Science Department, Horticulture Research International, Wellesbourne, Warwick, CV35 9EF, UK

ABSTRACT

Failure to control downy mildew of horticultural brassicas and lettuce due to phenylamide insensitivity was first reported in the UK in 1983. In both cases, there was evidence that a single insensitive clone was rapidly disseminated throughout production regions with transplanting material. The insensitive population of *Peronospora parasitica* appeared to have become phenotypically more diverse a few years later. Over a 10 year period, there have been cycles of increase and decline in three different phenylamide insensitive pathotypes of *Bremia lactucae* that became prevalent among the 12 that have been characterised. Decline has always been associated with widespread adoption of an integrated control strategy involving the deployment of cultivars carrying resistance genes against which the prevalent insensitive pathotype was avirulent.

INTRODUCTION

Downy mildews of cruciferous vegetables (caused by *Peronospora parasitica*) and lettuce (caused by *Bremia lactucae*) are economically damaging diseases in the UK and most other important production regions throughout the world (Channon, 1981; Crute, 1992a). During young plant propagation, both diseases can result in serious crop losses and a high level of control is demanded. Subsequently, disease in the mature crop adversely affects quality and may in extreme circumstances, lead to complete loss of marketable yield.

Exceptionally effective control of these diseases became possible in the late 1970s following the introduction of phenylamide fungicides, and in particular, products containing metalaxyl. However, in the UK, the high selection pressure imposed by sustained and frequent use of fungicides containing metalaxyl resulted in the emergence of pathogen variants with a high level of insensitivity 5 years after initial commercial use (Crute *et al.*, 1985). Field control failures due to insensitivity were subsequently reported from other countries (Table 1).

[a] Present address: AgrEvo UK Limited., Chesterford Park, Saffron Walden, Essex, CB10 1XL.

TABLE 1. Reports of phenylamide insensitivity in *B. lactucae* and *P. parasitica*.

Year	*Bl*	*Pp*	Reference
1983	UK	UK	Crute *et al.*, 1985
1987	France		Leroux *et al.*, 1988
	California, USA		Schettini *et al.*, 1991
1989	Florida, USA		Raid *et al.*, 1990

Bl = *Bremia lactucae*; *Pp* = *Peronospora parasitica*

Circumstances associated with the initial failure to control *B. lactucae* with phenylamide fungicides in UK production have been fully documented elsewhere (Crute, 1987; Crute *et al.*, 1987) as have details of an effective integrated control strategy based on continued fungicide use in combination with cultivars carrying an appropriate downy mildew resistance gene (*Dm* genes) or gene combination (McPherson & Crute, 1986; Crute, 1987; Crute 1992a, 1992b, 1992c). This paper provides an up-dated summary of studies conducted over a 10 year period on variation in the UK population of *B. lactucae* with respect to cultivar specific virulence, sensitivity to metalaxyl and sexual compatibility type (SCT).

Studies on the occurrence and distribution of phenylamide insensitive variants of *P. parasitica* associated with control failure in UK production of horticultural brassicas (all forms of *Brassica oleracea*) have not previously been documented. This paper provides information from studies conducted between 1983 and 1985, soon after the initial discovery of phenylamide insensitivity, together with further information gathered a few years later in 1989.

MATERIALS AND METHODS

A single mass-spore pathogen culture was derived from each disease sample obtained using methods previously described for the establishment and maintenance of *P. parasitica* and *B. lactucae* (Crute *et al.*, 1985, 1987; Moss *et al.*, 1994). All samples were obtained from crops being grown commercially in the UK.

Methods for determination of cultivar specific virulence and SCT type were as described fully elsewhere (Michelmore & Crute, 1982; Crute *et al.*, 1987; Moss *et al.*, 1994). The response of pathogen isolates to phenylamide fungicides was assayed on fungicide treated seedlings grown and inoculated under controlled conditions. Full details of the methods employed have been published previously (Crute *et al.*, 1985; Crute, 1987; Moss *et al.*, 1994). Variation for fungicide sensitivity was described by the incidence of sporulation and the time taken for it to occur on a batch of inoculated seedlings treated with a range of fungicide concentrations. Standard control isolates with a known response to fungicide were included in all assays. For each batch of treated seedlings, a value for the mean reciprocal latent period (time in days from inoculation to sporulation; 1/LP) was calculated and expressed as a percentage of this value obtained following inoculation of untreated seedlings. Reciprocals were used to allow seedlings on which the fungus did not sporulate to be readily included in calculations (1/LP = 0). On this basis, isolates that were completely insensitive to a particular concentration of fungicide had values close to 100 while those that were completely sensitive had values close to 0. A few isolates of both *P. parasitica* and *B. lactucae* consistently expressed differing degrees of intermediate response as indicated by values between these extremes. Isolates thought to be heterozygous at a locus (*P*) controlling response to phenylamide fungicides have previously been shown to exhibit an intermediate response phenotype (Crute & Harrison, 1988; Moss *et al.*, 1994).

RESULTS

Peronospora parasitica

Isolates of *P. parasitica* highly insensitive to phenylamide fungicides were first obtained from Lincolnshire, UK in 1983; a commercial propagator of brassica plants experienced failure to control the disease with metalaxyl. Subsequently, insensitive isolates were recovered from mature crops in the same region (Crute *et al.*, 1985). During 1984 metalaxyl insensitive isolates of the pathogen were also recovered from crops in Norfolk, Lancashire, Cambridgeshire, Cheshire, Worcestershire, Fife, Mid-Lothian and Borders. During 1985, phenylamide insensitive isolates from Jersey, Kent and Cardigan were also recovered. There was circumstantial evidence that a phenylamide insensitive pathotype had been rapidly disseminated with module-raised plants from which transplanted crops were produced. During 1985, 16 brassica propagation sites in Lincolnshire were surveyed between March and June for incidence of downy mildew; isolates obtained from sites where mildew was present were assayed for response to metalaxyl (Table 2). By April, there was downy mildew present at most sites and in excess of 60% of sites yielded insensitive isolates. This confirmed the prevalence of the problem and the conjecture that planting material provided an efficient means for dispersal of the pathogen.

TABLE 2. Incidence and response to metalaxyl of *P. parasitica* at 16 brassica propagation sites in Lincolnshire, UK (March - June 1985).

	March	April	May	June
Incidence[a]	38	56	50	63
Response to metalaxyl[b]	67	67	63	80

[a] Percentage of sites (sample size = 16) from which a viable sample of *P. parasitica* was obtained for testing.

[b] Percentage of those sites from which a viable *P. parasitica* sample was obtained where metalaxyl insensitivity was confirmed.

An interesting characteristic observed for the standard phenylamide insensitive isolate studied (referred to as P006) was the lack of complete cross-insensitivity between metalaxyl and cyprofuram (Table 3) (Crute *et al*, 1985; Moss *et al*., 1994). While the standard sensitive isolate (referred to as P005) was more sensitive to metalaxyl than cyprofuram, the opposite was the case for isolate P006. Cyprofuram retained significant activity at 10 μg/ml in assays of P006 while metalaxyl exerted no control at this concentration and exhibited little activity at 100 μg/ml. Isolate P006 was also shown to be of P2 SCT and was virulent on the a land-race cauliflower (Palermo Green) previously identified as expressing isolate specific resistance to *P. parasitica*.

Subsequently, 13 isolates from seven counties obtained during 1983-1985 were tested for SCT, virulence on Palermo Green and response to metalaxyl and cyprofuram. All 13 isolates were phenotypically identical to P0583 which strongly suggests that a single pathotype of clonal origin had become widely distributed during this period.

TABLE 3. Differential response of *P. parasitica* to metalaxyl and cyprofuram.

	μg/ml				
Isolate	0.01	0.1	1.0	10.0	100.0
P005	85[a]	0	0	0	0
Metalaxyl					
P006	107	106	105	101	80
P005	109	84	2	0	0
Cyprofuram					
P006	99	99	101	5	0

[a] Values are 1/LP as a % of untreated

In 1989, a small sample of seven phenylamide insensitive isolates was obtained from Lincolnshire and these were also subjected to phenotypic characterisation for SCT, virulence on Palermo Green and response to metalaxyl and cyprofuram (Table 4). Interestingly, five different phenotypes were evident among the seven isolates. Two isolates were identical to those derived from the 1983/85 sample indicating that the original clone was probably still prevalent in the population. Isolates avirulent on Palermo Green, either homothallic or of P1 SCT and not exhibiting differential sensitivity between cyprofuram and metalaxyl were represented among the other five isolates. This diversity of phenotypes is indicative of either sexual recombination or the selection, since 1983, of novel phenylamide insensitive mutants.

TABLE 4. Phenotypic characteristics of seven phenylamide insensitive isolates of *P. parasitica* obtained from Lincolnshire in 1989.

Phenylamide response	F(M>C)	I(M=C)	F(M=C)	F(M>C)	F(M>C)
"Palermo Green"	V	V	nt	V	A
SCT	P2	P1	nt	H	H
Number of isolates	2	1	1	2	1

V = virulent; A = avirulent; H = homothallic
F = fully insensitive; I = intermediate;
M>C = differential response to metalaxyl and cyprofuram (see Table 3);
M=C = equivalent response to metalaxyl and cyprofuram

Bremia lactucae

In a continuing study of variation within the UK population, over 520 isolates of *B. lactucae* have been assayed since 1983 for cultivar specific virulence, SCT and response to metalaxyl. Over this period, 12 phenotypically distinct metalaxyl insensitive pathotypes have been identified (Table 5) of which three have sequentially become widespread and caused significant problems for UK lettuce producers. While several of these pathotypes are fully insensitive, others express differing levels of intermediate sensitivity (Table 5). During 1992 and 1993, a pathotype of intermediate insensitivity, referred to as "1992A", became prevalent in several UK lettuce production regions. This indicates that field control failure is not necessarily always associated with the occurrence of isolates fully insensitive to phenylamide fungicides, an observation borne out by similar experiences in California (Schettini *et al.*, 1991).

A pathotype referred to as the "NL10 type" caused the initial failure of metalaxyl to control lettuce downy mildew; this pathotype was first located in Lancashire and between 1983 and 1986 it became distributed to all lettuce production regions in the UK (Crute *et al.*, 1987) (Table 6). Isolates of the "NL10 type" have not been recovered since 1988 when a single sample from Northern Ireland proved to be of this pathotype. The decline of the "NL10 type" was associated with encouragement to grow cultivars of lettuce carrying *Dm11* on which this pathotype is avirulent (Table 5). In 1987 the first occurrences of an insensitive pathotype referred to as the "NL15 type" were recorded; this pathotype was virulent on cultivars carrying *Dm11* (Table 5) and rapidly became prevalent throughout lettuce production regions during 1988 and 1989 (Table 6). The "NL15 type" was countered by encouragement to grow cultivars carrying either *Dm6* or *Dm16* on which it was avirulent. A phenylamide insensitive pathotype identical to the "NL15 type" was also reported to be prevalent in France during 1987/88 (Leroux *et al.*, 1988). Only a few isolates of this pathotype have been recovered from UK crops each year since 1990 indicative of the success of the control strategy. In 1992 several new insensitive pathotypes were identified mostly virulent on cultivars carrying *Dm16* which were by this time being favoured by growers. Of these new pathotypes, one, referred to as "1992A", has predominated (Table 6). This pathotype is avirulent on cultivars carrying *Dm3* and *Dm6* and it should therefore be possible readily to counteract it by choice of appropriate cultivars.

TABLE 5. Characteristics of 12 metalaxyl insensitive pathotypes of *Bremia lactucae* from UK lettuce crops.

Pathotype Name	Years Found	Metalaxyl Response	Virulence/avirulence *Dm*:3	6	11	16	18	SCT	Distribution
NL10	83-88	103[a]	+	+	-	-	-	B2	Widespread
86A	86	67	-	-	+	-	-	B1	Limited
NL15	87-92	105	+	-	+	-	-	B1	Widespread
87A	87	64	-	+	-	-	-	nt	Limited
87B	87	10	+	+	-	+	-	B2	Limited
89A	89	102	+	+	+	-	-	B2	Limited
89B	89+93	45	+	+	+	-	-	B1	Limited
92A	92-93	82	-	-	+	+	-	B2	Widespread
92B	92	102	+	+	+	-	-	H	Limited
92NI	92	93	+	-	(+)	+	-	B1	Limited
NL16	92	82	+	+	+	+	-	B1	Limited
92C	92	94	+	-	+	+	-	nt	Limited

[a] Values are 1/LP as a % of untreated; nt = not tested
+ = virulent; - = avirulent; (+) = partially virulent;
Dm = Downy mildew resistance gene.

TABLE 6. Incidence of occurrence of *B. lactucae* pathotypes among 522 isolates tested from 1983-1993.

Year	N[a]	NL10	86A	NL15	87A	87B	89A	89B	92A	92B	92NI	NL16	92C	SEN[b]
83	33	17												16
84	50	28												22
85	133	94												39
86	39	20	2											17
87	34	4		3	1	1								25
88	96	1		79										17
89	44			24			1	1						18
90	12			8										4
91	8			2										6
92	36			3					22	1	2	4	1	3
93	36			1				2	25					8
Tot	522	164	2	120	1	1	1	3	47	1	2	4	1	175

(Columns NL10 to SEN are headed "Pathotype Name".)

[a] Number of isolates tested; [b] All sensitive pathotypes

As illustrated in Table 6, since 1983, there have been three cycles of rapid increase and decline in three different phenylamide insensitive pathotypes of *B. lactucae*. There is little reason to doubt that rapid dissemination from the original area where a new pathotype emerges is associated with the movement of transplanting material. The first two pathotypes to become prevalent were countered by encouragement to grow appropriately resistant cultivars while continuing the use of fungicides containing metalaxyl to control virulent but sensitive components of the pathogen population. It remains to be seen how long this dynamic integrated control strategy can be successfully maintained but there are good prospects that the recently emerged "1992A" type can be similarly controlled once cultivars carrying the required *Dm* genes have been identified.

DISCUSSION

The studies reported in this paper clearly illustrate the impact of modern horticultural production practices on the rapid dissemination of new pathogen variants including those insensitive to fungicides. These observations assume a particular relevance with the deregulation of trade in plant material throughout countries of the European Union. These investigations have provided some important insights into pathogen population biology and have led to a successful strategy for lettuce downy mildew control and the continued effective use of phenylamide fungicides. However, several important biological questions remain to be answered.

What is the relative importance of sexual recombination and selection of novel mutants as the origins of new pathotypes?

How many genetic loci control response to phenylamides and is the observed phenotypic variation related to allelic variation?

Is it chance, or are there important phenotypic characters which determine if a new pathotype will become widespread or remain of limited commercial significance?

The answers to these questions are likely to emerge from inheritance and population studies employing techniques to identify selectively neutral DNA-based genetic markers.

ACKNOWLEDGEMENTS

The research reported in this paper was conducted with financial support from the Ministry of Agriculture, Fisheries and Food (England and Wales).

We wish to thank all the regional ADAS officers who collected and despatched samples from which pathogen isolates for testing were recovered and in particular Dr J. Wafford and Dr J.M.L. Davies for their prominent role in the study of *P. parasitica*.

NAM was in receipt of an SERC CASE studentship during the conduct of studies on *P. parasitica* and the interested involvement of Dr J.A. Lucas in this work is gratefully acknowledged.

REFERENCES

Channon, A.G. (1981) Downy mildew of brassicas. In: *The Downy Mildews*, D.M. Spencer (Ed), London, Academic Press, pp 321-339.

Crute, I.R. (1987) The occurrence, characteristics, distribution, genetics and control of a metalaxyl-resistant pathotype of *Bremia lactucae* in the United Kingdom. *Plant Disease*, **71**, 763-767.

Crute, I.R. (1992a) Downy mildew of lettuce. In: *Plant Diseases of International Importance, Diseases of Vegetables and Oil Seed Crops Vol II*, H.S. Chaude, U.S. Singh, A.N. Mukhopadhyay and J. Kumar (Eds), New Jersey, USA, Prentice Hall, pp 165-185.

Crute, I.R. (1992b) From breeding to cloning (and back again?): a case study with lettuce downy mildew. *Annual Review of Phytopathology*, **30**, 485-506.

Crute, I.R. (1992c) The role of resistance breeding in the integrated control of downy mildew (*Bremia lactucae*) in protected lettuce. *Euphytica*, **63**, 95-102.

Crute, I.R.; Harrison, J.M. (1988) Studies on the inheritance of resistance to metalaxyl in *Bremia lactucae* and on the stability and fitness of field isolates. *Plant Pathology*, **37**, 231-250.

Crute, I.R.; Norwood, J.M.; Gordon, P.L. (1985) Resistance to phenylamide fungicides in lettuce and brassica downy mildew. *Fungicides for Crop Protection: 100 years of Progress. British Crop Protection Council Monograph*, **No 31**, 311-314.

Crute, I.R.; Norwood, J.M.; Gordon, P.L. (1987) The occurrence, characteristics and distribution in the United Kingdom of resistance to phenylamide fungicides in *Bremia lactucae* (lettuce downy mildew). *Plant Pathology*, **36**, 297-315.

Leroux, P.; Maisonneuve, B.; Bellec, Y. (1988) Détection en France de souches de *Bremia lactucae* agent du mildiou de la laitue, résistantes au métalaxyl et à l'oxadixyl. *P.H.M. Revue Horticole*, **No 292**, 37-40.

McPherson, G.M.; Crute, I.R. (1986) Strategies for the control of lettuce downy mildew. *Aspects of Applied Biology*, **12**, 189-199.

Michelmore, R.W.; Crute, I.R. (1982) A method for determining the virulence phenotype of isolates of *Bremia lactucae*. *Transactions of the British Mycological Society*, **79**, 542-546.

Moss, N.A.; Crute, I.R.; Lucas, J.A. (1994) Laboratory production of oospores of *Peronospora parasitica* (crucifer downy mildew) and the recovery and characterisation of sexual progeny from crosses between isolates of different host specificity. *Plant Pathology* (In Press).

Raid, R.N.; Datnoff, L.E.; Schettini, T.; Michelmore, R.W. (1990) Insensitivity of *Bremia lactucae* to metalaxyl on lettuce in Florida. *Plant Disease*, **74**, 81.

Schettini, T.M.; Legg, E.J.; Michelmore, R.W. (1991) Insensitivity to metalaxyl in California populations of *Bremia lactucae* and resistance of California lettuce cultivars to downy mildew. *Phytopathology*, **81**, 64-70.

PHENYLAMIDE RESISTANCE IN POTATO LATE BLIGHT (*Phytophthora infestans*) IN THE UNITED KINGDOM IN 1993

N. J. BRADSHAW *, R. I. PRICE **

* ADAS Pwllpeiran Research Centre, St Agnes Road, Gabalfa, Cardiff. CF4 4YH

** ADAS Pwllpeiran Research Centre, Trawsgoed, Aberystwyth, Dyfed. SY23 4HT

T. LOCKE

ADAS Worcester, Whittington Road, Worcester. WR5 2LQ

ABSTRACT

A modified floating leaf disc technique was used to test isolates of *P. infestans* collected from primary sources of infection and commercial field crops of potatoes for sensitivity to phenylamide containing fungicides. The survey in 1993 detected resistance in 76.6% of all sites and 66.2% of all isolates tested. High levels of resistance were detected in primary infection sources (82% of sites and 72.8% of isolates tested). The implications of the results and interpretation of the data are discussed.

INTRODUCTION

Since the mid-1980's, ADAS has tested bulk-isolates of potato late blight (*Phytophthora infestans*) in England and Wales for sensitivity to phenylamide fungicides. Samples collected from blight infected field crops, private gardens and experimental plots were tested using a modified floating leaf disc method. Since the mid-1980's there has been an apparent increase in the level of phenylamide resistance from 31.7% of samples tested in 1986 to 81% in 1988 and 1989. A decline in resistance was recorded thereafter and in 1992, 46.3% of samples contained phenylamide resistant *P. infestans*. There is no published information on the phenylamide resistance status of primary sources of potato blight in England and Wales early in the season before fungicides have been applied and allowed to exert selection pressure.

The data presented here show the results of phenylamide resistance monitoring of field crops in England, Wales and Scotland in 1993 and includes incidence of phenylamide resistant *P. infestans* from primary infection sources such as untreated crops and waste heaps.

** Present address: Central Science Laboratory, Hatching Green, Harpenden, Herts., AL5 2BD

MATERIALS AND METHODS

Sample selection

Samples of potato blight were collected as actively sporulating leaf or stem lesion from infected crops of both early and maincrop potatoes or infected foliage found on discard heaps. Samples were packaged separately in clean, dry polythene bags which were lightly inflated and then sealed. They were then placed in a cool dark protective container to avoid damage during transit.

Phenylamide resistance tests

Up to five sporulating leaf or stem lesions were selected from samples from each site, each providing a spore population for a separate test. If necessary, sporulation was induced either by damp incubation at temperatures of 16-18°C or by re-activation following inoculation between potato tuber slices (cv King Edward).

Isolates of *P. infestans* were tested for sensitivity to phenylamide fungicides using metalaxyl (as Ridomil 25 WP). The test procedure used the modified floating leaf disc technique developed by Carter *et al.* (1982) and also described by Sozzi *et al.* (1992). Metalaxyl was used at concentrations of 2.0 mg/l and 100 mg/l and sporangial suspensions were not adjusted to a standard concentration because the quality of material received was too variable. Five leaf discs (14mm diameter) of cv King Edward, were used to test each isolate at each concentration of metalaxyl and in water alone. Each test was incubated for 9h at 15°C in the dark and 15h at 18°C during the day. All the potato tubers and leaf discs used in these tests were taken from plants grown from non-phenylamide treated seed tubers.

The leaf discs were examined microscopically after five days and the presence or absence of sporulation was recorded. No attempts were made to quantify sporulation or assess any tissue necrosis associated with lesion development. A site was defined as having resistance present if growth and sporulation of *P. infestans* was recorded at either 2.0 mg/l or 100 mg/l in at least one bulk-isolate. Isolates were defined as resistant if growth and sporulation was recorded at either 2.0 mg/l or 100 mg/l metalaxyl in at least one of the leaf discs at either concentration.

RESULTS

The results of the phenylamide sensitivity tests are given in the table. The incidence of phenylamide resistance has been classified according to the sample source, either untreated crop or discard heap or whether or not a phenylamide-containing fungicide had been used prior to sampling of a sprayed crop.

Samples of *P. infestans* were collected from a total of 145 sites. Phenylamide resistance was detected at 82.2% of sites which had not received any fungicide and also in 82.5% of samples from crops which had received at least one phenylamide fungicide. In sprayed crops, where phenylamides had not been used, 64.1% of sites contained resistant *P. infestans*.

Of the isolates tested (total 610), 69.3% from untreated crops and 76.3% from waste heaps contained phenylamide resistant *P. infestans*. This compares with 68.8% of isolates from phenylamide treated crops and 58.3% from non-phenylamide treated crops.

CONCLUSIONS

The results of the 1993 phenylamide resistance survey suggest an increase in the level of phenylamide resistant isolates of *P. infestans* collected from commercially grown fungicide treated potato crops from 46.3% in 1992 (Holmes, 1992) to 63.2% in 1993. Somewhat higher levels of phenylamide resistance were found in crops where phenylamide fungicides had been applied prior to sample collection compared with crops which had not been treated with a phenylamide fungicide. The incidence of phenylamide resistance in isolates of *P. infestans* from untreated crops or infected foliage found on waste heaps was also high at 69.3% and 76.3% of isolates tested respectively. The overall impression gained from these data is that a high level of phenylamide resistant *P. infestans* is present within the resident UK population. The apparent high incidence of resistance in important primary sources of infection is cause for concern and suggests that the resistant isolates are fit and able to overwinter in tubers that remain viable and survive the winter conditions. The phenylamide fungicides are marketed in mixtures with dithiocarbamates and therefore loss of performance due to resistence is difficult to judge in field crops.

This survey data should be viewed with caution and interpreted carefully. Williams & Gisi, (1992) draw attention to the difficulties encountered when carrying out a survey of this kind particularly in relation to late blight. They state that *'The nature and timing of the sampling, combined with the sensitivity testing used, can have a significant influence on the relevance of the sensitivity data obtained...... Testing of bulk populations (as described here) provide a qualitative answer that either resistance occurs or does not occur within the sample tested. A positive resistance reading is obtained with as little as 1% of the test population being resistant'.*

Undoubtedly, fungicide usage patterns are likely to play an important role in selecting for phenylamide resistance and FRAC guidelines should be followed (Urech & Staub, 1985). Nevertheless, it would be particularly beneficial to the industry if a quantitative assessment of phenylamide resistance in potato late blight programme could be made. The data presented here is almost certainly a worst case scenario, however even if as little as 1% of a test population is resistant a subsequent application of a phenylamide containing fungicide may leave the phenylamide resistant strains dominant (Clayton, 1993).

ACKNOWLEDGEMENTS

The authors gratefully acknowledge the financial support of the Ministry of Agriculture, Fisheries and Food and the UK Phenylamide Working Group of the Fungicide Resistance Action Committee.

Phenylamide resistance tests on *P. infestans* in 1993 - Proportion of sites and isolates where resistance was detected

Source or previous crop treatment	Number of sites (isolates) tested	Number and (%) of sites where phenylamide resistance* detected	Number and % Resistant isolates @ 100 mg/l	@ 2 mg/l + 100 mg/l
Untreated crop	35 (153)	29 (82.9)	72 (47.1)	106 (69.3)
Waste heap	22 (97)	18 (81.8)	64 (66.0)	74 (76.3)
Phenylamide ** treated crop	40 (170)	33 (82.5)	97 (57.1)	117 (68.8)
Non-phenylamide treated crop	39 (151)	25 (64.1)	71 (47.0)	88 (58.3)
Previous treatement not known	9 (39)	6 (66.7)	15 (38.5)	19 (48.7)
Totals	145 (610)	111 (76.6)	319 (52.3)	404 (66.2)

* Resistance detected at either 2.0 or 100 mg/l metalaxyl

** One or more treatments with a fungicide or mixture containing a phenylamide fungicide.

REFERENCES

Carter, G.A.; Smith, R.M.; Brent, K.J. (1982) Sensitivity to metalaxyl of *Phytophthora infestans* populations in potato crops in south-west England in 1980 and 1981. *Ann. appl. Biol* **100**, 433-441.

Clayton, R.C. (1993) Evaluation of different strategies for the development of phenylamide fungicides: effects on the development of phenylamide resistance in population of *Phytophthora infestans*. Proc., 12th Triennial Conference of the European Association for Potato Research 503-504.

Holmes, S.J. (1993) Potato Blight Control in 1993. In: *Proceeding of the Dupont Potato Seminar, NAC Stoneleigh.*

Sozzi, D.; Schwinn, F.J.; Gisi, U. (1992) Determination of the sensitivity of *Phytophthora infestans* to phenylamides: a leaf disc method. EPPO Bulletin 22, 306-309

Urech, P.A., Staub, T. (1985) The resistance strategy for acylalanine fungicides. EPPO Bulletin **15** 539-543.

Williams, R.J.; Gisi, U. (1992) Monitoring pathogen sensitivity to phenylamide fungicides: principles and interpretations. EPPO Bulletin **22**, 299-306.

TOWARDS THE ISOLATION OF GENES DETERMINING INSENSITIVITY TO PHENYLAMIDE FUNGICIDES FROM *PHYTOPHTHORA INFESTANS*.

H.S. JUDELSON, P. VAN WEST*

NSF Center for Engineering Plants for Resistance Against Pathogens, University of California, Davis, California 95616, USA.

R.C. SHATTOCK

School of Biological Sciences, University of Wales, Bangor, Gwynedd, LL57 2UW, UK.

* current address: Department of Phytopathology, Wageningen Agricultural University, P.O. Box 8025, 6700 EE Wageningen, The Netherlands.

ABSTRACT

A map-based (positional) cloning approach is being employed to isolate gene(s) from *Phytophthora infestans* that determine insensitivity to phenylamide fungicides. Crosses were performed to generate progeny segregating for the insensitivity phenotype. DNA markers linked to the gene are now being identified by screening the progeny for RAPD (random amplified polymorphic DNA) loci using bulked segregant analysis. So far, one potential linked marker has been identified. In the future, these markers will enable the molecular cloning of the gene(s) by chromosome walking.

INTRODUCTION

Insensitivity to the phenylamide, metalaxyl, occurred in potato late-blight (Phytophthora infestans) isolates soon after its commercial introduction in 1979. Outbreaks of late-blight were reported in metalaxyl-treated European crops of potato in 1980 and insensitive isolates are now widespread (Shattock *et al*; 1990). Recently it has been suggested that metalaxyl insensitivity, among other factors, may be contributing to the displacement of the "old" clonal populations of P.infestans by "new" genotypes (Fry *et al*., 1993, W.E. Fry, unpublished data).

Three phenotypes can be found among field isolates of late-blight, namely metalaxyl-sensitive, -insensitive and -intermediate. Where inheritance studies have been carried out by mating highly insensitive phenotypes with sensitive isolates, mostly intermediate phenotypes were obtained (Shattock, 1988). Patterns of segregation in some subsequent backcross and sib-matings suggest single gene control with sensitive and incompletely dominant insensitive alleles. In other cases, however, more complex patterns of segregation have been observed and are not consistent with simple inheritance (reviewed by Shaw, 1991).

A long-term goal of this project is to identify the molecular basis of insensitivity to the phenylamides in Phytophthora. The gene conferring insensitivity could be a mutated or amplified version of the target of the fungicide, or may be an entirely different locus. Several studies have suggested that phenylamides inhibit the synthesis of RNA by RNA polymerase I. One likely explanation for insensitivity is an altered target for the fungicide, and binding studies have indicated that extracts of insensitive isolates had a reduced affinity for phenylamides (Davidse, 1988). However, as the machinery of transcription in eukaryotes is complex the precise target of the phenylamides cannot be deduced from this data.

A map-based (positional) cloning approach is being employed to isolate gene(s) from *P. infestans* that determine insensitivity to phenylamide fungicides. This species has been chosen for these studies since it is amenable to both classical and molecular genetic analysis (Shaw and Shattock, 1991; Judelson *et al.*, 1991). Crosses were made to generate progeny segregating for insensitivity and DNA markers linked to the gene are now being identified by screening the progeny using RAPD (random amplified polymorphic DNA) loci (Williams et al., 1991).

CROSSES

A cross was initially made between metalaxyl-insensitive A1 isolate 88.1.1 (United Kingdom) and -sensitive A2 isolate E13a (Egypt). Subsequently a cross was made between metalaxyl -sensitive A1 isolate Ca65 (California) and A2 isolate 2.16 (a F1 progeny from cross 88.1.1 x E13a), an intermediate phenotype.

From cross Ca65 x 2.16, 100 single oospore progeny were characterized for mating type and metalaxyl sensitivity. A ratio of 5 : 1 for A1 : A2 mating types was observed. On rye agar with and without 10 μg/ml metalaxyl, most progeny could be classified as either sensitive or intermediate phenotypes in a ratio close to 1 : 1. Twenty three progeny were selected representing approximately equal numbers of parental and recombinant phenotypes for metalaxyl sensitivity and mating type.

Since *P. infestans* is bisexual, the progeny of crosses may potentially contain selfs of the parents, which would complicate the genetic analysis. To verify that the progeny used in this analysis were not selfs, their genotypes were examined by scoring RAPD loci, as described below. All of the progeny of the 88.1.1 x E13a cross and 22 of the 23 progeny of the Ca65 x 2.16 cross were confirmed to be outcrossed individuals.

SCREENING FOR RAPD MARKERS

RAPD loci linked to the insensitivity gene are being identified using polymerase chain reaction (PCR) assays using 10-mer oligonucleotide primers (Williams *et al.*, 1991). Due to the small size of the primers and the low annealing temperatures employed, the primers bind to many sites within genomic DNA. This results in the amplification of bands when the binding sites are nearby and in the opposite orientation. Sequence variation between isolates results in polymorphic banding patterns, which are scored as genetic loci.

Five each of the most metalaxyl -sensitive and -insensitive progeny of the Ca65 x 2.16 cross were selected and used to screen for RAPD polymorphisms linked to the insensitivity gene by bulked segregant analysis (Michelmore *et al.*, 1991). The bulked segregant analysis technique simplifies and accelerates the screening process. Individual 10-mer primers were used to prime amplification reactions using 2 ng of template DNA pooled from five insensitive or five sensitive progeny. This approach places bands (loci) unlinked to phenylamide insensitivity in both pools. Consequently, bands appearing only in one reaction will be enriched for those linked to the targeted gene (Fig. 1).

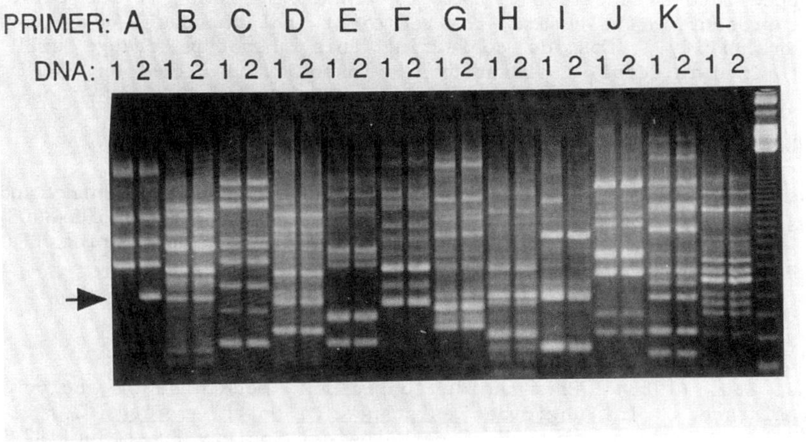

FIGURE 1: Screening for RAPD polymorphisms by bulked segregant analysis. Two bulks of DNA (1 and 2) were prepared that contained DNA pooled from five progeny of each genotype. This DNA was then used as template in PCR amplification reactions using different RAPD primers (A to L). Indicated in the left margin is the position of a polymorphism detected between bulks 1 and 2 using primer A. The right-hand lane contains a 123 bp ladder.

Preliminary RAPD reactions performed on the Ca65 and 2.16 parents indicated that each primer resulted in the amplification of about seven bands, with 0.75 polymorphic bands detected per primer. Approximately 1800 RAPD primers are available within our collection. Therefore, our assays are expected to test 1350 loci for linkage to the insensitivity gene. So far, approximately 520 primers have been tested by bulked segregant analysis of progeny DNA. Primers resulting in differences between the sensitive and insensitive pools were retested in amplification reactions using the DNA of individual progeny. One loosely linked marker, 19 cM from the insensitivity gene, has been identified, although the significance of this linkage (LOD score of 2.1; Ott, 1991) needs to be tested further by scoring additional progeny.

DISCUSSION

We have begun to construct a genetic map of the region containing gene(s) determining insensitivity to phenylamide fungicides. One potentially interesting linkage has been identified which needs to be studied further by scoring additional progeny. Additional RAPD primers will also be screened to identify additional linked markers. When a detailed genetic map of the region is available, a physical map will be constructed by long-range restriction mapping. This will reveal the maximum physical distance between the markers and indicate the feasibility of isolating the locus by chromosome walking.

In future, these markers will enable the molecular cloning of the genes by serving as probes for screening cosmid libraries. Cosmid clones absolutely linked to the insensitivity phenotype will be introduced into a sensitive isolate by transformation. If the targeted gene is present on the cosmid, the recipient isolate should be converted to insensitivity. Cosmids conferring this phenotype change will be characterized in detail to identify the insensitivity gene.

ACKNOWLEDGMENTS

We thank Wendy Jones for technical support in maintaining and storing cultures. This work was supported by USDA grant 93-3703-8908 to H.S.J. and NSF grant BIR-8920216 to the Center for Engineering Plants for Resistance Against Pathogens.

REFERENCES

Davidse, L.C. (1988) Phenylamide fungicides: mechanism of action and resistance. In Fungicide Resistance in North America, C.J. Delp (Ed), St Paul: American Phytopathological Society Press, pp. 63-65.

Fry, W.E.; Goodwin, S.B.; Dyer, A.T.; Matuszak, J.M.; Drenth, A.; Tooley, P.W.; Sujkowski, L.S.; Koh, Y.J.; Cohen, B.A.; Spielman, L.J.; Deahl, K.L.; Inglis, D.A.; Sandlan, K.P. (1993) Historical and recent migrations of *Phytophthora infestans*: chronology, pathways, and implications. Plant Disease 77, 653-661.

Judelson, H.S.; Tyler, B.M.; Michelmore, R.W. (1991) Transformation of the oomycete pathogen, *Phytophthora infestans*. Molecular and Plant Microbial Interactions, 4, 602-607.

Michelmore, R.W.; Paran, I.; Kesseli, R.V. (1991) Identification of markers linked to disease-resistance genes by bulked segregant analysis: a rapid method to detect markers in specific genomic regions using segregating polymorphisms. Proceedings of the National Academy of Sciences, USA, 88, 9828-9832.

Ott, J. (1991) Analysis of Human Genetic Linkage. Baltimore: John Hopkins University Press.

Shattock, R.C. (1988) Studies on the inheritance of resistance to metalaxyl in *Phytophthora infestans*. Plant Pathology, 37, 4-11.

Shattock, R.C.; Shaw, D.S.; Fyfe, A.M.; Dunn, J.R.; Loney, K.H.; Shattock, J.A. (1990) Phenotypes of *Phytophthora infestans* collected in England and Wales from 1985 to 1988 : mating type, response to metalaxyl and isoenzyme analysis. Plant Pathology, 39, 242-248.

Shaw, D.S. (1991) Genetics. In: *Phytophthora infestans*, The Cause of Late-Blight of Potato, D.S. Ingram and P.H. Williams (Eds), London: Academic Press, pp. 131-170.

Shaw, D.S.; Shattock, R.C. (1991) Genetics of *Phytophthora infestans* : the Mendelian approach. In: *Phytophthora*, J.A. Lucas, R.C. Shattock, D.S. Shaw and L.R. Cooke (Eds), Cambridge: Cambridge University Press, pp. 218-230.

Williams, J.G.K.; Kubelik, A.R.; Livak, K.J.; Rafalski, J.A.; Tingey, S.V. (1991) DNA polymorphisms amplified by arbitrary primers are useful as genetic markers. Nucleic Acids Research, 18, 6531-6535.

MINIMISING PHENYLAMIDE RESISTANCE - A SUCCESSFUL STRATEGY

R. A. COLLIER, S. J. LE BOUTILLIER

Department of Agriculture and Fisheries, P.O. Box No. 327, Howard Davis Farm, Trinity, Jersey, JE4 8UF, Channel Islands

ABSTRACT

In an area where inoculum pressure is high due to the intensive cultivation of potatoes phenylamide resistance has been contained between 14% (28%) to 39% (62%) with strategic use of the relevant fungicides. Annual monitoring of all potato blight outbreaks from 1985 to 1993 has revealed no dramatic change in the susceptibility of populations of ***Phytophthora infestans*** to the phenylamide fungicides.

INTRODUCTION

The inoculum pressure and risk of resistance developing to the phenylamide fungicides by ***Phytophthora infestans***, the fungus causing late blight of potatoes, is intensified in Jersey where potatoes are grown nearly 12 months of the year. This includes glasshouse, polythene tunnel, early mulched, maincrop and autumn crops being harvested from November through to October.

The cultivar grown International Kidney - "Jersey Royal" is also extremely susceptible to blight infection. Due to marketing strategy and the history of the product there is no scope to change to a cultivar with higher resistance to blight.

Graph 1. First Outbreaks of Potato Blight Recorded in Jersey 1985 - 1993

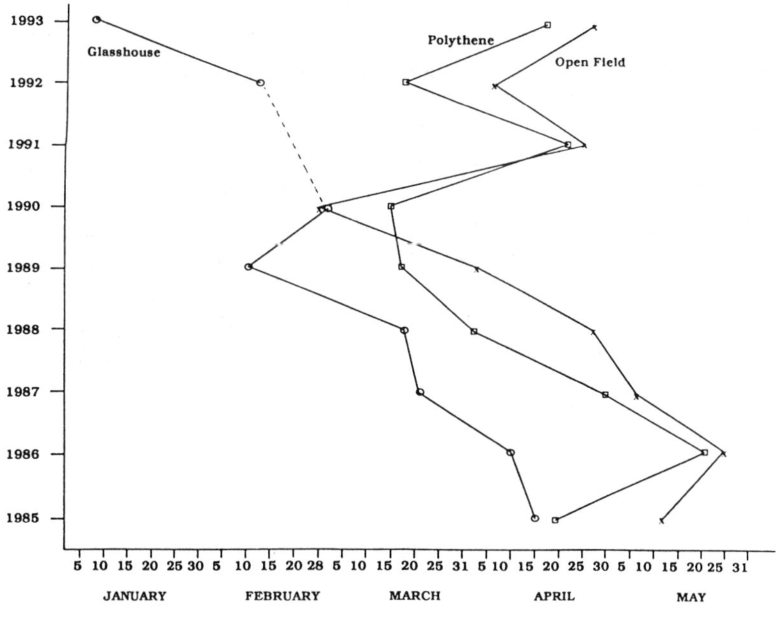

METHOD

Laboratory assays using the leaf disc method (Sozzi, et al., 1992), of survey samples collected by the Department's Inspection Service have been carried out for 9 years (1985-1993). Isolates have been tested from crops grown under protection (glasshouse or polythene tunnel) and field crops, either polythene covered or open field. The increase in use of polythene mulches has encouraged earlier development of late blight symptoms in the field crop. In 1989, after an exceptionally mild winter, blight symptoms were first recorded in a field crop on 28th February as soon as the polythene mulch was lifted. More usually the blight epidemic occurs throughout April and early May.

RESULTS

The level of resistance found in Jersey has remained relatively steady over the period of testing. Even in bad blight years, eg. 1987-89, the level of resistance did not exceed 39% (62%). More recently in a moderate season resistance levels have dropped to 14% (28%) of fieldsamples tested. (Table 1, Graph 2).

However, in 1993 of the samples tested, 86 were sensitive at 2 ppm and 100 ppm, and 30 resistant at 100 ppm, with 11 "partially resistant" ie. sensitive at 100 ppm, but resistant at 2 ppm. The relevance of this category remains a point of discussion in the context of the sampling procedure available.

Table 1. Potato Blight - Phenylamide Group Resistance

Year	Total	Sensitive (%)	Partial Resistance %	Resistance (%)	Total
1985	95	36 (38%)	22 (23%)	37 (39%)	59 (62%)
1986	259	117 (45%)	49 (19%)	93 (36%)	142 (55%)
1987	272	136 (50%)	46 (17%)	90 (33%)	136 (50%)
1988	412	164 (40%)	102 (25%)	146 (35%)	248 (60%)
1989	523	271 (52%)	99 (19%)	153 (29%)	252 (48%)
1990	34	18 (53%)	7 (21%)	9 (26%)	16 (47%)
1991	87	55 (63%)	15 (17%)	17 (20%)	32 (37%)
1992	176	126 (72%)	25 (14%)	25 (14%)	50 (28%)
1993	127	86 (68%)	11 (8%)	30 (24%)	41 (32%)

The pattern of usage of phenylamide products has varied little over the years. Between 23% to 28% of fields tested had used systemic products at the beginning of their spray programme.

The disease is statutorily controlled under the Blight Disease (Jersey) Order, 1982, which allows for a fairly high degree of supervision in the growing of the potato crop. This has included the development of a comprehensive strategy for the use of phenylamide fungicides, which will be adhered to by the growers. The consequences of failure to control the disease is the issue of a Statutory Scorching Order by the Department of Agriculture Inspectors, for the focus of infection to be destroyed and removed. It is considered that this policy has reduced the potential risk of a build-up of resistant inoculum.

Graph 2. Potato Blight
Phenylamide Group Resistance Results Jersey 1985 - 1993
(figures above columns are the total tested)

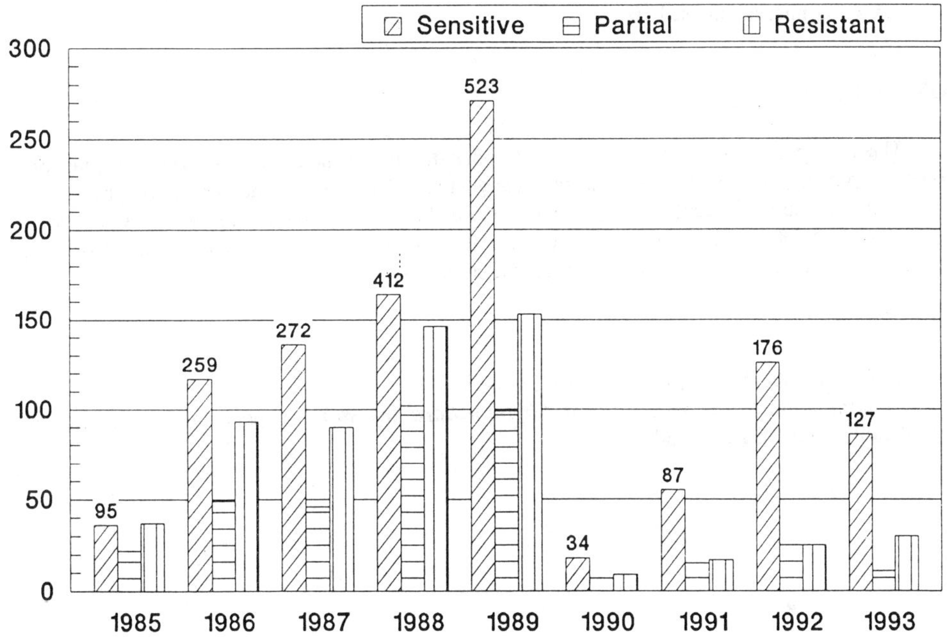

Jersey Strategy for the Use of Systemic Fungicides 1994

1. **NOT** to be applied to crops grown under permanent protection ie. glass or polythene tunnels.

2. **NOT** to be applied to crops grown for seed or non export ware.

3. A maximum of 3 sprays may be applied to any single crop, the last of which must be before 30th June.

4. **NOT** to be applied curatively.

5. Crops under polythene covers must be treated immediately on removal of film provided no blight infection present.

6. Only to be used on field grown export crop of Jersey Royals.

7. Spraying interval advised 10-14 days.

CONCLUSIONS

The field performance of the fungicides under discussion have remained good despite some growers apparently losing confidence in the phenylamides for a time. The occasional problems of lack of disease control identified by the Extension Pathologist appear to be more related to application technology, and timing. These are the crucial factors to be addressed. Continued monitoring of the pathogen population is regarded to be essential.

REFERENCES

SOZZI, D, SCHWINN, F.J. and GISI, U (1992)
Determination of the sensitivity of ***Phytophthora infestans*** to phenylamides: a leaf disc nethod EPPO Bulletin 22 306-309

PHENYLAMIDE RESISTANCE IN *PHYTOPHTHORA INFESTANS* IN NORTHERN IRELAND

L.R. COOKE, R.E. SWAN

Plant Pathology Research Division, Department of Agriculture for Northern Ireland, Newforge Lane, Belfast, Northern Ireland, BT9 5PX.

ABSTRACT

In Northern Ireland, after three successive years (1989-91) in which the proportion of isolates of *Phytophthora infestans* containing phenylamide-resistant strains declined from 90% to 42%, there was an increase to 68% in 1992 followed by a decrease in 1993 (to 54%). The crop area treated with phenylamides is now much smaller than in the period 1985-88 and most growers now only use one or two applications per season. The higher incidence of phenylamide-resistant strains in 1992 appears to be related to weather very favourable to blight in July-August of that year, which resulted in a larger *P. infestans* population being exposed to selection by phenylamides.

INTRODUCTION

Resistance to phenylamide fungicides has been monitored in the Northern Ireland *Phytophthora infestans* population since it was first detected in 1981 (Cooke, 1981). In the late 1980's resistant strains were detected in over 80% of isolates tested, but between 1988 and 1991, the proportion declined (Cooke & Penney, 1992). This paper reports the results of surveys of the occurrence of phenylamide-resistant *P. infestans* up to 1993.

MATERIALS & METHODS

Samples of infected potato foliage and data on sample site, potato cultivar, fungicide usage and disease incidence were obtained by members of the Department of Agriculture's Potato Inspection Service, as previously described (Cooke & Penney, 1992). At the end of each season, Inspectors supplied details of fungicide usage for all seed potato crops in their areas.

Isolates were maintained on detached leaves of glasshouse-grown potato plants cvs King Edward or Bintje and sporangial/zoospore suspensions prepared as previously described (Cooke, 1986). Isolates, tested using the floating leaf disc technique (Cooke, 1986), were designated resistant if they sporulated on 100 mg metalaxyl/litre-treated discs and sensitive if they sporulated on untreated discs but not on any metalaxyl-treated disc. Isolates which failed to grow on at least four out of six untreated discs were re-tested.

RESULTS

Incidence of phenylamide resistance

The proportion of *P. infestans* isolates containing phenylamide-resistant strains, having declined from 90% in 1988 to 42% in 1991, increased to 68% in 1992 (Figure 1), but further decreased to 54% in 1993.

FIGURE 1. The proportion of Northern Ireland isolates of *Phytophthora infestans* containing phenylamide-resistant strains, 1981-1993.

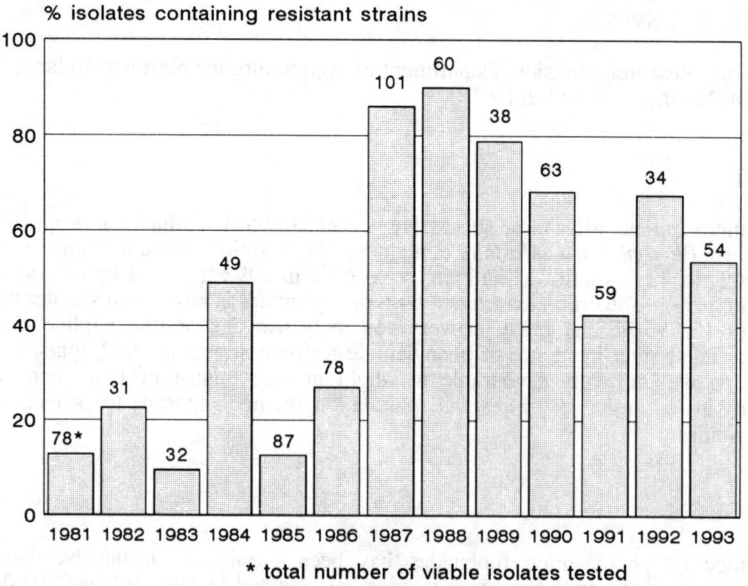

As in previous years, isolates obtained from crops which had received phenylamide applications tended to have a greater proportion of isolates containing resistant strains. However, a substantial proportion of isolates containing resistant strains came from crops where only protectant fungicides had been used (Table 1).

TABLE 1. Fungicide usage on potato crops sampled for *P. infestans* phenylamide resistance survey, 1991-93

Crop treatment	% crops yielding isolates containing resistant strains		
	1991	1992	1993
protectant only*	29	57	43
phenylamide	65	88	63
all crops**	42	68	54

* no usage of phenylamide-containing products

** includes untreated crops and those where fungicide used is not known

Usage of phenylamide-containing products

Potato Inspectors' estimates of usage of products containing phenylamides on seed potato crops indicated that in the mid-1980's 50-70% of the crop area received phenylamides, but after 1987 usage declined (Figure 2). For the past four years the reported area treated has again increased. However, the majority of growers are now using only one or two applications per season (Table 2), whereas in the mid-1980's anecdotal evidence suggests that programmes of five or more sprays were common.

FIGURE. 2. Estimated seed potato crop area treated with phenylamides, 1983-1993.

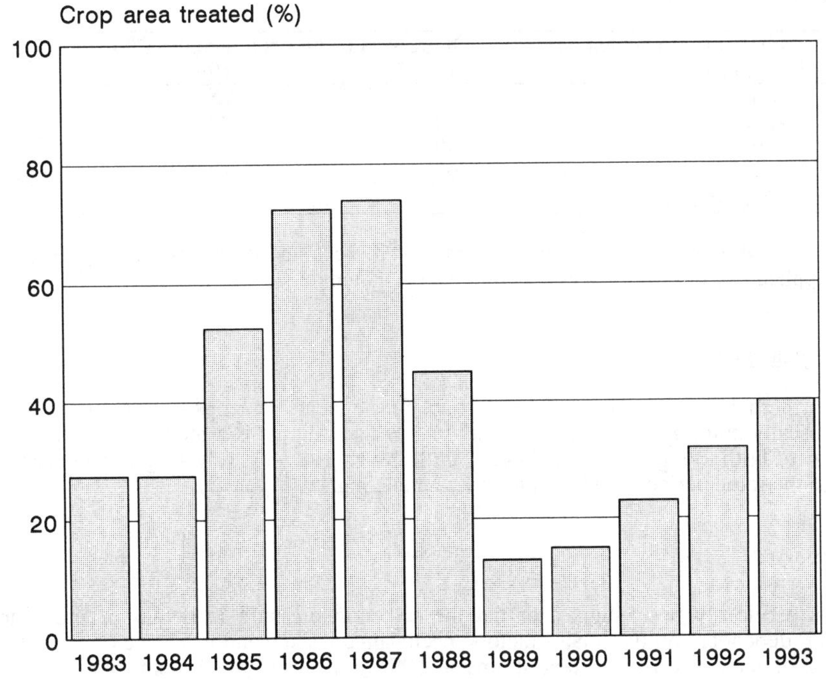

FIGURE 3. Total monthly rainfall for July-August, 1981-1993, Co. Londonderry.

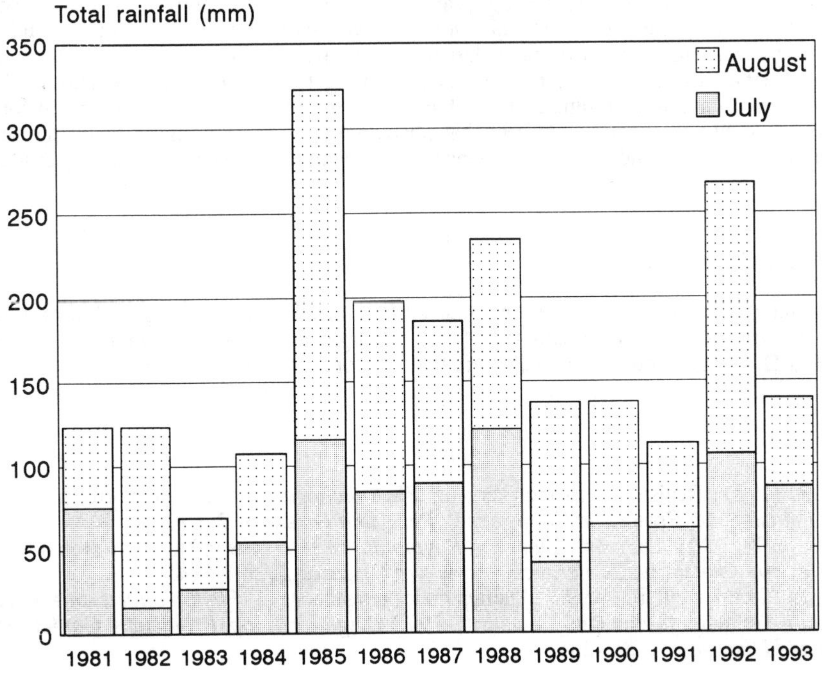

TABLE 2. Number of applications used by growers applying phenylamides, 1992-93

Number of phenylamide applications	Of growers using phenylamides (%) 1992*	1993*
one	34	25
two	24	27
three	30	23
four	9	18
five or more	4	7

* Of all seed growers, in 1992 and 1993, 26% and 44%, respectively applied one or more sprays containing a phenylamide.

Meteorological data

In Northern Ireland, although foliage blight first appears in crops during June, infection does not usually build up until July. Rainfall in July and August is a major factor in affecting the severity of both foliage and tuber blight. In the years 1989-91, July-August were relatively dry, but, in the same period in 1992, rainfall was much higher (Figure 3).

DISCUSSION

It has previously been suggested that the selection pressure in favour of phenylamide resistance is dependent on two major factors, phenylamide usage and incidence of *P. infestans*. The increased incidence of resistant strains in 1992 is not adequately explained by greater phenylamide usage alone; although the crop area treated was slightly larger than in 1991, most growers only used one or two applications. However, the July-August rainfall in 1992 was much greater than in the three preceding years. This produced conditions which favoured blight, so that a large population of *P. infestans* was exposed to fungicide treatment, and also encouraged growers to use more phenylamide-containing products. In the rather drier conditions of 1993, the proportion of isolates containing resistance again decreased. With several new non-phenylamide fungicide formulations having now received UK approval for the control of potato blight, it seems unlikely that phenylamide usage will again reach the levels attained in the mid 1980's and that the incidence of phenylamide-resistant strains may gradually decline.

ACKNOWLEDGEMENTS

The authors are very grateful to members of the Department of Agriculture's Potato Inspection Service for supplying samples of potato blight and crop information. Mr. D.G. Wilson, Miss D. Carlisle and Miss C. Allen are thanked for technical assistance.

REFERENCES

Cooke, L.R. (1981) Resistance to metalaxyl in *Phytophthora infestans* in Northern Ireland. *1981 British Crop Protection Conference - Pests and Diseases*, **2**, 641-649.

Cooke, L.R. (1986) Acylalanine resistance in *Phytophthora infestans* in Northern Ireland. *1986 British Crop Protection Conference - Pests and Diseases*, **2**, 507-514.

Cooke, L.R.; Penney, R.E. (1992) Phenylamide resistance in *Phytophthora infestans* in Northern Ireland - a changing situation. *1992 Brighton Crop Protection Conference - Pests and Diseases*, **1**, 201-206.

STRATEGIES FOR PHENYLAMIDE DEPLOYMENT: EFFECTS ON PHENYLAMIDE RESISTANCE IN POPULATIONS OF *PHYTOPHTHORA INFESTANS*.

R.C. CLAYTON[1] , R.C. SHATTOCK

School of Biological Sciences, University of Wales, Bangor, Gwynedd, LL57 2UW

[1] Present Address, Potato Marketing Board Experimental Station, Sutton Bridge, Lincs., PE12 9YB

ABSTRACT

A simple, rapid method for measuring the proportion of phenylamide-resistant sporangia within mixed populations of *Phytophthora infestans* was developed. Different strategies for the deployment of phenylamides were simulated on potato crops grown in polythene tunnels and on single leaflets in Petri-dishes, each of which were inoculated with selected mixed populations of *P.infestans*. The percentage of resistant phenotypes in a mixed population was higher after six weeks where phenylamides were used at the beginning of a spray programme compared with simulations where phenylamides were used following several applications of non-systemic compounds. An increase in phenylamide resistance in the absence of selection pressure was also demonstrated using some U.K. isolates.

INTRODUCTION.

Phenylamide fungicides were introduced for the control of *Phytophthora infestans* in the late 1970s. Their systemicity offered the grower a degree of freedom from the usual constraints imposed by weather conditions and rapid plant growth. Phenylamide-resistant isolates of *P.infestans* however, rapidly appeared in Europe (Carter *et al*, 1982; Davidse, 1981; Dowley & O'Sullivan, 1981) reaching levels of over 60% of all isolates in the U.K. and The Netherlands (Davidse *et al*, 1989; Shattock *et al*, 1990).

Monitoring changes in the level of phenylamide resistance has traditionally been a macro-dynamic affair. A rapid method for measuring the proportion of resistant individuals within a population is presented in this paper and has been used to quantify the effects of fungicide deployment on the development of phenylamide-resistance.

MATERIALS AND METHODS

% Phenylamide-resistance assay.

Petri-dishes containing Rye-A agar amended with technical grade metalaxyl (to 100μg ml-1) and Ryfamycin, Ampicillin and Nystatin were spot inoculated with 10μl droplets of a sporangial suspension (200 sporangia ml-1) from a test population (15 spots per dish; 4

dishes per test). After incubation at 18°C in the dark for seven days, the number of spots supporting sporulation were counted and % Phenylamide-resistance (%PR) was calculated.

Strategies for phenylamide deployment.

Crops of potato (*Solanum tuberosum*) cv. Maris Piper were grown in polythene tunnels and were inoculated with a 10%PR sporangial suspension of *P.infestans* when plants had met between rows. The phenylamide-resistant and -sensitive isolates used were selected after components of aggressiveness had been characterised (Tooley *et al*, 1986) so that the effects of isolate aggressiveness on the development of phenylamide-resistance could be quantified.

After visual confirmation of infection, crops were treated with fungicides in accordance with the strategies shown in Table 1. Sporangial samples were collected in sterile deionised water at 20°C at weekly intervals and %PR was measured.

Table 1. Simulation of strategies for phenylamide deployment in small crops grown in polythene tunnels.

weeks after inoculation	Strategy 1	Strategy 2
1	mancozeb[a]	Fubol
2	mancozeb	Fubol
3	Fubol[b]	Fubol
4	Fubol	mancozeb
5	mancozeb	mancozeb
6	mancozeb	mancozeb

a applied at 50% recommended rate with a Cooper CP15 Knapsack sprayer
b applied at 40% recommended rate

RESULTS AND DISCUSSION

The assay developed to measure %PR was cheap, rapid and accurate with calculated %PR correlating highly with %PR in prepared standard sporangial mixtures (R^2>0.99). .

The data presented in Table 2 show that when phenylamide compounds are used early in the season, taking advantage of their systemicity, then %PR can reach 100% within a 4 week period whereas, when mancozeb alone was used early in the season (a strategy followed in The Netherlands), then only 40%PR was reached after 6 weeks.

When isolates were selected such that the resistant isolate was more aggressive than the sensitive isolate, the proportion of resistant sporangia in samples reached 80% regardless of fungicide treatment. This suggests that quantifying isolate aggressiveness is an equally important step in the risk assessment process.

This study suggests that the use of phenylamides, for late-blight control later in the season and for a shorter period of time, may be useful in reducing the rate of development of resistance to phenylamides.

Table 2. Development of phenylamide-resistant genotypes (%PR) in mixed populations of *P. infestans* under different strategies for phenylamide deployment.

			strategy[a]			
Weeks after inoculation	1[b]	1[c]	2[b]	2[c]	untreated[b]	untreated[c]
1	10±0.2[d]	32±2.4	20±1.6	38±4.2	10±0.0	10±0.2
2	11±0.6	53±1.4	42±0.6	72	7±1.2	30±2.4
3	16±1.2	54±3.2	66±1.4	100	3±0.0	41±3.6
4	42±2.4	98	93±3.6	100	5±0.0	60±1.4
5	40±1.6	100	100	100	1±0.0	60±1.4
6	41±1.6	100	100	100	-	72±3.6

[a] Strategies in Table 1
[b] mixed isolates with similar fitness indices
[c] mixed isolates with a more aggressive resistant isolate
[d] SEM

ACKNOWLEDGEMENTS

R.C.Clayton was supported by a MAFF Postgraduate Scholarship.

REFERENCES

Carter G.A., Smith R.M. & Brent K.J. (1982) Sensitivity to metalaxyl of *Phytophthora infestans* populations in S.W. England in 1980 and 1981. *Annals of Applied Biology* **100** 433-441.

Davidse L.C., Henken J., van Dalen A., Jespers A.B.K. & Mantel B.C (1989) Nine years practical experience with phenylamide resistance in *P.infestans* in The Netherlands. *Netherlands Journal of Plant Pathology* **95** 197-213

Davidse L.C.,Looijen D., Turkensten L.J. & van der Wal D. (1981) Occurrence of metalaxyl resistant strains of *Phytophthora infestans* in Dutch potato fields. *Netherlands Journal of Plant Pathology* **87** 65-68

Dowley L.J. & O'sullivan E. (1981) Metalaxyl resistant strains of *Phytophthora infestans* (Mont) de Bary in Ireland. *Potato Research* **24** 417-421

Shattock R.C., Shaw D.S., Fyfe A.M., Dunn J.R., Loney K.H. & Shattock J.A.. (1990) Phenotypes of *Phytophthora infestans* collected in England & Wales from 1985- 1988; mating type, response to metalaxyl and isozyme analysis. *Plant Pathology* **39** 242-248.

Tooley P.W., Sweigard J.A. & Fry W.E (1986) Fitness and virulence of *Phytophthora infestans* isolates from sexual and asexual populations . *Phytopathology* **76** 1209-1212.

[illegible]

[illegible]

[illegible]

[illegible]

ACKNOWLEDGEMENTS

[illegible]

REFERENCES

[illegible]

METALAXYL RESISTANCE IN FINNISH STRAINS *OF PHYTOPHTHORA INFESTANS*

A.O. HANNUKKALA

Agricultural Research Centre, Institute of Plant Protection,
FIN-31600 Jokioinen, Finland

ABSTRACT

Metalaxyl resistance of Finnish *Phytophthora infestans* strains was monitored in 1990-93. Strains isolated from single lesions on potato leaflets were tested with a floating leaf disc method. In 1990-91 samples were collected from 30 potato fields in major potato growing areas. In 1992-93 isolates were obtained from unsprayed potato (cv. 'Bintje') plots at seven experimental stations. In 1990-92 20-30 % of strains were sensitive to metalaxyl, 20 % sporulated in 0.1-10 ppm and 60-70 % of isolates sporulated at 100 ppm concentration of the fungicide. In 1993 only 2 % of strains were sensitive and 80 % highly resistant to metalaxyl in spite of anti-resistance strategies and decreasing use of Ridomil MZ.

INTRODUCTION

Potato late blight (*Phytophthora infestans*) is the major disease on potato in Finland. Metalaxyl was registered for potato late blight control in 1985 as a mixture with mancozeb (Ridomil MZ). Since then the fungicide has been widely used in potato production. Ridomil MZ has been recommended for the first two sparays, whereafter contact fungicides, maneb or mancozeb should be applied.

Though development of metalaxyl resistance was already recognized (Staub & Sozzi, 1984) when Ridomil MZ was registered in Finland, the experience in other countries was not taken seriously enough. During the first years after registration of Ridomil MZ it was frequently used curatively. The spraying season in Finland is usually 4-6 weeks and in many cases two Ridomil MZ applications were enough to control late blight.

Since 1989 several farmers have frequently failed in late blight control. In most cases the reason for failure has certainly been wrong timing of applications. However, the presence of metalaxyl resistant strains of *Phytophthora infestans* was also suspected, especially as the existence of the problem had been reported throughout the world (Davidse *et al.*, 1989, Olofsson, 1989, , Deahl *et al.*, 1991, Dowley & O'Sullivan, 1991, Staub, 1991). The monitoring of metalaxyl resistance was initiated in 1990. Much attention has been paid to the educaton of farmers to follow the internationally accepted anti-resistance strategies (Staub, 1991, Williams & Gisi, 1992).

MATERIALS AND METHODS

In 1990-91 *P. infestans* was sampled from commercial potato fields where different spraying programs had been used. Approximately 50 % of the samples originated from fields where Ridomil MZ had been included in the spraying program; 25 % from fields where only protectants had been applied and 25% of the samples were collected from unsprayed home gardens.

In 1992-93 monitoring fields were situated at seven locations at the Experimental Stations of Finnish Agricultural Research Centre and Potato Research Centre. Potato cv. 'Bintje' was planted in 10-15 rows, 10-20 m long. No fungicides were applied.The first sample was taken when the first visible blight symptoms were detected, the second 10 and the third 20 days later. Also some samples were taken from commercial potato fields and home gardens.

The sampling was done by consultants in the potato processing industry and by the staff of experimental stations. Each sample consisted of 20-30 leaflets containing a single lesion (Ø < 1 cm). Every leaflet was taken from a different plant, the distance between plants was more than 2 m. Each leaflet was placed between two sheets of blotting paper in a plastic bag, moistened with a few drops of water and the sample was mailed to the institute of Plant Protection.

Infected leaflets usually produced enough sporangia for resistance testing during mailing. If the sporulation was poor they were removed to plastic Petri- dishes (Ø 9 cm) on moistened filter paper and incubated in moisture chamber at 95-100% r. h. and +15°C for 2-4 days. If blight had destroyed leaflets during mailing, spores were collected from the plant tissue with a paint brush and transferred to fresh potato leaves. Leaves were then incubated for 5 days in a moisture chamber.

A floating leaf disc test (Anonymous, 1982, Sozzi *et al.*, 1992) was used for the routine monitoring of metalaxyl resistance. Leaf discs were obtained from leaves of potato cv. 'Bintje' grown in a greenhouse for 4-5 weeks. Sporangia produced on infected leaflets were washed into distilled water. Spore concentration of the inoculum was adjusted to 100 000 sporangia/ml and 20 μl of the suspension was pipetted in the centre of each leaf disc. Six leaf discs for each strain and metalaxyl concentration were inoculated.

After inoculation leaf discs were kept at +8-10°C for 12 hours to stimulate zoospore release. Sporulation of the pathogen was tested on leaf discs placed in distilled water and 0.01, 0.1, 1.0, 10.0 and 100.0 ppm concentrations of metalaxyl (Ridomil 25 WP). In 1992-93 the lowest metalaxyl concentration was abandoned because it never gave any control in tests.

In addition 60 strains of *P. infestans* isolated from tubers were tested. The pathogen was isolated using the method of Tantius *et al.* (1986). Sporangia for resistance testing were produced on detached potato leaves (cv. 'Bintje'). Some single spore isolates derived from field strains were also tested for their resistance to metalaxyl. 40 field strains that gave resistant reaction were tested also by a semi-quantitative test (Nuninger *et. al.*, 1992).

RESULTS AND DISCUSSION

The resistance to metalaxyl was present in all sampled fields and monitoring plots. In 1990-92 the proportion of sensitive strains varied between 20 and 30%. In 1993 only 2% of the strains proved to be sensitive (Table 1). The frequency of metalaxyl resistant strains is suprisingly high compared to the resistance situation in other Scadinavian (Olofsson 1989, personal communication) and European countries (Davidse *et al.*, 1989, Staub, 1991).

TABLE 1. The proportion of metalaxyl resistant, intermediate and sensitive strains of *Phytophthora infestans* in Finland in 1990-93. Sensitive sporulating at 0-0.01 ppm, intermediate 0.1-10 ppm and resistant at 100 ppm metalaxyl.

Year	sensitive	intermediate	resistant	sample size
	% of strains tested			
1990	19	5	76	85
1991	21	18	61	305
1992				
leaves	28	18	54	440
tubers	48	12	40	60
1993	2	15	83	404

The Leaf disc method tends to overestimate the importance of resistance since populations containing 1% of resistant spores usually gives resistant reaction (Sozzi *et al.*, 1992). All single spore isolates derived from resistant field strains gave resistant reaction in this study. The semi-quantitative test indicated nearly 100% proportion of resistant sporangia in the field strains tested also.

Today farmers are very doubtful of the efficacy of Ridomil MZ. It is extremely difficult though to show in farm conditions whether the failure of the fungicide is due to resistance or false timing of the application. The majority of potato growers gave up Ridomil sprayings last summer. However the proportion of resistant strains in 1993 was higher than ever before. In addition the proportion of resistant strains was very high at the beginning of the epidemic at several locations (Table 2). This is contradictory to the observations from previous years and the reports from other countries (Davidse *et al.*, 1989, Olofsson 1989, Deahl *et al.*, 1991, Dowley & Sullivan, 1991, Staub, 1991).

TABLE 2. The proportion of resistant *P. infestans* strains at 7 locations at the oneset of the epidemic, 10 and 20 days later in 1993.

Location	first symptoms	+10 days resistant strains %	+20 days	first symptoms observed
Lammi	100	100	frost	24 Aug
Jokioinen	100	83	76	16 Jul
Kokemäki	92	77	100	02 Aug
Partala	40	84	68	03 Aug
Laukaa	61	76	100	01 Aug
Ylistaro	90	92	88	16 Aug
Ruukki	100	83	frost	17 Aug

The results from Finnish fungicide registration experiments at three locations indicate that since 1989 the marketable yield in spraying programs containing Ridomil MZ has been at the same level as in programs with contact fungicides only. In 1993 most contact fungicides gave better control and higher yields than Ridomil MZ -programme.

For the next season a newly registered mixture of propamocarb-HCl and mancozeb (Tattoo) will partly replace Ridomil MZ. However it is important to continue monitoring resistance to metalaxyl and to investigate why the resistance situation in Finland differs so much from many other countries.

ACKNOWLEDGEMENTS

The author thanks Ciba Geigy company for training the methods for monitoring metalaxyl resistance and the staff on Experimental Stations for collecting the blight samples from the fields.

REFERENCES

Anonymous (1982) Recommended methods for detection and measurement of resistance of plant pathogens to fungicides. Method for fungicide resistance in late blight of potato - FAO Method No. 30. *FAO Plant Protection Bulletin* **30**, 69-71.

Davidse, L. C.; Henken, J.; VanDalen, A.,; Jaspers, A. P. K.; Mantel, B. C. (1989) Nine years of practical experience with phenylamide resistance in *Phytophthora infestans* in the Netherlands. *Netherlands Journal of Plant Pathology* **95**, 197-213.

Deahl, K. L.; Inglis, D. A.; DeMuth, S. P. (1991) Testing for resistance to metalaxyl in *Phytophthora infestans* isolates from Northwestern

Washington. *American Potato Journal* **68**, 605.
Dowley, L. J.; O'Sullivan, E. (1991) Changes in the distribution of phenylamide-resistant strains of *Phytophthora infestans* (Mont.) de Bary in Ireland 1985-89. *Potato Research* **34**, 67-69.
Nuninger, C.; Bolton, N.; De Wijs, J. J.; Staub, T. (1992) Determination of the sensitivity of Phytophthora infestans to phenylamides: a semi-quantitative test method.*Bulletin OEPP/EPPO Bulletin* **22**, 309-311.
Olofsson, B. (1989)Metalaxylkänsligheten hos svenska stammar av potatisbladmögel (*Phytophthora infestans*). *Växtskyddsnotiser* **53**, 139-143.
Sozzi, D.; Schwinn, F. J.; Gisi, U. (1992) Determination of the sensitivity of *Phytophthora infestans* to phenylamides: a leaf disk method. *Bulletin OEPP/EPPO Bulletin* **22**, 306-309.
Staub, T. (1991) Fungicide resistance: Practical experience with Anti-resistance strategies and the role of integrated use. *Annual review of Phytopathology* **29**, 421-442.
Staub, T.; Sozzi, D. (1984) Fungicide resistance: A continuing challenge. *Plant Disease* **68**, 1026-1031.
Tantius, P. H.; Fyfe, A. M., Shaw, D. S.; Shattock, R. C. (1986) Occurrence of the A2 mating type and self-fertile isolates of *Phytophthora infestans* in England and Wales. *Plant Pathology* **35**, 578-571.
Williams, R. J.; Gisi, U. (1992) Monitoring pathogen sensitivity to phenylamide fungicides: principles and interpretation. *Bulletin OEPP/EPPO Bulletin* **22**, 299-306.

Session 3
Benzimidazoles and Dicarboximides

Chairman D SLAWSON

Session Organisers D SLAWSON and J T FLETCHER

EFFECT OF FUNGICIDE MIXTURES AND ALTERNATIONS ON DICARBOXIMIDE RESISTANCE DYNAMICS

G. W. MOORMAN, R. J. LEASE

Department of Plant Pathology, The Pennsylvania State University, University Park, PA, 16802

ABSTRACT

Botrytis cinerea populations initiated from spores, 5% of which were resistant to the dicarboximide fungicide vinclozolin, were cycled at 10 day intervals over a 30 day period on excised *Pelargonium* leaf tissue that had been sprayed once with single or mixtures of fungicides. Vinclozolin, whether used alone or in alternations or in mixtures with chlorothalonil, maintained the selection pressure in favour of resistance as a consequence of its long residual activity. It was concluded that once resistance is present, dicarboximide fungicide use should cease in greenhouses, particularly when disease pressure is high.

INTRODUCTION

Botrytis cinerea infects a wide range of greenhouse plants as a result of conidiospore germination and penetration or hyphal growth from infected tissue that has fallen onto healthy tissue. Spores do not necessarily germinate immediately and can remain dormant on plant tissue for weeks before germinating and penetrating (Salinas *et al.*, 1989). The few fungicides effective against *Botrytis* in greenhouses are used repeatedly to protect crops, particularly when disease pressure is high. When populations of *B. cinerea* are repeatedly exposed to dicarboximide fungicides, selection for resistance can be rapid (Gullino *et al.*, 1983, 1984; Locke and Fletcher, 1988; Northover and Matteoni, 1986). The presence of significant numbers of dicarboximide resistant *Botrytis* spores has been documented to occur in greenhouses where dicarboximides have never been used (Northover and Matteoni, 1986; Moorman and Lease, unpublished). It is presumed that this is a result of the movement of apparently healthy plants, especially vegetatively propagated plants such as *Pelargonium* harbouring fungicide-resistant spores, from greenhouses where dicarboximides were employed.

It appears that carrying dicarboximide resistance is not significantly detrimental to the overall fitness of the fungus (Faretra et al., 1989; Gullino *et al.*, 1982; Moorman and Lease, 1992; Vali and Moorman, 1992). While the dicarboximide resistant portion of the *Botrytis* population in the greenhouse persists even after use of this class of fungicides ceases (Locke and Fletcher, 1988), there can be fluctuations in the fungicide resistant:sensitive ratio in the spore population during the year. Katan and Ovadia (1985) noted that in greenhouses during the summer dormant season, when few plants are grown in Israel, the dicarboximide resistance proportion in the population declined and was relatively low early in the subsequent growing season. This allowed the successful use of dicarboximides once or twice late in the growing season. Although this practice fostered an increase in the proportion of resistant spores, it was postulated that conditions which were not defined, would suppress the resistant portion of the population during the following summer thereby allowing growers to apply dicarboximides late the next season. A similar decline in the dicarboximide resistance frequency was found to occur during the winter dormancy of strawberries grown in greenhouse-like plastic tunnels (Hunter *et al.*, 1987). Thus, there appear to be as yet undefined environmental selection pressures either against the resistant or in favour of the sensitive portion of the population.

Since populations with high proportions of resistance can be correlated with the frequency of fungicide use (Northover, 1988), the goal of management practices has been to limit the exposure of the fungus to the resistance selection pressure of the fungicide by applying fewer sprays per season or mixing or alternating dicarboximides with fungicides that have different modes of action. The effectiveness of alternations depends on selection pressures against the slightly less fit resistant portion of the population to reduce the frequency of resistance. The efficacy of mixtures depends on the dicarboximide to reduce or eliminate the sensitive portion of the population while the partner chemical holds in check the resistant portion. However, Gullino and Garibaldi (1987) and Vali and Moorman (1992) demonstrated that once dicarboximide resistance is present, dicarboximides should not be used. Neither mixtures nor alternations with non-dicarboximide fungicides significantly delayed the build-up of the dicarboximide-resistant portion of the population. In our previous work (Vali and Moorman, 1992), methods to cycle a population of *Botrytis* on plant tissue involved inoculation as soon as the spray had dried. Spores subsequently developing on this treated tissue were harvested and assayed to determine the percentage of resistant spores and were also used to inoculate newly treated plant tissue. Those experiments tested the situation where the fungus was continually exposed to freshly applied active ingredient. We have since modified the procedures to determine the effects of various chemicals when inoculation is done 0,

10, or 20, or 30 days after spraying in order to simulate the situation where inoculum lands on the treated tissue after the activity of the fungicide has begun to decline. In this way, the effect of the declining residual efficacy on the selection for resistance in the population was examined.

METHODS

To identify the combined effects of fungicide use patterns and declining residual activity on the population dynamics of resistance in *Botrytis*, the previously described *Pelargonium* leaf disk technique (Vali and Moorman, 1992) was modified slightly. Briefly, 12 wk old seed geraniums were sprayed only once with water, a fungicide, or a mixture of fungicides. Leaf disks were excised the day following treatment (day 0) and inoculated with a population of *Botrytis* spores, 5% of which were resistant to vinclozolin. After a 10 day incubation period, the number of infected disks was recorded as a measure of disease and the resistant:sensitive ratio in the spore population was ascertained by plating some spores from each treatment on fungicide-amended (20 μg vinclozolin/ml) and fungicide-free agar and counting the number of germinating spores. Spores were then used to inoculate leaf disks excised from the remaining plants that had been treated at the beginning of the experiment. In this manner, tissue was inoculated 0, 10, 20, and 30 days after spraying with spores harvested from the previously inoculated leaf disks. Fungicide-free tissue was inoculated with spores from each treatment on each inoculation date to be certain that the pathogenicity of the populations being carried in each treatment did not change. The experiment was done twice.

RESULTS AND DISCUSSION

As in previous research (Vali and Moorman, 1992), the use of vinclozolin alone and in a mixture with another fungicide favoured the rapid increase in the proportion of resistant spores (Table 1). The long residual activity of vinclozolin maintained the selection pressure favouring resistance over the 30 day period after treatment as indicated by the level of disease that developed (Table 2) and the high percent of resistant spores in all treatments involving vinclozolin. The 10 day interval between exposures to vinclozolin afforded in the alternation with chlorothalonil did not allow a significant decline in the percentage of resistant spores. Thus, these experiments indicate that under conditions of high disease pressure, once resistance is present in the population, dicarboximides should not be used in the greenhouse.

Beever *et al.* (1991) postulated that with a particular fungicide use pattern, the frequency of resistance will stabilize. This theory is based on the assumption that the selection for resistance exerted by the fungicide is balanced by the selection against the slightly less environmentally fit resistant part of the population. The problem indicated by the experiments reported here is that the selection pressure of dicarboximides is so strong, the high residual activity so prolonged, and the differences in fitness between resistant and sensitive strains of *Botrytis* so slight (Moorman and Lease, 1992a , 1992b; Vali and Moorman, 1992) that the frequency of resistance may stabilize at a level too high for dicarboximides to be effective.

It remains to be determined whether there is a proportion of resistance to dicarboximides in *Botrytis* populations at which dicarboximides can still be useful in greenhouses. While some (Beever and Brien, 1983; Locke and Fletcher, 1988) report good control despite a resistance frequency of 38%, others (Gullino and Garibaldi, 1987) report relatively poor control associated with a 10% frequency. If an acceptable resistance level can be identified, along with specifications of the conditions or treatments that force the resistant portion of the population below that level, it may be possible to use dicarboximide fungicides to a limited extent, despite the presence of resistance. The grower must clearly understand, however, that the resistance frequency will sharply increase after treatment. Where growers are unable or unwilling to use dicarboximides judiciously, it may be wisest to avoid dicarboximide use and recommend the application of mixtures of fungicides to which *B. cinerea* is not resistant and which have been demonstrated to provide the initial protection and long residual activity approaching that obtained with dicarboximides (Moorman and Lease, 1992a, 1992b, 1993).

Table 1. Mean percent of *Botrytis cinerea* spores carrying resistance to vinclozolin (dicarboximide) fungicide harvested from leaf disks inoculated the indicated number of days after spraying and incubated 10 days. Five percent of the initial population of spores used on day 0 were resistant to vinclozolin.

Treatment	Days after spraying			
	0	10	20	30
No fungicide (water)	9	2	1	1
vinclozolin[1]	99	99	98	95
chlorothalonil[2]	4	13	44	24
mancozeb[3]	59	98	99	96
vinclozolin[1] + chlorothalonil[2]	91	95	99	99
Alternation of chlorothalonil[2] (c) and vinclozolin[1] (v)	(c) 4	(v) 96	(c) 98	(v) 99

[1] vinclozolin, 0.6 g AI/litre (Ornalin 50% WP)
[2] chlorothalonil, 1.2 g AI/litre (Daconil 2787, 4.17F)
[3] mancozeb, 0.5 g AI/litre (Dithane F-45)

Table 2. Mean percent disease based on the number of leaf disks infected per 50 inoculated with *Botrytis cinerea* in each of two experiments.

Treatment	Days after spraying			
	0	10	20	30
No fungicide (water)	91	97	66	100
vinclozolin[1]	62	83	62	95
chlorothalonil[2]	30	17	22	47
mancozeb[3]	52	60	64	90
vinclozolin[1] + chlorothalonil[2]	10	28	19	61
Alternation of chlorothalonil[2] (c) and vinclozolin[1] (v)	(c) 30	(v) 33	(c) 14	(v) 96

[1] vinclozolin, 0.6 g AI/litre (Ornalin 50% WP)
[2] chlorothalonil, 1.2 g AI/litre (Daconil 2787, 4.17F)
[3] mancozeb, 0.5 g AI/litre (Dithane F-45)

REFERENCES

Beever, R. E.; Brien, H. M. R. (1983) A survey of resistance to the dicarboximide fungicides in *Botrytis cinerea*. *New Zealand Journal of Agricultural Research*, **26**, 391-400.

Beever, R. E.; Pak, H. A.; Laracy, E. P. (1991) An hypothesis to account for the behavior of dicarboximide-resistant strains of *Botrytis cinerea* in vineyards. *Plant Pathology*, **40**, 342-346.

Faretra, F.; Pollastro, K. S.; DiTonno, A. P. (1989) New natural variants of *Botryotinia fuckeliana* (*Botrytis cinerea*) coupling benzimidazole-resistance to insensitivity toward the N-phenylcarbamate diethofencarb. *Phytopathologia Mediterranea*, **28**, 98-104.

Gullino, M. L.; Romano, M. L.; Garibaldi, A. (1982) Characterization of dicarboximide-resistant strains of *Botrytis cinerea* Pers. naturally occurring in Italy. *Mededelingen van de Faculteit Landbouwwetenschappen Rijksuniversiteit Gent*, **47**, 781-791.

Gullino, M. L.; Romano, M. L.; Garibaldi, A. (1983) The influence or different spray programmes on the behavior of dicarboximide-resistant strains of *Botrytis cinerea* Pers. in greenhouse conditions. *Rivista de Patologia Vegetale*,**19**, 59-65.

Gullino, M. L.; Romano, M. L.; Garibaldi, A. (1984) Fungicide resistance on tomato in Italian greenhouses. *1984 British Crop Protection Conference*, **2**, 447-451.

Gullino, M. L.; Garibaldi, A. (1987) Control of of *Botrytis cinerea* resistant to benzimidazoles and dicarboximides with mixtures of different fungicides. *Mededelingen van de Faculteit Landbouwwetenschappen Rijksuniversiteit Gent*, **52**, 895-900.

Hunter, T.; Brent, K. J.; Carter, G. A.; Hutcheon, J. A. (1987) Effects of fungicide spray regimes on incidence of dicarboximide resistance in grey mould (*Botrytis cinerea*) on strawberry plants. *Annals of Applied Biology*, **110**, 515-525.

Katan, T.; Ovadia, S. (1985) Effect of chlorothalonil on resistance of *Botrytis cinerea* to dicarboximides in cucumber glasshouses. *European and Mediterranean Plant Protection Organization Bulletin*, **15**, 365-369.

Locke, T.; Fletcher, J. T. (1988) Incidence of benomyl and iprodione resistance in isolates of *Botrytis cinerea* in tomato crops in England and Wales. *Plant Pathology*, **37**, 381-384.

Moorman, G. W.; Lease, R. J. (1992a) Residual efficacy of fungicides used in the management of *Botrytis cinerea* on greenhouse-grown geraniums. *Plant Disease*, **76**, 374-376.

Moorman, G. W.; Lease, R. J. (1992b) Benzimidazole and dicarboximide-resistant *Botrytis cinerea* from Pennsylvania greenhouses. *Plant Disease*, **76**, 477-480.

Moorman, G. W.; Lease, R. J. (1993) Control of gray mold on greenhouse geraniums. *Fungicide & Nematicide Tests*, **48**, 390.

Northover, J. (1988) Persistence of dicarboximide-resistant *Botrytis cinerea* in Ontario vineyards. *Canadian Journal of Plant Pathology*, **10**, 123-32.

Northover, J.; Matteoni, J. A. (1986) Resistance of *Botrytis cinerea* to benomyl and iprodione in vineyards and greenhouses after exposure to the fungicides alone or mixed with captan. *Plant Disease*, **70**, 398-402.

Salinas, J.; Glandorf, D. C. M.; Picavet, F. D.; Verhoeff, K. (1989) Effects of temperature, relative humidity, and age of conidia on the incidence of spotting on gerbera flowers by *Botrytis cinerea*. *Netherlands Journal of Plant Pathology*, **95**, 51-64.

Vali, R.J.; Moorman, G. W. (1992) Influence of selected fungicide regimes on frequency of dicarboximide-sensitive strains of *Botrytis cinerea*. *Phytopathology*, **76**, 919-924.

Locke, T.; Fletcher, J. T. (1988) Incidence of benomyl and iprodione resistance in isolates of *Botrytis cinerea* in tomato crops in England and Wales. *Plant Pathology* **37**, 381-384.

Moorman, G. W.; Lease, R. J. (1992a) Residual efficacy of fungicides used in the management of *Botrytis cinerea* on greenhouse grown geraniums. *Plant Disease* **76**, 374-376.

Moorman, G. W.; Lease, R. J. (1992b) Benzimidazole- and dicarboximide-resistant *Botrytis cinerea* from Pennsylvania greenhouses. *Plant Disease* **76**, 477-480.

Moorman, G. W.; Lease, R. J. (1991) Control of gray mold on greenhouse geraniums. *Fungicide and Nematicide Tests* **46**, 300.

Northover, J. (1987) Persistence of dicarboximide-resistant *Botrytis cinerea* in Ontario vineyards. *Canadian Journal of Plant Pathology* **10**, [illegible].

Northover, J.; Matteoni, J. A. (1986) Resistance of *Botrytis cinerea* to benomyl and iprodione in vineyards and greenhouses after exposure to the fungicides alone or mixed with captan. *Plant Disease* **70**, 398-402.

Salinas, J.; Glandorf, D. C. M.; Picavet, F. D.; Verhoeff, K. (1989) Effects of temperature, relative humidity, and age of conidia on the incidence of spotting on gerbera flowers caused by *Botrytis cinerea*. *Netherlands Journal of Plant Pathology* **95**, 51-64.

Vali, R. J.; Moorman, G. W. (1992) [illegible] of selected fungicide regimes on frequency of dicarboximide-resistant [illegible] strains of *Botrytis cinerea*. *Phytopathology* [illegible].

RESISTANCE OF *HELMINTHOSPORIUM SOLANI* TO THIABENDAZOLE IN RELATION TO DIFFERENT STRATEGIES OF FUNGICIDE USE DURING SEED POTATO PRODUCTION.

P.J. BURGESS, F.J. FORBES

SAC Aberdeen, 581 King Street, Aberdeen, AB9 1UD

S.J.P OXLEY, S.A. BOWEN

SAC Edinburgh, King's Buildings, West Mains Road, Edinburgh ED9 3JG

ABSTRACT

A single application of thiabendazole to seed tubers resulted in 54% of *Helminthosporium solani* isolates from daughter tubers being resistant (IC50 >100 mg/l) to the fungicide. A further application of thiabendazole to these tubers resulted in 24% of isolates from the progeny crop being resistant at the Aberdeen site and 32% at the Edinburgh site. Application of thiabendazole in a mixture with imazalil appeared to delay the formation of resistant isolates.

INTRODUCTION

Thiabendazole has been widely used since the mid 1970's in the UK as a treatment for both ware and seed potatoes. Applied to tubers immediately after harvest the incidence of silver scurf (*Helminthosporium solani*), skin spot (*Polyscytalum pustulans*), gangrene (*Phoma foveata*) and dry rot (*Fusarium* spp.) in stored ware or seed tubers is reduced (Hide and Cayley 1983; 1985; 1987). When used as a seed tuber treatment, applied either as a spray during storage or as a dust pre-planting the incidence stem canker and black scurf (*Rhizoctonia solani*), silver scurf and skin spot in the growing crop is reduced (Hide *et al.,* 1980). However, since 1985 resistance to thiabendazole has been found in isolates of both *H. solani* and *P. pustulans* (Hide et al., 1988; Carnegie and Cameron, 1992; Burgess et al., 1993).

Other fungicides are available for the treatment of seed potato tubers. However, the availability of treatments for stored ware potatoes is limited to thiabendazole in the UK. Thus thiabendazole remains widely used despite problems with resistance in economically important pathogens.

The present study aims to determine the effect of fungicide applications to seed tubers on the level of control of silver scurf and the incidence of resistance to thiabendazole over three generations of seed production.

MATERIALS AND METHODS

Field trials

On 15 September 1990, a stock of seed potatoes (cv. Pentland Squire, VTSC2 Scottish Seed Potato Classification Scheme) was machine harvested at Tillycorthie Farm, Aberdeenshire and obtained for use in this study. This seed stock had been multiplied from supplied clones supplied by The Scottish Office Agriculture and Fisheries Department (SOAFD) at Tillycorthie Farm and had been treated with 2-Aminobutane during previous storage seasons. However, no other fungicide seed treatments had been applied to the seed tubers of this stock, or others, during multiplication at Tillycorthie Farm.

The crop was sized over square riddles and a quantity of the seed fraction (35-55 mm diameter tubers) was used for these trials. The next day liquid fungicides were applied as a hydraulic spray (2.2 litres diluted product applied per tonne) as tubers passed on a roller table. 2-Aminobutane was applied as a fumigant some four to five weeks later.

The fungicides applied were :-

1. Untreated (UT). No fungicide applied to tubers.
2. Thiabendazole (TBZ). 90 ml/t (40.5 mg AI kg^{-1}) 'Storite Flowable' (MSD Agvet, Hoddeston Herts.)
3. Thiabendazole + imazalil (TBZ+IM). 100 ml/t (30 + 10 mg AI kg^{-1}) 'Extratect' (MSD Agvet, Hoddeston, Herts.)
4. Imazalil (IM). 100 ml (10 mg AI kg^{-1}) 'Fungazil 100SL' (Rhone Poulenc Agriculture Ltd, Brentwood Essex)
5. 2-Aminobutane (2AB). 200 mg 2AB kg^{-1} (Chemical Spraying Co. Ltd., Perth)

Half the tubers of each treatment were stored at Tillycorthie Farm with the remainder transported to Boghall Farm, The Scottish Agricultural College, Edinburgh. At each site, the tubers were stored under commercial storage conditions in closed paper potato sacks. The temperature was maintained at *c.* 4°C throughout the storage period at both sites.

The tubers were planted as unreplicated large plots at each site in May 1991 (Table 1). The husbandry of plots was as recommended for the production of a specialist seed potato crop under local conditions. The plots were individually machine harvested and graded before treatment with the same fungicide as applied to the mother tubers. Samples of tubers were removed before treatment for resistance testing (Aberdeen only) and disease assessment. Treatments were applied by hydraulic spray at Aberdeen. At Edinburgh an electrostatic sprayer (Microstat, Horstine Farmery) was used to apply fungicide treatments (1.1 l of diluted product was applied per tonne). Treated seed tubers were stored in closed paper bags at each site under commercial storage conditions (*c.* 4°C) until planting as replicated block field trials in May 1992 (Table 1). Husbandry of the plots at each site was as recommended for a specialist seed potato crop.

After harvest in September 1992 samples from each plot were removed for disease assessment and resistance testing. Further fungicide treatments were applied and tubers were stored and planted as a

TABLE 1. Trial details and timetable of operations at Aberdeen and Edinburgh sites.

Growing season	Operation	Aberdeen	Edinburgh
	Harvest of initial seed stock	15/9/90	-
	Treatments applied	16/9/90	-
1st Growing season	Trial Planted	10/5/91	1/5/91
	Site	Udny Green Aberdeenshire	Boghall Farm Midlothian
	Plot size	45m x 2 drills	20m x 8 drills
	Replication	one	one
	Harvest	26/9/91	4/10/91
	Treatments applied	27/9/91	21/10/91
2nd growing season	Trial planted	25/5/92	7/5/92
	Site	Udny Green, Aberdeenshire	Boghall Farm, Midlothian
	Plot size	7.5m x 4 drills	7 m x 4 drills
	Replication	five	six
	Harvest	26/9/92	9/10/92
	Treatments applied	9/10/92	13/10/92
3rd growing season	Trial planted	7/5/93	5/5/93
	Site	Longside, Aberdeenshire	Boghall Farm, Midlothian
	Plot size	5m x 4 drills	6m x 4 drills
	Replication	six	six
	Harvest	18/10/93	21/10/93

replicated block field trial in May 1993. The husbandry of each site was as recommended for the production of a ware potato crop. Samples of tubers were removed after harvest for disease assessment.

The yield and number of tubers in each plot was measured for each replicated field trial.

Disease assessment

At harvest samples of tubers from each plot were washed and the percentage surface area infected with silver scurf was estimated by visual assessment. The results are presented as the % incidence (% tubers infected with silver scurf) and the % severity (mean % of the surface area infected with silver scurf). Assessments were also conducted on samples of tubers stored (as harvest) in closed paper sacks at 4°C for *ca*. 6 months. Samples of 100 (1991), 50 (1992) or 25 (1993) tubers were assessed from each plot.

Sensitivity of *H. solani* to thiabendazole

Samples of tubers were stored, in sealed paper bags, from harvest at 8°C until isolations of *H. solani* were made between 2 and 4 months later. Isolates were taken from each treatment at the Aberdeen site after harvest 1991 and from both the Aberdeen and Edinburgh sites after harvest 1992. Infected tubers were washed under tap water then incubated in humid conditions at 15°C for 14 days. After this period, conidia of *H. solani* could be removed using a sterile needle. These were streaked onto malt extract agar containing streptomycin and penicillin (2700 and 626 mg/l respectivley). The plates were incubated at 20°C for 7 days, after which time single spore isolates of *H solani* were selected and transferred onto buffered (pH 7) plates of malt extract agar and incubated (20°C) for 21 days before inoculation onto test media. Only one isolate was obtained from each infected tuber. Each isolate was tested for sensitivity to TBZ (as 'Storite Flowable') at 0 mg/l, 5 mg/l and 100 mg/l, using the method of Hide *et al.* (1988). The results are expressed as the concentration of TBZ required to reduce the colony diameter by 50% when compared to colonies growing on agar without TBZ (IC50).

Residues of thiabendazole present on tubers

Samples for residue analysis were stored at 4°C until analysis was commenced. Analysis for TBZ was by the method of Martindale (1988). Sub-samples of whole tubers were extracted with acetone/methanol and the extracts subjected to mild acid hydrolysis. After solvent extraction clean-up, the extracts were made alkaline and TBZ extracted into dichloromethane. After concentration, TBZ was determined by high performance liquid chromatography using a 25 cm CPS column and fluorescence detection.

RESULTS

Residues of thiabendazole on tubers after treatment

Residues of TBZ ranged from 1.01 to 6.20 mg/kg (Table 2). This represents between 2.5 and 16.6% of the target dose. The residues obtained after harvest in 1992 were generally less than those obtained in 1991. Residues at Aberdeen and Edinburgh sites were similar despite the different application methods used at each site (hydraulic and electrostatic respectively).

TABLE 2. Results of analysis of tuber samples for residues of thiabendazole after treatment in 1991 and 1992.

Year	Site	Treatment	Thiabendazole (mg/kg)	Percentage of target dose
1991	Aberdeen	TBZ	4.00	9.9
		TBZ+IM	4.20	14.0
	Edinburgh	TBZ	6.20	15.3
		TBZ+IM	5.00	16.6
1992	Aberdeen	TBZ	2.33	5.8
		TBZ+IM	1.77	5.9
	Edinburgh	TBZ	1.01	2.5
		TBZ+IM	1.61	5.3

Silver scurf on progeny tubers

Aberdeen

After storage of tubers harvested in 1991 the only treatment that appeared to reduce silver scurf were those that included imazalil (Table 3). However, there was no replication at this stage of the trial. After harvest 1992, silver scurf levels were generally higher. Treatment IM significantly ($P<0.05$) reduced both the incidence and severity of disease when compared to TBZ and UT treatments. There was no difference in the amount of silver scurf observed on TBZ and UT tubers. Application of TBZ+IM resulted in intermediate levels of disease control. There was a large and significant increase in the amount of disease following 2AB treatment. The levels of disease present on tubers did not increase during storage and there were no significant differences between treatments after storage. After harvest 1993 only low levels of disease were observed.

TABLE 3. Percentage incidence and severity of silver scurf at harvest and after storage at 4°C, Aberdeen site

Treatment	After storage 1991-1992		Harvest 1992		After storage 1992-1993		Harvest 1993	
	Incidence	Severity	Incidence	Severity	Incidence	Severity	Incidence	Severity
UT	72	6.0	50	14.9	20	7.5	4	1.1
TBZ	88	9.8	54	15.4	45	12.7	13	2.1
TBZ+IM	26	2.1	38	11.0	27	7.8	13	1.6
IM	60	4.2	17	5.3	32	10.9	11	1.6
2AB	80	8.5	73	33.6	42	14.5	4	0.2
LSD ($P<0.05$)	-	-	11.6	9.01	NS[1]	NS	NS	NS

[1] NS Not significant

Table 4 Percentage incidence and severity of silver scurf at harvest and after storage at 4°C, Edinburgh site

Treatment	After storage 1991-1992		Harvest 1992		After storage 1992-1993		Harvest 1993	
	Incidence	Severity	Incidence	Severity	Incidence	Severity	Incidence	Severity
UT	60	3.3	0	0	19	3.1	77	9.4
TBZ	26	1.3	0	0	16	2.2	78	10.9
TBZ+IM	50	7.1	0	0	14	1.7	67	5.9
IM	30	2.4	0	0	8	0.9	69	7.2
2AB	34	2.5	0	0	1	2.1	62	5.1
LSD ($P<0.05$)	-	-	-	-	6.1	0.99	18.5	3.60

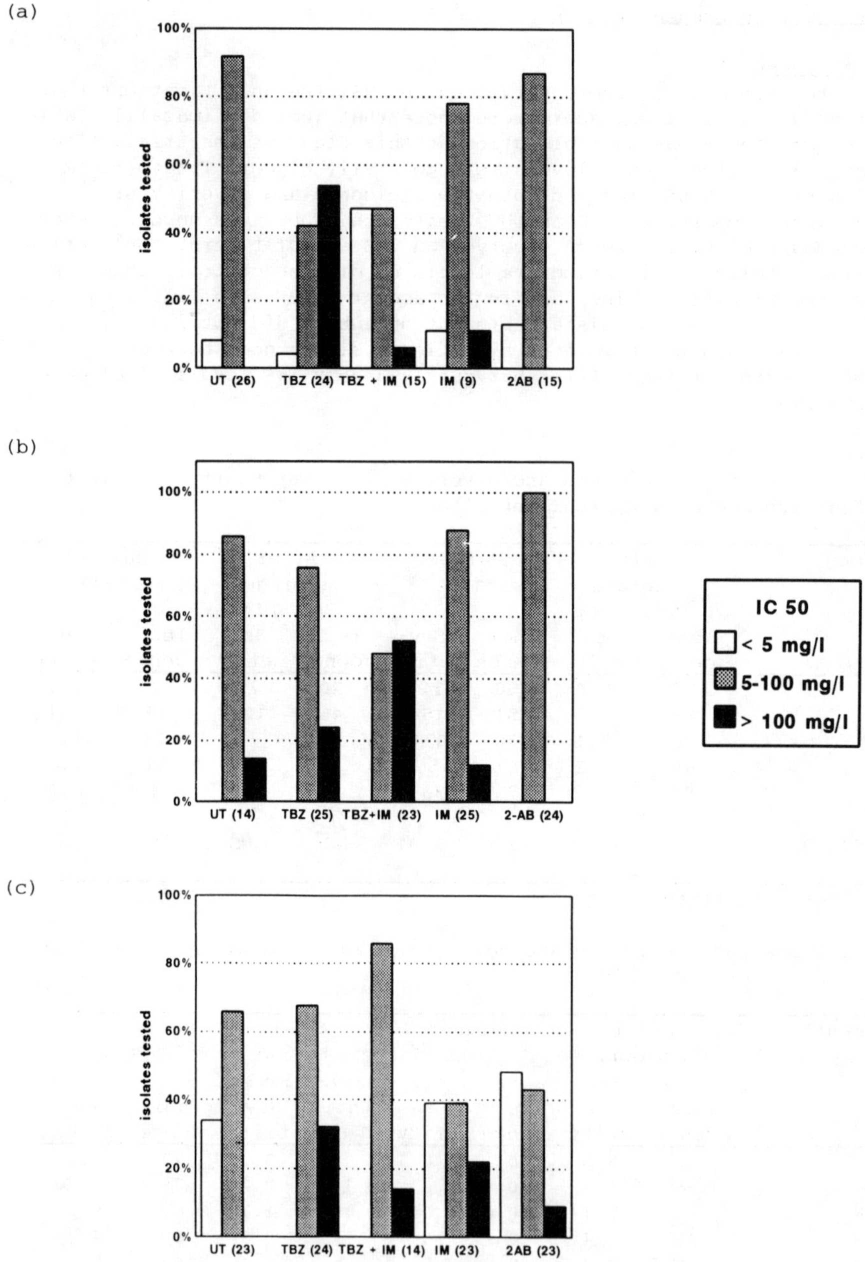

FIGURE 1. Percentage of isolates from each treatment with an IC 50 of < 5 mg/l, 5-100 mg/l or > 100 mg/l. (a) Aberdeen after harvest 1991 (b) Aberdeen after harvest 1992 and (c) Edinburgh after harvest 1992. Number of isolates tested given in parenthesis.

Edinburgh

At the Edinburgh site disease levels have remained generally low throughout the trial (Table 4). After storage 1991-1992, TBZ treatment appeared to result in least disease (no replication at this stage of the trial). At harvest 1992, no silver scurf was observed. However, during the subsequent storage period, disease did develop but remained at low levels. After harvest 1993, higher levels of disease were observed. There was least disease where mother tubers had been treated with TBZ+IM or 2AB. Treatment with TBZ resulted in no difference from the UT control.

Sensitivity of *H solani* isolates to thiabendazole

Aberdeen site after harvest 1991

After a single growing season, most isolates from untreated tubers were of intermediate sensitivity to TBZ (IC50 5-100 mg/l) (Figure 1a). After treatment of mother tubers with TBZ 54% of isolates were fully resistant (IC50 >100 mg/l). The sensitivity of isolates from 2AB and IM treatments were similar to the untreated control. After application of TBZ+IM a single isolate (6%) was found to be fully resistant to TBZ.

Aberdeen site after harvest 1992

After harvest 1992, no isolates (from a total of 107 tested) fully sensitive to TBZ were recovered from tubers of any treatment (Figure 1b). Isolates fully resistant to TBZ were recovered from TBZ (24% of isolates), TBZ+IM (52%) and IM (12%) treated tubers. The proportion of isolates with an IC50 >100 mg/l had decreased in TBZ treated tubers but increased in those treated with TBZ+IM when compared to the previous season. A small proportion (12%) of isolates from imazalil treated tubers had an IC50 >100 mg/l.

Edinburgh site after harvest 1992

In contrast to results from the Aberdeen site a proportion of isolates from all treatments in which TBZ had not been included were fully sensitive to TBZ (IC50 <5 mg/l) (Figure 1c). However, resistant isolates (IC50 >100 mg/l) were obtained from all treatments except the UT control. The range of sensitivities was greater for IM and 2AB treatments. Treatment with TBZ resulted in 32% of isolates being fully resistant to TBZ. The proportion was reduced to 14% when TBZ+IM was applied.

Yield and tuber numbers

Assessments of yield and tuber number were made only after harvest of replicated field trials (1992 and 1993), the results are given in Table 5. At Aberdeen in 1993, both TBZ+IM and IM treatments significantly ($P<0.05$) increased the total yield when compared to the UT control. However, TBZ applied alone did not increase the yield. At Edinburgh (1993), treatment IM significantly reduced the total yield when compared to the UT control. However, the yield of other treatments was not significantly different from the UT control. At this site in both 1992 and 1993, significantly ($P<0.05$) less tubers were produced following treatment with IM than TBZ.

TABLE 5. Yield (t/ha) and tuber number (1000's/ha) at Aberdeen and Edinburgh sites in 1992 and 1993.

Treatment	Aberdeen				Edinburgh			
	1992		1993		1992		1993	
	t/ha	000's/ha	t/ha	000's/ha	t/ha	000's/ha	t/ha	000's/ha
UT	43.9	449	54.6	458	58.6	367	49.6	349
TBZ	47.2	457	57.0	447	61.5	393	52.5	395
TBZ+IM	45.9	437	60.8	445	59.4	365	51.6	368
IM	48.7	479	61.4	432	58.0	343	45.7	344
2AB	41.0	425	59.1	432	55.0	329	51.2	388
LSD (P>0.05)	5.48	51.8	5.13	49.7	3.70	37.70	3.56	35.1

DISCUSSION

In these trials a large proportion of *H. solani* isolates, across all treatments and at both sites were found to be intermediate in their sensitivity (IC50 5-100 mg/l) to thiabendazole. Although no isolates were taken from the initial mother crop (1990 harvest), the majority of isolates (92%) from the untreated plot after one season were intermediate in their sensitivity to TBZ. The remaining isolates were fully sensitive to TBZ (IC50 <5 mg/l). Hide *et al.* (1988) also reported isolates of intermediate sensitivity. This result would appear to confirm that there is a large range of sensitivity to TBZ within populations of *H. solani*, which is unusual in fungal resistance to benzimidazole fungicides (Hide *et al.*, 1988).

A single treatment of mother tubers with TBZ resulted in 54% of isolates being fully resistant to the fungicide at the Aberdeen site. This is a similar proportion to that reported in previous studies (Hide *et al.*, 1993). Treatment with a mixture of TBZ and imazalil resulted in fewer resistant isolates than treatment with TBZ alone. The increased sensitivity to the fungicide was also reflected in the degree of silver scurf control observed after treatment with the mixture of TBZ and imazalil.

In the second year of TBZ treatment, a smaller proportion of isolates were resistant to TBZ (24% at Aberdeen and 32% at Edinburgh). This result contrasts with that of Hide *et al.* (1993) who recorded almost complete resistance to TBZ after two treatments. Treatment of tubers with a mixture of TBZ and imazalil resulted in a higher proportion of resistant isolates at Aberdeen but fewer at Edinburgh.

Residues of TBZ on tubers after treatment were recorded after treatment in 1991 and 1992. In both years, the residues achieved were low (2.5% to 16.6% of target dose). These low residues may be the reason that the levels of resistance did not increase in the second season of these trials. Hide *et al.* (1993) dipped tubers in differing concentrations of fungicide and concluded that resistance was most likely to develop where higher concentrations of TBZ had been used. The work reported in this paper would seem to support this conclusion. The residues achieved on tubers in this study are similar to those achieved in commercial practice. Thus over the two years of these trials the degree of resistance observed may be similar to that present in commercial practice.

The higher than expected degree of sensitivity to TBZ could also be explained by an inherent inability of the *H. solani* population to become resistant to TBZ. Alternatively, it could be due to introduction of 'wild-type' *H. solani* from outside sources, although efforts were made throughout this trial series to reduce such sources of contamination. Some cross-contamination of treatments would appear to have occurred in these experiments with resistant isolates being isolated from treatments to which no TBZ had been applied. Cross-contamination would also occur in commercial practice and thus this study may give a realistic picture of the restsistance situation in commercail stores.

TBZ treatment of mother tubers resulted in no appreciable control of disease, despite the fact that not all the isolates obtained from daughter tubers were resistant to the fungicide. In general, treatments containing imazalil performed better. Similar results were reported by Hall and Hide (1992).

The results reported here indicate that resistance to TBZ by *H. solani* is perhaps not as widespread as previously considered due to the fact that in commercial practice good coverage of tubers with fungicides is often not achieved. However, it does appear from these results that the level of silver scurf control achieved with TBZ is not sufficient to justify its inclusion in seed production. Based on these results and those of Hide *et al.* (1993), the use of formulations of TBZ and imazalil reduce the build-up of resistance but does not prevent it developing entirely.

ACKNOWLEDGEMENTS

This research was supported in part by the Potato Marketing Board. The trials form part of a project funded jointly by the Potato Marketing Board, MSD Agvet, Schering Agriculture and Rhone Poulenc Agriculture. Residue analysis was conducted by Dr N Stephen, SAC Aberdeen.

REFERENCES

Burgess, P.J.; Wale, S.J.; Oxley, S.; Lang, R.W. (1993). Fungicide treatment of seed potatoes: Strategies for the control of silver scurf (*Helminthosporium solani*). *Proceedings Crop Protection in Northern Britain 1993, Dundee*, 319-324.

Carnegie, S.F.; Cameron, A.M. (1992). Resistance to thiabendazole in isolates of *Polyscytalum pustulans* (skin spot) and *Fusarium solani* var. *coeruleum* (dry rot) in Scotland. *Plant Pathology*, **41**, 606-610.

Hall, S.M.; Hide, G.A. (1992) Fungicide treatment of seed tubers infected with thiabendazole-resistant *Helminthosporium solani* and *Polyscytalum pustulans* for controlling silver scurf and skin spot on stored progeny tubers. *Potato Research*, **35**, 143-148.

Hide, G.A.; Cayley G.R. (1983). Effects of delaying fungicide treatment on the incidence of gangrene in stored potato tubers. *Annals Applied Biology*, **102**, 107-115.

Hide, G.A.; Cayley G.R. (1985). Effects of delaying fungicide treatment on the incidence of *Fusarium* dry rot in stored potato tubers. *Annals Applied Biology*, **107**, 429-438.

Hide, G.A.; Cayley G.R. (1987). Effects of delaying fungicide treatment and of curing and chloropropham on the incidence of skin spot on stored potato tubers. *Annals Applied Biology*, **110**, 617-627.

Hide, G.A.; Cayley G.R.; Read, P.J.; Fraser, J.H. (1980). Treatment of seed and ware tubers with thiabendazole for control of storage diseases. *Annals of Applied Biology*, **96**, 119-131.

Hide, G.A.; Hall, S.M. (1993) Development of resistance to thiabendazole in *Helminthosporium solani* (silver scurf) as a result of potato seed tuber treatment. Plant Pathology, **42**, 707-714.

Hide, G.A.; Hall, S.M.; Boorer, K.J. (1988) Resistance to thiabendazole in isolates of *Helminthosporium solani*, the cause of silver scurf disease of potatoes. *Plant Pathology*, **37**, 377-380.

Martindale R.W. (1988) Determination of residues of a range of fungicides, anti-sprouting agents and (organochlorine and organophosphorus) insecticides in potatoes by gas-liquid and high performance liquid chromatography. *Analysist,* **113,** 1229-1233

BENZIMIDAZOLE AND DICARBOXIMIDE RESISTANCE IN PATHOGENS OF STORED APPLES AND PEARS

A.M. BERRIE[1], I. KOOMEN[2]

ADAS, Boxworth, Cambridge, CB3 8NN, UK

ABSTRACT

Between 1988 and 1993, during the storage period for apples and pears (December-April), regular visits were made to commercial fruit packhouses in England when fruit was being graded. Samples were taken of the major fungal pathogens causing rotting: *Botrytis cinerea*, *Monilinia fructigena*, *Nectria galligena*, *Gloeosporium* spp., *Penicillium expansum* and *Phytophthora syringae*. Tests for fungicide response were conducted on the cultures established. More than 50% of *B. cinerea* isolates and over 95% of *P. expansum* isolates were consistently found to be resistant to benzimidazole fungicides. No resistance was detected in *M. fructigena* or *N. galligena* isolates. Resistance of *Gloeosporium* spp. to benzimidazole fungicides was detected (approx. 25% of isolates) in the last three years of the study. Resistance of *B. cinerea* to dicarboximide fungicides was detected, but at a low incidence. A very low incidence of *P. syringae* isolates resistant to metalaxyl was also detected. Strategies for minimising the development of fungicide resistance in stored fruit pathogens are discussed.

INTRODUCTION

The harvesting period for most British apples and pears is restricted to the period September to mid October. Efficient storage is therefore essential in order to allow the fruit industry to regulate its supply of fruit onto the UK market for most of the year and enable it to compete with imports of high quality fruit from other EC countries and outside Europe. Losses due to post-harvest rots can seriously affect profitability and their effective control is an integral part of efficient storage. In the 1960's, losses in store were more than 30 per cent from some orchards as a result of infection by *Gloeosporium* spp. (Preece, 1967). Fungicidal control was applied as pre-harvest sprays, principally of protectant fungicides, such as captan. More recent surveys of rotting in stored fruit (Berrie, 1989, 1992a) initially in fungicide-treated fruit (1979-91) and latterly in untreated fruit (1991-93), have indicated that while economic losses (> 2 per cent) occur in most seasons, they are not as high as those recorded previously. This is partly due to improvements in fruit mineral composition and in storage techniques that increase the ability of the fruit to resist fungal attack,

[1] Present address: Horticulture Research International, East Malling, West Malling, Kent ME19 6BJ, UK

[2] Present address: IPO-DLO, Postbus 9060, 6700 GW Wageningen, The Netherlands

but also to the introduction of post-harvest fungicide dips/drenches of initially benzimidazoles in the 1970's and later of metalaxyl (in combination with carbendazim as 'Ridomil mbc 60 wp'). Benzimidazole fungicides were introduced mainly for control of *Gloeosporium* spp. and metalaxyl for the control of *Phytophthora syringae* which emerged as a problem in the 1970's following changes in orchard cultural practices (Edney, 1978; Upstone, 1978). The main fungi responsible for losses in stored apples and pears are shown in Table 1 and the fungicides used to control these rots are given in Table 2.

TABLE 1. Principle fungal species responsible for losses due to rotting in apple (cvs Cox and Bramley) and pear in the U.K.

Apple cv. Cox	Apple cv. Bramley	Pear
Botrytis cinerea	*Monilinia fructigena*	*Botrytis cinerea*
Monilinia fructigena	*Penicillium expansum*	*Monilinia fructigena*
Nectria galligena		
Gloeosporium spp.		
Phytophthora syringae		
Penicillium expansum		

TABLE 2. Fungicides used or previously used as post-harvest dips/drenches on apples and pears in the UK and the fungi controlled.

Active ingredient	Chemical Product	B	M	N	Gl	Ps	Pe
benomyl[1]	'Benlate'	+r	+	+	+	-	+r
carbendazim	e.g. 'Bavistin', 'Derosal'	+r	+	+	+	-	+r
captan	'PP Captan 83'	-	-	-	-	+	-
iprodione[2]	'Rovral Flo'	+	+	-	-	-	-
metalaxyl + carbendazim	'Ridomil mbc 60 wp'	+r	+	+	+	+	+r
thiophanate methyl	'Mildothane'	+r	+	+	+	-	+r
vinclozolin[3]	'Ronilan'	+	+				

Key to fungi

B = *B. cinerea*, M = *M. fructigena*, N = *N. galligena*, Gl = *Gloeosporium* spp., Ps = *Phytophthora syringae*, Pe = *Penicillium* spp.

Notes

\+ Fungi controlled
+r Sensitive isolates only controlled
[1] No longer recommended
[2] Off-label approval on pears only
[3] No longer recommended and use withdrawn in 1990

Between 1979 and 1993 surveys of rotting in stored fruit were conducted, mainly in south-east England. The main objective of these studies was to identify the fungi responsible for rotting in store and to monitor the resistance of these fungi to benzimidazole, dicarboximide and acylalanine fungicides. The results of the surveys of rotting and the earlier study of the incidence of fungicide resistance have been reported elsewhere (Berrie, 1989, 1992a). The purpose of this paper is to report the results of the study on fungicide resistance carried out between 1988 and 1993.

MATERIALS AND METHODS

Source of fungal isolates (1988-93)

Sixteen commercial packhouses and three fruit co-operatives in the south-east and east of England were visited regularly (weekly to some) during the storage period between December and April. Observations were made on fungal rots of apple cvs. Cox and Bramley's Seedling and pear cv. Conference during grading of fruit out of store, by examining one hundred rejected rotted fruit of each cultivar and identifying the rots visually. Samples of fruits with *Botrytis cinerea*, *Monilinia fructigena*, *Nectria galligena*, *Gloeosporium* spp., *Phytophthora syringae* and *Penicillium expansum* were collected. In 1990-93 visits to commercial packhouses in Gloucestershire, Somerset and Herefordshire were also included. Up until 1991 observations on rotting in store were mainly on fruit that had been treated with a fungicide post-harvest. Between 1991 and 1993 samples of rots were also obtained from a more specific survey of rotting in stored Cox apples and Conference pears which were treated or untreated with a fungicide post harvest. In 1991/92 and 1992/93 this survey covered a total of 47 and 76 Cox orchards and 24 and 8 Conference orchards respectively, located mainly in the south-east and east of England, but also from Herefordshire.

Tests for fungicide resistance

Fungal isolates of *B. cinerea*, *M. fructigena* and *P. expansum* were obtained by culturing from rotted fruit onto Potato Dextrose agar (PDA) amended with 100 ppm streptomycin to inhibit bacterial growth. *N. galligena* and *Gloeosporium* spp. were cultured on PDA amended with 2 ppm vinclozolin to inhibit growth of *B. cinerea* and *Penicillium*; and *P. syringae* was cultured on PDA amended with 2 ppm vinclozolin or V8 agar. Fungicide resistance tests were carried out using PDA amended with benomyl, vinclozolin or metalaxyl, all at 2 or 20 ppm. Inoculated plates were assessed after 2 or 7 days depending on the test fungus. Isolates showing no colony growth were regarded as sensitive; those showing normal growth were considered resistant.

RESULTS AND DISCUSSION

The results of the fungicide resistance tests are summarised in Tables 3 (apple) and 4 (pear). The numbers of isolates tested varied from year to year depending on disease incidence and the ease of obtaining uncontaminated isolates for testing.

TABLE 3. Percentage of fungal isolates obtained from stored apples with resistance to 20 ppm benomyl, vinclozolin and metalaxyl (1988-93).

Fungus	Cultivar	Fungicide	1988/89	1989/90	1990/91	1991/92	1992/93
B. cinerea	Cox	benomyl	87 (47)	58 (205)	83 (60)	55 (156)	65 (134)
B. cinerea	Cox	vinclozolin	0 (47)	1 (209)	2 (49)	-	-
B. cinerea	Bramley	benomyl	16 (6)	40 (5)	50 (14)	54 (13)	-
B. cinerea	Bramley	vinclozolin	16 (6)	0 (5)	7 (14)	-	-
P. expansum	Cox	benomyl	97 (30)	92 (99)	88 (49)	75 (32)	-
P. expansum	Bramley	benomyl	100 (24)	100 (26)	79 (38)	-	-
P. syringae	Cox	metalaxyl	0 (2)	-	-	7 (42)	0 (22)
P. syringae	Bramley	metalaxyl	0 (5)	-	-	-	0 (7)
M. fructigena	Cox	benomyl	0 (2)	0 (36)	0 (64)	0 (41)	0 (13)
M. fructigena	Cox	vinclozolin	0 (2)	0 (36)	0 (29)	-	-
M. fructigena	Bramley	benomyl	0 (16)	0 (9)	0 (33)	0 (8)	-
M. fructigena	Bramley	vinclozolin	0 (14)	0 (9)	0 (8)	-	-
N. galligena	Cox	benomyl	0 (69)	0 (33)	0 (28)	0 (151)	0 (180)
N. galligena	Bramley	benomyl	0 (16)	0 (2)	0 (2)	0 (4)	-
Gloeosporium spp	Cox	benomyl	-	0 (13)	27 (11)	41 (58)	27 (102)

Figures in brackets = number of isolates tested

TABLE 4. Percentage of fungal isolates obtained from stored pears with resistance to 20 ppm benomyl, vinclozolin and metalaxyl (1988-93).

Fungus	Fungicide	1988/89	1989/90	1990/91	1991/92	1992/93
B. cinerea	benomyl	80 (99)	74 (31)	82 (60)	68 (97)	71 (38)
B. cinerea	vinclozolin	3 (87)	3 (31)	0 (72)	-	-
P. expansum	benomyl	-	89 (9)	100 (5)	-	-
P. syringae	metalaxyl	-	-	-	0 (1)	0 (2)
M. fructigena	benomyl	0 (3)	0 (4)	0 (2)	0 (7)	0 (4)
M. fructigena	vinclozolin	0 (3)	0 (4)	0 (1)	-	-
N. galligena	benomyl	0 (11)	-	0 (4)	0 (9)	0 (1)
Gloeosporium spp.	benomyl	-	100 (1)	0 (1)	20 (5)	0 (5)

Figures in brackets = number of isolates tested

Fungicide resistance in *B. cinerea*

B. cinerea from Conference pear consistently accounted for 70-80% of rotting over the period of the study and is the main fungus at which post-harvest fungicide treatment is targeted (Berrie, 1989; 1994). Consequently large numbers of isolates were tested, over 70% of which on average were resistant to benomyl (Table 1). The incidence of benomyl-resistant *B. cinerea* was higher (90%) in the early 1980s (Table 5), falling to less than 60% resistance in 1987 before increasing again to over 70%. This incidence reflects fungicide product use. Up until 1983, benzimidazole fungicides were principally used as post-harvest drenches. The high incidence of resistance resulted in significant losses (> 2% on average; 10% losses in some badly affected orchards) which jeopardised the long-term storage of pears (up to June in some years). Vinclozolin was introduced as an alternative treatment in 1984 and was rapidly taken up by pear growers as the principle fungicide treatment. The temporary suspension of approval for vinclozolin in 1990 resulted in a switch back to benzimidazole fungicides and a consequent increase in losses due to *B. cinerea.* In 1993 the granting of an off-label approval for use of iprodione as a post-harvest treatment has again resulted in a major shift in fungicide use. Dicarboximide-resistant *B. cinerea* (Table 4 and 5) has been detected, but at a very low incidence to date, such that good control of rotting is still achieved. The influence of fungicide treatment in selecting resistant isolates of *B. cinerea* is shown in Table 6 where over 70% of isolates from benzimidazole-treated fruit were resistant compared to less than 50% in untreated fruit.

TABLE 5. Incidence of resistance to benomyl and vinclozolin of *B. cinerea* from apple and pear 1980-93.

	% resistance to 20 ppm benomyl		% resistance to 20 ppm vinclozolin	
Year	Cox apple	Conference pear	Cox apple	Conference pear
1980/81		90.3		
1982/83	75.0	90.9		
1983/84	64.3	66.7		
1984/85	70.0	63.6		
1987/88	75.6	56.3	9.8	14.1
1988/89	87.2	79.8	0	3.4
1989/90	58.0	74.2	0.5	3.2
1990/91	83.3	81.7	2.0	0
1991/92	55.1	68.0	-	-
1992/93	64.9	71.1	-	-

B. cinerea from Cox apple is of increasing importance as a cause of rotting (Berrie, 1994) although post-harvest fungicide treatment is mainly targeted at control of *Gloeosporium* spp., *Phytophthora syringae, Nectria galligena* and *Monilinia fructigena.* Consequently, benzimidazole fungicides are the main products used as vinclozolin is ineffective against *Gloeosporium* and *Nectria*. Over 50% of *B. cinerea* isolates were resistant to benomyl and this incidence has remained consistent over the period of study (Tables 3 and 5). The incidence of resistance to dicarboximide was lower than in pear indicating the low or nil usage of these products as drenches for apple.

Fungicide resistance in *M. fructigena*

M. fructigena consistently causes losses in most seasons in both apple and pear, although incidence in pear has declined in recent years (Berrie, 1992a). The fungus is controlled by both benzimidazole and dicarboximide fungicides. Over the ten years of the study, no isolates of *M. fructigena* resistant to dicarboximide or benzimidazole fungicides have been detected. Similar results have been found elsewhere, although resistance of the closely related *Monilinia fructicola*, which causes a brown rot of stone fruit, to both these fungicide groups has been frequently recorded (Penrose *et al.*, 1979; 1985; Ritchie, 1982). A possible explanation for this could be that in *M. fructicola*, which is not found in the UK, the sexual stage is common and forms an important part of the annual disease cycle. In *M. fructigena* occurrence of the sexual stage is rare and the fungus survives asexually in cankers or mummified fruit (Byrde & Willetts, 1977).

Fungicide resistance in *P. expansum*

Many different species of *Penicillium* have been recorded as fruit rots of apple and pear (Rosenberger, 1990), but in the UK *P. expansum* appears to be the most frequently isolated species. The rot commonly occurs on apple and pear every season although actual losses are very low. Almost all isolates tested were found to be resistant to benomyl (Table 3 and 4), which is consistent with results from elsewhere (Kim *et al.* 1989; Rosenberger *et al.*, 1991). In 1991-93, the survey of rotting in treated and untreated fruit has indicated a higher incidence of *P. expansum* in drenched fruit (Berrie, 1993). The use of benzimidazole fungicides as drenches appears to be an effective way of spreading *P. expansum*.

Fungicide resistance in *Gloeosporium* spp.

In the UK three species of *Gloeosporium* have been recorded during the ten years of the survey - *G. album* (*Pezicula alba*), *G. perennans* (*Pezicula malicorticis*) and *G. fructigenum* (*Glomerella cingulata*). Individual species were not identified during the fungicide resistance study. In the 1960's, *Gloeosporium* was the most important cause of rotting in Cox, however the incidence over the ten years of the survey has been very low or absent until the last two seasons. This higher incidence, although actual losses were very low, is probably associated with the poor mineral composition of fruit and consequent poorer quality of cv. Cox apparent in 1992/93 (Sharples, 1980). Resistance of *Gloeosporium* spp. to benzimidazole fungicides has been frequently recorded in Europe (Palm, 1986; van der Scheer and Remijnse, 1988), as a result of reliance on pre-harvest fungicide sprays for storage rot control. The occurrence of such isolates in the UK was thought to be rare or absent. Due to the low incidence of disease, very few isolates were tested until the last three seasons (1990-93). Twenty-five per cent or more of isolates tested were found to be resistant to 20 ppm benomyl (Table 3). Reasons for this are not clear, as the use of benzimidazole fungicides as orchard sprays has been avoided in favour of post-harvest drenching. It is possible that resistance has gradually built-up since the introduction of the benzimidazole fungicides, or that the natural level variation in the population has become evident in the last three seasons when the incidence of *Gloeosporium* was higher and more tests were carried out. Tests to date suggest that either could be true. Table 6 compares the incidence of benomyl resistant isolates from treated and untreated fruit samples. In 1991/92 a higher incidence of resistance is apparent in treated fruit, whereas in 1992/93 the incidence in the two treatments is similar. The situation may become clearer after studies in the current season, when data on the incidence of resistant isolates from fruit treated pre-harvest with carbendazim sprays will be compared to that in untreated fruit. In addition, individual *Gloeosporium* spp. will be identified.

TABLE 6. Percentage resistance to 20 ppm benomyl of *B. cinerea* or *Gloeosporium* spp. from fruit treated or untreated with a benzimidazole fungicide post harvest.

	% resistant isolates			
	B. cinerea		*Gloeosporium*	
Year	Treated	Untreated	Treated	Untreated
1991/92	68.1	46.3	72.7	11.5
1992/93	73.2	47.2	21.9	27.4

Fungicide resistance in *Nectria galligena*

The incidence and importance of *Nectria* fruit rot varies according to the occurrence of *Nectria* canker. In orchards where canker is present, losses due to rotting can be severe (up to 30%) following wet summers (Berrie, 1992a). Only benzimidazole fungicides are effective in rot control (Berrie, 1992b) although they appear to be less effective when applied as post-harvest treatments. In culture, *Nectria galligena* is variable, isolates being very pale, almost white in colour, to dark orange with similar variation in fluffiness and growth rate. No isolates have been detected which are resistant to 20 ppm benomyl (Table 3 and 4), during the ten years of the survey (Berrie, 1989) or from other studies on control of *Nectria* canker, even after the application of 48 sprays of carbendazim to an orchard over four seasons (Berrie, 1992b). However, isolates do vary in sensitivity to benomyl, some being sensitive to 2 ppm and some being resistant to 2 ppm but sensitive to 20 ppm (Table 7). This would appear to be a feature of natural variation in the population with incidence varying from season to season, and not apparently influenced by post-harvest treatment (Table 8).

Fungicide resistance in *P. syringae*

Phytophthora fruit rot is mainly a problem of cv. Cox, although occasionally it can be serious on Conference pears and Bramley apples, associated with wet weather at harvest. The occurrence of the rot is thus sporadic and particularly associated with

TABLE 7. Sensitivity of *Nectria galligena* isolates to benomyl (1988-93).

		% isolate sensitive to benomyl									
		1988/89		1989/90		1990/91		1991/92		1992/93	
Crop	Cultivar	2 ppm	20 ppm	2 ppm	20 ppm	2 ppm	20 ppm	2 ppm	20 ppm	2 ppm	20 ppm
Apple	Cox	42	58	76	24	96	4	33	68	68	32
Apple	Bramley	38	63	100	0	100	0	0	100	-	-
Pear	Conference	9	91	-	-	50	50	22	78	-	-

TABLE 8. Sensitivity to benomyl of *Nectria galligena* isolates from fruit treated or untreated post harvest with benzimidazole fungicide.

	% isolates sensitive to benomyl			
	1991/92		1992/93	
Post harvest Treatment	2 ppm	20 ppm	2 ppm	20 ppm
Treated	33	67	57	44
Untreated	13	87	69	31

modern intensive orchard systems. The rot is effectively controlled by the use of metalaxyl (in combination with carbendazim as 'Ridomil mbc 60 wp'). Numbers of isolates tested have been low, but a low level of resistance to metalaxyl at 20 ppm was detected in 1991/92 (Table 3). This is of particular concern as at present there are no other effective products available for use on fruit.

STRATEGIES FOR FUNGICIDE USE TO CONTROL STORAGE ROTS

Up until the introduction of post-harvest treatments in the 1970's, control of storage rots had relied on the use of pre-harvest sprays of captan or benzimidazole fungicides. Once the efficacy of post-harvest treatments was established, use of benzimidazole fungicides in the orchard was actively discouraged, to preserve the efficacy of these treatments for storage rots. A similar policy was later adopted for metalaxyl. In Europe, post-harvest treatments were never developed and in some countries their use was prevented by law. Repeated pre-harvest use of benzimidazoles for control, in particular of *Gloeosporium* spp., has resulted in resistant isolates developing and becoming widespread. In the UK, it would appear that the adopted strategy has at least delayed the appearance of benzimidazole-resistant *Gloeosporium*.

To develop a strategy of fungicide use for resistance management of storage rots is difficult when the availability of effective products is limited. Preserving the products for post-harvest use only was and is a sound strategy, however this is now threatened by the future of post-harvest treatments, not for scientific or environmental reasons but because of consumer concerns over such uses. Alternative, less effective, fungicides such as captan are available as pre-harvest fungicide sprays to control *Nectria*, *Gloeosporium* and possibly *Monilinia*. However, for control of *Botrytis* and *Phytophthora*, alternative fungicide options are limited and much less effective. For control of *Phytophthora* fruit rot, an alternative orchard treatment is the use of metalaxyl + mancozeb (as an off-label approval) as an orchard spray to the soil and low hanging fruit (Harris, 1979; Edney & Chambers, 1981). Should post-harvest treatments be withdrawn, such a use of metalaxyl would become routine in high risk *Phytophthora* orchards (modern intensive Cox orchards) and would probably greatly accelerate the development of resistance in both *P. syringae* (fruit rot) and *P. cactorum* (the cause of crown rot and collar rot in apple).

For fungi, such as *Penicillium*, chemical control offers few solutions and attention must be focused on cultural procedures, enhancing the ability of the apple to resist rotting and attention to hygiene. The possibility of using chlorine as a water disinfectant treatment should also be considered.

Any strategy for control of storage rots must be based on an integrated approach, relying on orchard cultural treatments, good storage conditions and methods, such as ensuring the correct fruit mineral composition, that will enhance the ability of the fruit to resist rotting. Such cultural methods can be combined with a system of determination of rot risk to ensure that fungicide treatment is only used when necessary, and preferably applied post-harvest, when all the factors relating to rot-risk are known. The continued availability of post-harvest treatments is, therefore, essential both to ensure the long-term and economic storage of UK apples and pears, and to preserve the effectiveness of the limited range of available fungicide products.

ACKNOWLEDGEMENTS

The authors wish to thank the Ministry of Agriculture, Fisheries and Food, and the Apple and Pear Research Council for financial support. Thanks are also due to Dr. T. Locke and Mr. D. Lockley for providing information on resistance from the south-west and West Midlands and to Dorothy Baines, Sandra Reynolds, Geoff Thorpe and other staff at ADAS Wye for carrying out the fungicide resistance tests.

REFERENCES

Berrie, A.M. (1989). Storage rots of apple and pear in South-East England 1980-1988 incidence and fungicide resistance. *Integrated control of pome fruit diseases, Vol II, IOBC Bulletin XII/6*, 229-239.

Berrie, A.M. (1992a). Alternative strategies for the control of post harvest rots in apples and pears. *Brighton Crop Protection Conference - Pest and Diseases 1992*, **1**, 301-310.

Berrie, A.M. (1992b). Comparison of fungicide sprays for the control of canker (*Nectria galligena* Bres.) in Apples cvs. Cox's Orange Pippin and Spartan. *Acta Phytopathologica et Entomologica Hungarica*, **27**, 103-109.

Berrie, A.M. (1993). Strategies for rot control. *East Malling Research Association Members Day Report, May 1993.*

Berrie, A.M. (1994). The importance of *Botrytis cinerea* as a storage rot of apple cv. Cox and pear cv. Conference. *Norwegian Journal of Agricultural Science* (In Press).

Byrde, R.J.W.; Willetts, H.J. (1977). The brown rot fungi of fruit: their biology and control. Pergamon Press, Oxford.

Edney, K.L. (1978). The infection of apples by *Phytophthora syringae*. *Annals of Applied Biology*, **88**, 31-36.

Edney, K.L.; Chambers, D.A. (1981). The use of metalaxyl to control *Phytophthora syringae* rot of apple fruits. *Plant Pathology*, **30**, 167-170.

Harris, D. (1979). The suppression of *Phytophthora syringae* in orchard soil by furalaxyl as a means of controlling fruit rot of apple and pear. *Annals of Applied Biology*, **91**, 331-336.

Kim, N.Y.; Kim, K.H.; Lee, C.U. (1989). Spore germination/mycelial growth and pathogenicity of benomyl-resistant *Penicillium expansum* causing apple blue mould rot. *Korean Journal of Plant Pathology, Vol 5*, 344-348.

Palm, G. (1986). Die Aktvelle bedeutung der Fruchtfäule erreger an der Niederelbe und möglichkeiten ihver bekämpfung. *Mitt. O.V.R. Alten Landes*, **1**, 14-21.

Penrose, L.J.; Davis, K.C.; Koffmann, W. (1979). The distribution of benomyl-tolerant *Sclerotinia fructicola* (Wint) Rehm in stone fruit orchards in New South Wales and comparative studies with susceptible isolates. *Australian Journal Agricultural Research*, **30**, 307-319.

Penrose, L.J.; Koffmann, W; Nicholls, M.R. (1985). Occurrence of vinclozolin resistance in *Monilinia fructicola*. *Plant Pathology*, **34**, 228-234.

Preece, T.F. (1967). Losses of Cox's Orange Pippin apples during refrigerated storage in England, 1961-65. *Plant Pathology*, **16**, 176-180.

Ritchie, D.F. (1982). Effect of dichloran, iprodione, procymidone and vincozolin on the mycelial growth, sporulation, and isolation of resistant strains of *Monilinia fructicola*. *Plant Pathology*, **66**, 484-486.

Rosenberger, D.A. (1990). Blue mould. In: *Compendium of apple diseases*, A.L. Jones and H.S. Aldwinckle (Eds), American Phytopathological Society, pp. 54-55.

Rosenberger, D.A.; Wicklow, D.T.; Korjagen, V.A.; Rondinaro, S.M. (1991). Pathogenicity and benzimidazole resistance in *Penicillium* species recovered from flotation tanks in apple packing houses. *Plant Disease*, **75**, 712-715.

Sharples, R.O. (1980). The influence of orchard nutrition on the storage quality of apples and pears grown in the UK. In: *Mineral Nutrition of Fruit Trees*, D. Atkinson; J.E. Jackson; R.O. Sharples and W.M. Waller (Eds), Butterworth, 1980, 17-28.

Upstone, M.G. (1978). *Phythophthora syringae* fruit rot of apples. *Plant Pathology*, **27**, 24-30.

Van der Scheer, H.A.Th.; Remijnse, W. (1988). Aanpassing betrjding *Gloeosporium* schimmels genast. *De Fruiteelt*, **29**, 16-17.

Pepper, [illegible] Kaufmann, W. Nichols, M.R. (1983). [illegible] Plant Pathology, [illegible] 228-234.

Preece, T.F. (19[illegible]). Losses of Cox's Orange Pippin apples during refrigerated storage in England, [illegible]. Plant Pathology, 16, 176-180.

[illegible] (198[illegible]). Effect of dichloran, iprodione, procymidone and vinclozolin on the mycelial growth, sporulation and isolation of resistant strains of Botrytis cinerea. Plant Pathology, [illegible].

Rosenberger, D.A. (1990). Blue mold. In Compendium of apple and pear diseases (A.L. Jones and H.S. Aldwinckle, Eds), American Phytopathological Society, pp. 54-55.

Rosenberger, D.A., Wicklow, D.T., Korjagin, V.A., Rondinaro, S.M. (1991). Pathogenicity and benzimidazole resistance in Penicillium species recovered from flotation tanks in apple packing houses. Plant Disease, [illegible]

[illegible] (1980). The influence of orchard nutrition on the storage quality of apples and pears grown in the [illegible]. In Mineral Nutrition of Fruit Trees [illegible]

[illegible]

[illegible]

SELECTION IN POPULATIONS OF THE EYESPOT FUNGUS IN CONTINUOUS WHEAT BY REPEATED APPLICATIONS OF CARBENDAZIM AND PROCHLORAZ

G.L. BATEMAN

Institute of Arable Crops Research, Rothamsted Experimental Station, Harpenden, Herts., AL5 2JQ

ABSTRACT

The effects of repeated applications of carbendazim, prochloraz or a mixture of the two fungicides, on eyespot, grain yield and population structure of the eyespot fungus (Pseudocercosporella herpotrichoides) were compared in nine years of consecutive winter wheat crops. Failure of control by carbendazim occurred within two years because of its selection for resistance in the fungus population. Prochloraz was moderately effective in most years, but selected for the R-type of the eyespot fungus and, slowly, for carbendazim sensitivity in the R-type; discriminating concentration tests showed carbendazim-resistant isolates to be slightly more sensitive than carbendazim-sensitive isolates to prochloraz. Prochloraz applied with carbendazim was sometimes more effective than prochloraz applied alone. Determination of EC50s after eight years showed R-type isolates from plots treated with prochloraz, either alone or with carbendazim, to be slightly less sensitive to prochloraz than were isolates from untreated plots, but no resistance to prochloraz was found.

INTRODUCTION

Changes in population structure of the eyespot fungus (Pseudocercosporella herpotrichoides) and in the performance against eyespot of the fungicides carbendazim and prochloraz, applied each year, were monitored from 1984 in a field experiment on continuous winter wheat. The detailed results of population and disease monitoring up to 1989 have been reported (Bateman et al., 1990; Bateman & Fitt, 1991).

This paper summarises previous findings and reports the most recent results on the long-term effects of the fungicides on eyespot and pathogen population structure. Data on fungicide sensitivity in fungal populations are used to attempt to explain variability in the performance of the fungicides when used separately and as mixtures.

METHODS

A field experiment was established in a third successive crop of winter wheat in 1984. Details of the design and treatments are reported elsewhere (Bateman et al., 1990). Some plots had inoculum applied artificially to create large populations with known initial proportions of carbendazim-sensitivity and resistance and of the R-type and W-type of the fungus. However, results only from the uninoculated plots are described here; the background population in these in the summer of 1984 was 85% W-type and 15% R-type, and 97% carbendazim-sensitive and 3% carbendazim-resistant. The fungicides, carbendazim, prochloraz or prochloraz + carbendazim, were applied in November or December and March or April in each year. Cv. Avalon was grown each year up to 1990, and cv. Mercia subsequently. The site was shallow tine-cultivated after each harvest to prevent the burial of inoculum, except in 1989 and 1990 when it was ploughed to control grass weeds.

Samples were taken in July for eyespot assessments and fungus isolations, using methods described previously (Bateman et al., 1990; Bateman & Fitt, 1991). Grain yields were measured in all years except 1987.

Sensitivity to carbendazim was determined in each year, except 1990 and 1991, by a single discriminating concentration (1 mg l^{-1}) in agar. Sensitivity to prochloraz was determined in 1988 by two discriminating concentrations (2 and 5 mg l^{-1}) and in 1992 by dosage-response tests (0.0008-0.4 mg l^{-1}) in which EC50s were estimated by fitting logistic curves, using log-transformed concentrations, and compared by Wald tests.

RESULTS

Effects of fungicides on eyespot and grain yields

Carbendazim was moderately effective against eyespot for one year only, after which resistance developed causing control failure (Fig. 1). Prochloraz was usually partially effective but failed to decrease eyespot in 1988. Prochloraz and carbendazim together decreased disease more than prochloraz in six years out of nine, the difference being significant in 1988 and 1992.

Yields were greater when carbendazim was applied with prochloraz than when prochloraz was applied alone in seven years out of eight (Fig. 2). Although these differences were not significant, a maximum difference of 1.41 t ha^{-1} occurred in 1985, before the selection of carbendazim-resistance, and the average difference was 0.63 t ha^{-1}.

Effects of fungicides on populations of *P. herpotrichoides*

Carbendazim selected rapidly for resistance (Fig. 3) but did not select for the W-type or R-type (Fig. 4). Prochloraz selected for the R-type and, when applied alone, for sensitivity to carbendazim in this type. After four years, in 1988, discriminating concentration tests showed carbendazim-sensitive R-type isolates to be less sensitive to prochloraz than were those that were carbendazim-resistant, regardless of whether or not they were from plots treated with prochloraz (Table 1); some

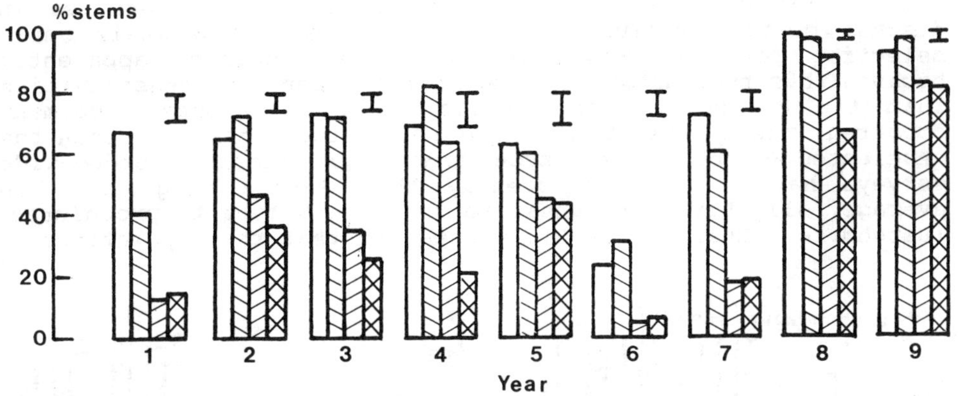

FIG. 1. Effects of fungicides applied each year on percentage of stems with eyespot in July in successive wheat crops, 1985-1993. ☐, untreated; ▧, carbendazim; ▨, prochloraz; ⊠, carbendazim + prochloraz. Vertical bars, SED (24 DF).

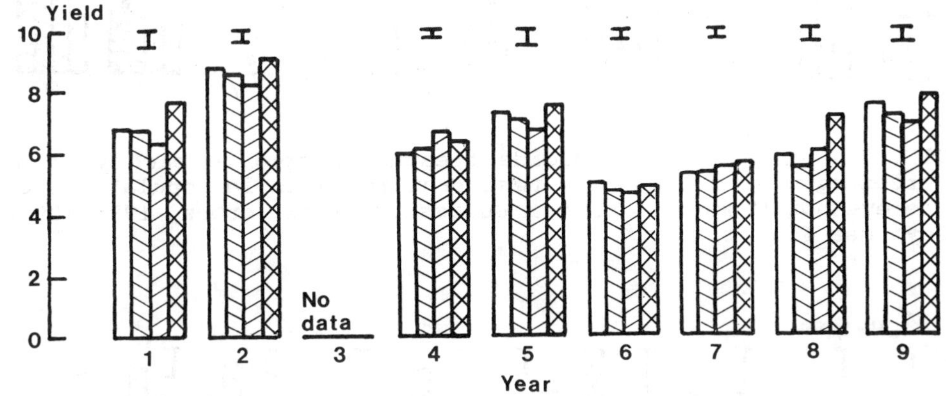

FIG. 2. Effects of fungicides applied each year on grain yields (t ha^{-1}) in successive wheat crops, 1985-1993. Key: see Fig.1.

plots yielded too few isolates to allow a full comparison of levels of sensitivity to prochloraz in relation to previous treatments. After eight years, in 1992, dosage-response tests to determine EC50s for prochloraz showed that R-type isolates from prochloraz-treated plots were less sensitive ($P<0.05$) than those not exposed to prochloraz (Table 2). There was no evidence on this occasion of differences between carbendazim-sensitive and carbendazim-resistant isolates. The range of EC50s for prochloraz was greater in R-type isolates (0.006-0.2 mg l^{-1}) than in W-type isolates (0.010-0.074 mg l^{-1}).

DISCUSSION

The most rapid selection in populations of the eyespot fungus was for carbendazim-resistance. The build-up from 3% to almost 100% resistant isolates in the population resulted in disease control failure within two years. Moderately rapid selection for the R-type by prochloraz occurred apparently because this type naturally has a greater range of sensitivities than the W-type. Where selection by prochloraz treatment occurred, the R-type fungi sampled were less sensitive than the unselected R-type or the W-type fungi to prochloraz. Europe-wide surveys have shown consistently that R-type fungi are, on average, slightly more sensitive than the W-type to prochloraz (Birchmore & Russell, 1990), suggesting that the populations

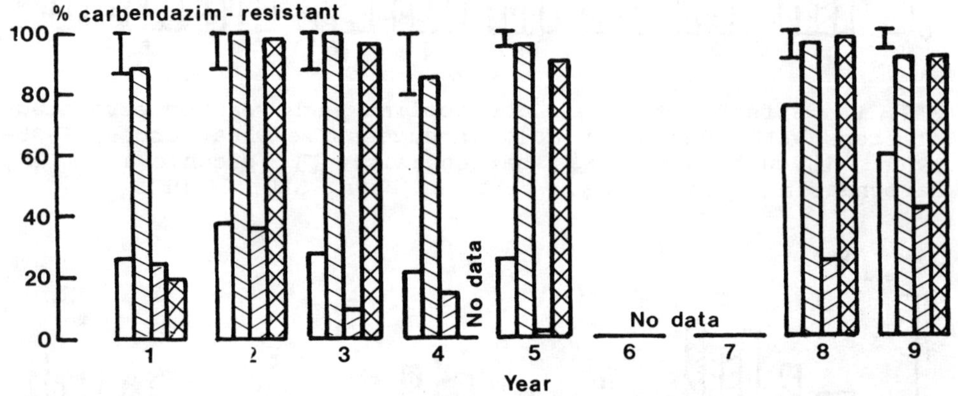

FIG. 3. Effects of fungicide treatments applied each year on percentages of carbendazim-resistant isolates in populations of the eyespot fungus in July in successive wheat crops, 1985-1993. Key: see Fig. 1.

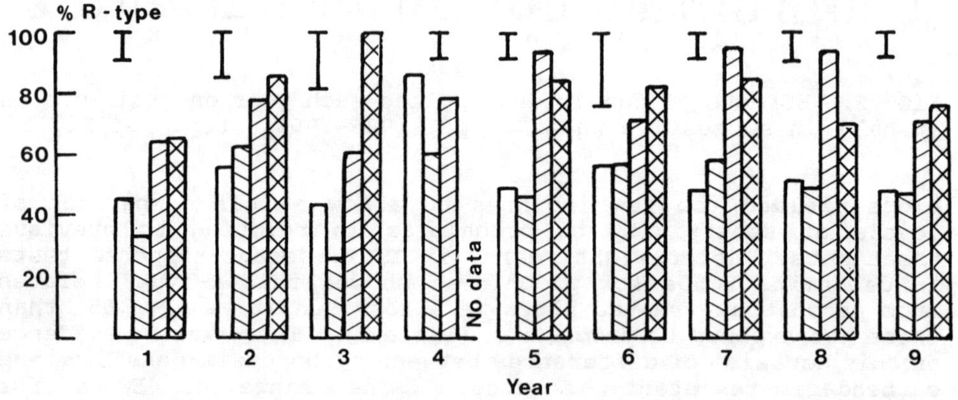

FIG. 4. Effects of fungicide treatments applied each year on the percentages of R-type isolates in populations of the eyespot fungus in July in successive wheat crops, 1985-1993. Key: see Fig. 1.

sampled in those surveys were largely unselected by prior treatments with prochloraz. Our results showed selection for decreased sensitivity to prochloraz and less control by that fungicide. Decreased sensitivity to prochloraz has been reported

TABLE 1. Sensitivity to prochloraz of isolates of Pseudocercosporella herpotrichoides taken in 1988, determined by discriminating concentration tests in agar plates.

Test	Carbendazim-sensitivity	No. of isolates	Percentage showing growth* 2 mg l^{-1}	5 mg l^{-1}
1	Sensitive	22	90.9	13.6
	Resistant	50	32.0(2.0)	2.0(2.0)
2	Sensitive	36	91.7(91.7)	25.0(25.0)
	Resistant	12	16.7	0.0
3	Sensitive	60	61.8	28.1
	Resistant	29	17.2	3.4

* The numbers in brackets are the percentages that were from prochloraz-treated plots.

TABLE 2. Sensitivity to prochloraz (EC50, back-transformed from $\log_{10}$) of isolates of the eyespot fungus taken in July 1992 from wheat in plots treated each year since 1985 with different fungicides.

Treatment to plots	W-type	R-type
None	0.037	0.029
Carbendazim	0.023	0.016
Prochloraz	No data	0.067
Carb. + proc.	0.039	0.088

also in northern France, with evidence of some decreased performance in field trials (Migeon et al., 1992), but selection was not detected elsewhere in small population samples (Birchmore et al., 1992).

The experiment consistently produced evidence of selection against carbendazim-resistance by prochloraz in the R-type fungus. The elimination of this by carbendazim probably accounts for the sometimes better eyespot control by the mixture of fungicides than by prochloraz alone. There was convincing evidence in 1988 that this resulted from greater tolerance to prochloraz in the carbendazim-sensitive fungus; this strain may be more fit and able to cope with the fungicide. This was not confirmed by a different testing procedure in 1992, perhaps because either the carbendazim-resistant strain had been further selected for tolerance to prochloraz in the intervening four years, or the high discriminating concentrations used in 1988 were better able to detect small differences than were the EC50 tests used in 1992. Evidence for the former explanation came in 1993 when carbendazim failed to add significantly to the performance of prochloraz, although the relative performances of the fungicides may differ from year to year as environmental conditions change.

Wheat in continuous cultivation, and the consequent need for annual applications of fungicides to control stem base disease, are becoming increasingly uncommon in Europe as attempts are made to reduce the surplus of grain. Even so, applying mixtures of fungicides may be a way of preventing slow selection of decreased sensitivity to a fungicide as reported here. Mixed formulations of prochloraz and benzimidazoles, however, are not currently an important part of the armoury against eyespot. Alternating fungicides were not tested in this experiment. The interactions between prochloraz and carbendazim, fungicides with different modes of action and selectivity, may be relevant to diseases in other crops grown continuously in which fungicides can be used separately or as mixtures.

ACKNOWLEDGEMENTS

This work was sponsored by the Ministry of Agriculture, Fisheries and Food, and the Pesticides Safety Directorate.

REFERENCES

Bateman, G.L.; Fitt, B.D.L. (1991) Effects of fungicides on eyespot in five successive crops of winter wheat in plots with different initial or fungicide-selected populations of the eyespot fungus, Pseudocercosporella herpotrichoides. Crop Protection, **10**, 479-484.

Bateman, G.L.; Fitt, B.D.L.; Creighton, N.F.; Hollomon, D.W. (1990) Changes in populations of Pseudocercosporella herpotrichoides in successive crops of winter wheat in relation to initial populations and fungicide treatments. Crop Protection, **9**, 135-142.

Birchmore, R.J.; Ashman, P.I.; Stanley, S.; Russell, P.E. (1992) Effects of crop history on sensitivity to prochloraz of Pseudocercosporella herpotrichoides isolates from cereals in western Europe. Brighton Crop Protection Conference - Pest and Diseases 1992, **1**, 183-188.

Birchmore, R.J.; Russell, P.E. (1990) Long-term monitoring of Pseudocercosporella herpotrichoides populations for sensitivity to prochloraz and carbendazim. Brighton Crop Protection Conference - Pests and Diseases 1990, **3**, 1153-1158.

Migeon, J.L.; Mathon, M.P.; Chudzicki, A.M. (1992) Piétin-verse du blé (Pseudocercosporella herpotrichoides): étude de la sensibilité au prochloraze dans le nord de la France. Mededelingen van de Faculteit Landbouwwetenschappen Rijksuniversiteit Gent, **57**, 205-212.

STRATEGIES TO CONTROL DICARBOXIMIDE-RESISTANT BOTRYTIS STRAINS IN GRAPES

G. Lorenz, R. Becker, K. Schelberger

BASF AG, Landwirtschaftliche Versuchsstation,
D-67114 Limburgerhof, Federal Republic of Germany

ABSTRACT

Dicarboximides replaced benzimidazoles for *Botrytis* control in most situations in the late 1970's and early 1980's. Meanwhile their initially high activity has, at least in part, been lost due to the development of resistance, but their use remains crucial in most crops, since good alternatives are lacking.

Therefore, in long-term trials on grapes, attempts were made to develop a strategy to maintain effective *Botrytis* control by dicarboximides despite the widespread occurrence of dicarboximide-resistant stains. This aim could be achieved by limiting the number of applications to two or at the most three per season or growing cycle and thus reducing the selection pressure. Additionally, the use of combinations with a conventional fungicide (e.g. thiram) is highly recommended since they stabilize the performance of the dicarboximides. As is proved by the results presented, this resistance management strategy works well due to the special characteristics of dicarboximide resistant strains, whose percentage in a given population varies depending on the selection pressure exerted throughout the year.

INTRODUCTION

The dicarboximides iprodione, vinclozolin and procymidone were introduced into the market between 1975 and 1977 for the control of grey mould, *Botrytis cinerea*, in grapes (Beetz and Löcher, 1979). This was shortly after the benzimidazoles could not be used any longer for this indication due to resistance problems.

The first dicarboximide-resistant field strains appeared after about three years of intensive use of these products in 1978 (Holz, 1979; Lorenz and Eichhorn 1980; Schuepp *et al.*, 1982; Leroux *et al.*, 1982). Due to the lack of good alternative fungicides and the fact, that inspite of resistance, no total loss of control occurred, dicarboximide use was continued. As a consequence, the proportion of resistant strains in the pathogen population increased considerably.

Initial studies and research projects on dicarboximide resistance management were started in Germany as early as 1979 (Löcher *et al.*, 1985). The aims of these projects were to determine possible strategies to maintain effective *Botrytis* control by reducing the number of treatments in order to reduce the selection pressure and by using combinations with different fungicides, such as thiram.

In all trials, the population dynamics of dicarboximide-resistant *Botrytis* strains were followed closely by constant and regular monitoring work.

The results of these studies (Löcher *et al.*, 1987; Lorenz and Löcher, 1988) clearly showed that:

1. Due to the special characteristics of dicarboximide-resistant strains (i.e. lower fitness/pathogenicity) their proportion within a population varies greatly depending on the selection pressure throughout the year (Pommer and Lorenz 1982 and 1987). Although they do not disappear completely from a population, their amount may drop to a certain minimum level during periods without selection pressure, but increase again rapidly when a new selection pressure is exerted.

2. With dicarboximides a successful resistance management has to be based on a reduction of the selection pressure and on prolonged times without selection pressure. In this respect, it could be shown that two treatments properly timed, i.e. shortly before berry touch (development stage 32/33) and at the beginning of ripening/softening of berries (stage 34/35), are adequate for satisfactory control, and the effect was not improved by three applications (see Löcher et al. 1987).

3. It is, however necessary for vinclozolin to be supported by an other fungicide with activity against *Botytis* to stabilize control, though in this case the use of combinations is no measure to reduce the selection pressure. This is probably due to the fact that those fungicides (i.e. Thiram) presently available as combination partners for dicarboximides are themselves comparably weak botryticides.

Even though after eight years of trial work results seemed very clear and as a consequence now ready mixes of dicarboximides with thiram are registered and used in France, it was decided to continue these trials in two sites in different regions. This was due to the fact, that at this time (1988) still no new botryticides, besides diethofencarb, were being developed which led to the question how long a resistance strategy based on specific properties of resistant strains (i.e. low degree of resistance, reduced fitness) could be used or whether eventually changes of these characterisics might create further problems in *Botrytis* control.

MATERIAL AND METHODS

The two trial sites were situated at Filzen in the Mosel area and at Wiesbaden in the Hessian area. In both trials, the same plots were used for the same treatments each year and each year the influence of two to three applications of vinclozolin alone or the combination of vinclozolin and thiram on *Botrytis* control and the population dynamics of resistant *Botrytis* strains were evaluated.

The Filzen trial was started already in 1980 and thus represents the most complete history of trial results with vinclozolin and combinations. The Wiesbaden trial was started in 1989 to survey a different region.

Materials and methods used are similar to those already described by Löcher *et al.* (1987).

The fungicides used were wettable powders containing either 50 % vinclozolin or 10 % vinclozolin and 64 % thiram. The spraying dates were determined according to the Eichhorn and Lorenz-scale, which describes the development stages (see Löcher *et al.*, 1987). Metiram was used for downy mildew control in both trials.

Each experimental plot contained 25 vines and covered an area of about 50 m^2. The number of replicates was two to four. To evaluate *Botrytis* attack, six samples, each of 100 bunches of grapes per treatment were classified in six categories. From the data obtained, the % disease intensity was calculated (Löcher *et al.*, 1987).

To determine the sensitivity of *Botrytis* strains to vinclozolin, diseased plant material was sampled several times per year from five to eight sites in each plot. Wood samples were taken in February, inflorescences and leaves were collected in June/July, and grapes and leaves in October shortly before harvest. 10 - 24 isolates per plot were prepared and tested. The procedures and methods used for the preparation of samples, and the isolation and testing of the *Botrytis* strains are the same as those described by Löcher *et al.* (1987).

To check the degree of resistance, each year altogether 50 *Botrytis* isolates from both trials (October sampling) were tested in an agar plate test using a concentration range of vinclozolin. ED_{50}-values were calculated by interpolation analysis from dose-response curves.

RESULTS

Table 1 contains all the data available from the Filzen trial concerning *Botrytis* control and population dynamics of resistant strains for both treatments in relation to the untreated plot for the years 1980 up to 1993, as well as the respective number of applications per year and the time between last application and evaluation. The data from 1980 up to 1987 have already been extensively discussed and published (Lorenz and Löcher 1988) and will, therefore, not be further considered here. An essential fact is that the same pattern repeats itself in the following years from 1988 to 1993.

A comparison of the control values obtained with vinclozolin alone with those of the combination over the years shows, that the performance of the combination in general is slightly to significantly better and results are more stable than with vinclozolin alone. This seems to be independent of the number of treatments and the disease intensity. On the other hand, the relevance of control data becomes slightly doubtful for those years where the time between last application and evaluation stretched beyond six weeks due to the late onset of *Botrytis* attack especially during the last seven years.

As far as the population dynamics are concerned, the data in Table 1 show that the proportion of resistant strains still varies greatly according to the time of year. From 1984 onwards, after only two applications in 1983, a general decrease of resistant strains in all plots could be observed, especially at the sampling dates of February and July. During the following years, in which only two to three applications late in the season were made, the resistant population established itself at a fairly low level in the untreated plots. Due to the prolonged period without selection pressure, seasonal variations became more pronounced, with the lowest levels of resistant strains normally observed during July.

With respect to the selection pressure exerted by the different treatments, data are consistent throughout the years. Independent of the number of treatments and the initial percentage of resistant strains, vinclozolin alone, as well as the combination, cause in general the same increase in the resistant population, even though there sometimes seems to be a slight advantage in this respect for the combination. Three and even two applications are sufficient for the resistant strains to obtain maximum levels (80 - 100 %) again.

The respective data for the Wiesbaden trial are presented in Table 2 and essentially show the same results and facts, though the amount of resistant strains in this area is higher and the yearly variation less pronounced than in the Filzen trial.

The examination of 50 isolates per year (Ocober sampling) from both trials and the calculation of their ED_{50}-values so far has not revealed the presence of strains with an increased degree of resistance. ED_{50}-values of resistant strains still lie between 3 and at maximum 10 ppm a.i. as was the case during the first years of these investigations.

CONCLUSIONS

The special properties of dicarboximide-resistant strains (low degree of resistance, low fitness and low competitive ability) obviously explain the fact, that inspite of the occurrence and rapid spread of resistant strains, no total loss of control occurred. The latter fact and the lack of good alternative botryticides encouraged the development of anti-resistance strategies in the early 1980's, which ensured the further use of these products. The data presented here from 13 years of trial work in the Mosel region (resistance started there), backed up by further results from five years trial work in an other region, clearly show that neither the properties of resistant strains nor the situation in the field have changed so far. That means, that the Fungicide Resistance Action Committee (FRAC) recommendations as regularly published in the Groupement International des Associations Nationales de Fabricants de Produits Agrochimiques (GIFAP) resistance newsletter are still valid.

TABLE 1. The effect of vinclozolin alone and in combinations on *Botrytis* control and on population dynamics of dicarboximide-resistant strains of *Botrytis cinerea* in Müller-Thurgau grapes in the Filzen/Mosel region during the years 1980 - 1993

Year Nr. of appl.	Treatment	% resistant strains			% Disease intensity	weeks from last application to evaluation
		Febr.	July	Oct.	Oct.	
1980 5	untreated vinclozolin vinclozolin + chlorothalonil	0 30 30	43 88 88	42 92 92	27 16 11	7
1981 5	untreated vinclozolin vinclozolin + chlorothalonil	27 52 35	50 71 66	70 100 100	30 29 12	5
1982 4	untreated vinclozolin vinclozolin + chlorothalonil	66 - 100	10 - 85	0 - 50	13 - 4	5
1983 2	untreated vinclozolin vinclozolin + chlorothalonil	40 60 50	50 * *	20 95 90	7 2 3	5
1984 3	untreated vinclozolin vinclozolin + thiram	0 20 0	20 0 50	20 80 90	39 20 13	7
1985 3	untreated vinclozolin vinclozolin + thiram	11 20 20	20 14 20	11 20 30	8 3 3	5
1986 3	untreated vinclozolin vinclozolin + thiram	20 10 #	* * #	43 93 #	62 58 #	6

* = no sampling
\# = no treatment

TABLE 1. (continued)

Year Nr. of appl.	Treatment	% resistant strains Febr.	July	Oct.	% Disease intensity Oct.	weeks from last application to evaluation
1987	untreated	35	27	20	28	6
2	vinclozolin	50	13	80	17	
3	vinclozolin + thiram	33	47	93	15	
1988	untreated	47	30	13	31	7
	vinclozolin	53	40	87	30	
3	vinclozolin + thiram	60	10	67	12	
1989	untreated	53	25	60	17	8
2	vinclozolin	53	46	100	12	
3	vinclozolin + thiram	47	18	33	11	
1990	untreated	73	57	33	21	6
2	vinclozolin	80	67	100	18	
3	vinclozolin + thiram	73	36	87	18	
1991	untreated	47	10	33	7	7
2	vinclozolin	60	30	80	5	
3	vinclozolin + thiram	40	0	53	7	
1992	untreated	40	33	40	18	6
	vinclozolin	87	13	73	19	
2	vinclozolin + thiram	33	33	67	12	
1993	untreated	13	14	40	17	9
	vinclozolin	53	13	80	42	
2	vinclozolin + thiram	33	13	40	18	

* = no sampling
# = no treatment

TABLE 2. The effect of vinclozolin alone and in combinations on *Botrytis* control and on population dynamics of dicarboximide-resistant strains of *Botrytis cinerea* Wiesbaden 1989 - 1993

Year	Treatment	% resistant strains			% Disease intensity	weeks from last application to evaluation
		Febr.	July	Oct.	Oct.	
1989	untreated	87	62	73	19	5
	vinclozolin	100	57	80	16	
	vinclozolin + thiram	83	29	100	11	
1990	untreated	60	43	93	41	7
	vinclozolin	87	50	100	50	
	vinclozolin + thiram	80	60	86	33	
1991	untreated	77	37	73	46	5
	vinclozolin	100	29	79	19	
	vinclozolin + thiram	75	0	83	20	
1992	untreated	67	45	79	21	8
	vinclozolin	83	53	90	23	
	vinclozolin + thiram	67	54	90	20	
1993	untreated	83	46	75	17	5
	vinclozolin	88	37	79	10	
	vinclozolin + thiram	88	45	83	10	

* = no sampling
# = no treatment

REFERENCES

Beetz, K.-J; Löcher, F. (1979) Botrytisbekämpfung im Weinbau - Versuchserbebnisse aus den Jahren 1973 - 1978. *Weinberg und Keller,* **25**, 236 - 249.

Holz, B. (1979) Über eine Resistenzerscheinung von *Botrytis cinerea* an Reben gegen die neuen Kontaktfungizide im Gebiet der Mittelmosel. *Weinberg und Keller,* **26**, 18 - 25.

Leroux, P.; Lafon, R; Gredt, M. (1982) La résistance du *Botrytis cinerea* résistentes aux benzimidazoles et aux imides cycliques situation dans les vignobles Alsaciens, Bordelais et Champenais. *OEPP/EPPO Bulletin,* **12**, 137 - 143.

Löcher, F.J.; Brandes, W.; Lorenz, G.; Huber, W.; Schiller, R.; Schreiber, B. (1985) Entwicklung einer Strategie zur Erhaltung der Wirksamkeit von Dicarboximiden bei Auftreten von resistenten Botrytis-Stämmen an Reben. *Gesunde Pflanzen,* **37**, 502 - 507.

Löcher, F.J.; Lorenz, G.; Beetz, K.J. (1987) Resistance management strategies for dicarboximide fungicides in grapes: results of six years' trial work. *Crop Protection,* **6**, 139 - 147.

Lorenz, D.H.; Eichhorn, K.W. (1980) Vorkommen und Verbreitung der Resistenz von *Botrytis cinerea* gegen Dicarboximid-Fungizide im Anbaugebiet der Rheinpfalz. Wein-Wissenschaft, **35**, 199 - 210.

Lorenz, G.; Löcher, F. (1988) Strategies to control dicarboximide resistant strains of *Botrytis cinerea*. Proc. Brit. Crop Protect. Conf., Pests and Diseases Vol. 3, 1107 - 1115.

Pommer, E.H.; Lorenz, G. (1982) Resistance of *Botrytis cinerea* to dicarboximide fungicides - a literature review. *Crop Protection,* **1**, 221 - 230.

Pommer, E.H.; Lorenz, G. (1987) Dicarboximide fungicides. In: *Modern Selective Fungicides*, Lyr, H. (Ed.) Longman Group UK Ltd., London, and VEB Gustav Fischer Verlag, Jena, pp, 91 - 106.

Schüepp; H.; Küng, M.; Siegfried, W. (1982) Dévelopment des souches de *Botrytis cinerea* résistentes aux dicarboximides dans les vignes de la Suisse alémanique. *OEPP/EPPO Bulletin,* **12**, 157 - 161.

THE EPIDEMIOLOGICAL AND GENETICAL BASIS OF BENZIMIDAZOLE AND DICARBOXIMIDE RESISTANCE IN *MONILINIA FRUCTICOLA* ON STONE FRUIT

R.E. GAUNT, P.A.G. ELMER and N. SANOAMUANG.

Department of Plant Science, P.O. Box 84, Lincoln University, Canterbury, New Zealand.

ABSTRACT

Examples are provided of the persistence of benzimidazole-resistant strains of *Monilinia fructicola* in commercial orchards and the lack of persistence of dicarboximide-resistant strains. The difference is explained by the apparent lack of fitness of the dicarboximide-resistant strains in host tissues, and their inability to survive between seasons. The implications for management of these resistant types is discussed.

INTRODUCTION

Brown rot is potentially the most damaging disease of stone fruit globally. In New Zealand the commonest causal agent, *Monilinia fructicola* (Wint) Honey, has been exposed to intensive fungicide use in many production systems. Strains of the pathogen resistant to the benzimidazole (Anon, 1979) and dicarboximide (Elmer & Gaunt, 1986) fungicides have been reported. The development of resistant sub-populations is affected by both epidemiological and genetical factors, which influence the selection and persistence of these strains in the production environment. Several aspects of these factors have been studied and these are discussed in this paper in relation to other pathosystems and pesticides.

PERSISTENCE

Disease control failures associated with benzimidazole resistance were first reported in New Zealand in 1979. Thereafter, use of these products was discontinued on most properties. Recently, the frequency of benzimidazole-resistant strains was determined in commercial properties where the products had not been applied for at least the last nine years. Resistant strains were still present in the *M. fructicola* populations, sometimes at high frequencies. This confirmed similar reports from Australia (Penrose *et al,* 1990) and the USA (Zehr *et al,* 1991).

Dicarboximide resistance was first reported in 1986, though the frequency and level of resistance was lower than for the benzimidazole fungicides. The persistence of resistant strains was studied in commercial orchards with different fungicide management strategies (Elmer & Gaunt, 1993). From these studies it was concluded that dicarboximide-resistant strains were maintained in the population only if the products continued to be used frequently, and especially when they were used during both the flowering and the pre-harvest periods. When dicarboximide selection pressure was withdrawn, the frequency of resistant strains in the pathogen population decreased.

EPIDEMIOLOGICAL FACTORS

Pathogenicity, virulence and fitness

For resistant strains to remain in the population, it is assumed that they must show equal levels of pathogenicity, virulence and fitness on the variety of substrates used during the life cycle of the pathogen. If one or more of these factors is reduced relative to the sensitive strains, the resistant strains will be selected against and will decline in the population except when the selection pressure of the fungicide is present. It has been demonstrated that some strains with high levels of resistance to the dicarboximide fungicides, and some with low levels of resistance to the benzimidazole fungicides, were pathogenic but were less virulent and were less fit compared to sensitive strains or those with high levels of resistance to benzimidazole fungicides (Elmer & Gaunt, 1994). For example, the rate of spore production on flower tissue was less with some dicarboximide-resistant strains than sensitive strains, though the durations of the latent period and incubation period were similar. On the other hand, the rate of spore production in some strains was similar to the sensitive strains, thus demonstrating the potential for selection within the resistant sub-population.

Competition

The population dynamics of sub-populations of sensitive and resistant strains may be influenced by the competitive use of substrates. This may be especially true of a pathogen such as *M. fructicola* which has extended periods of saprophytic behaviour interspersed with brief biotrophic and necrotrophic behaviour. Competition between strains on flower and fruit tissues was investigated, and it was found that the characteristics of individual strains was not necessarily a good predictor of the outcome of competition for substrates. Some strains, with equal apparent fitness individually, did not compete well with the sensitive strains in mixed culture in host tissues.

Survival

M. fructicola has clearly defined periods of survival associated with the production of twig cankers, latent infections in immature fruit and mummies from mature fruit. The relative survival of sensitive and resistant strains in host tissues has been studied (Sanoamuang & Gaunt, 1991). Some dicarboximide- and some low-level benzimidazole-resistant strains survived less well in twig cankers and in mummified fruit compared to the sensitive strains. In some dicarboximide resistant strains, this apparent reduced survival was the only factor affected of those measured, thus possibly explaining the lack of persistence observed in the field.

GENETICAL FACTORS

The production of sexual spores on apothecia by *M. fructicola* is observed in the field, especially in moist climates (Batra & Harada, 1986). The production of sexual progeny in the laboratory has met with limited success, but we have developed a reliable method based on the work of Willetts & Harada (1984). The production of sexual progeny provides an opportunity for genetic analysis of the inheritance of resistance to fungicides.

Level of resistance

The benzimidazole-resistant strains were classified into low and high level types. The high level types were distinct, with discretely separated EC_{50} values, but low level types were defined somewhat arbitrarily. The dicarboximide-resistant strains isolated from the field mostly had discretely separated EC_{50} values, unlike the situation reported for *Botrytis cinerea* (Beever *et al*, 1989). Most strains appeared to have some degree of stability, though reversion to greater dicarboximide sensitivity occurred when some dicarboximide-resistant strains were cultured in fruit tissue for nine generations.

Segregation ratios

Sexual progeny were derived from benzimidazole-resistant and sensitive strains (Sanoamuang *et al*, 1991). Progeny from sensitive strains were always sensitive, whereas segregation occurred from resistant strains. Segregation ratios from high level resistant types were not significantly different from 1:1 ratios when both apothecial populations and ascospore sets were examined (Table 1). Low level resistant types segregated at ratios significantly different from 1:1 and 9:7 ratios. Analysis of patterns of distribution of ascospores within the asci was consistent with the hypothesis that high level resistance is controlled by a single gene and that conversion towards resistance occurred occasionally.

CONCLUSIONS

Persistence and survival

The characteristics of benzimidazole-resistant strains was consistent with the observed persistence of these strains in the field. In contrast, it could be expected that the dicarboximide-resistant strains would not persist. Spatio-temporal analyses suggested that dispersal occurred during the season over relatively short distances, and that within-row spread was greater than between-row spread. There was no correlation between seasons at specific tree sites for the presence of dicarboximide-resistant strains. This and other evidence (Ellis *et al*, 1988) suggested that dispersal may occur by splash and insect transfer as well as by wind movement, and that mutations for resistance occurred and were selected each season. This may be a more significant factor in resistance management than the survival of resistant strains between seasons (Milgroom *et al,* 1989).

Resistance management

The findings from our research have been incorporated into recommended resistance management strategies, promoted by the New Zealand Committee for Pesticide Resistance (Elliot *et al*, 1988, Prince *et al*, 1989). These are being updated currently and will be modified for the next growing season based on recent experiences in New Zealand and elsewhere.

ACKNOWLEDGMENTS

We thank Rhone and Poulenc New Zealand Ltd., Rhone and Poulenc UK Ltd., New Zealand Summerfruit Council, New Zealand Fruitgrowers Federation Ltd, and the Hawkes Bay Fruitgrowers Association for providing a postgraduate scholarship (P.A.G.E.) and research funds. The New Zealand Ministry for External Relations and Trade also provided a postgraduate scholarship (N.S.). We thank Ms K. Boyd-Wilson and Mrs J. Pay for technical support.

REFERENCES

Anon., (1979) Disease and pest resistance to pesticides. The New Zealand Fruitgrowers Federation Ltd. *Growers Guide Bulletin,* **12**.

Batra, L.R.; Harada, Y. (1986) A field record of apothecia of *Monilinia fructigena* in Japan and its significance. *Mycologia,* **78(6)**, 913-917.

Beever, R.E.; Laracy, E.P.; Pak, H.A. (1989) Strains of *B. cinerea* resistant to dicarboximide and benzimidazole fungicides in New Zealand vineyards. *Plant Pathology,* **38**, 427-43.

Elliot, G.S.; Moore, M.S.; Wearing, C.H. (1988) The New Zealand Committee on pesticide resistance 1988 summary. *Proceedings 41st New Zealand Weed and Pest Control Conference*, **41**, 288-292.

Ellis E.C.; Penman D.R.; Gaunt R.E., (1988) Thrips as potential vectors of brown rot of stonefruit. *Proceedings of the 41st New Zealand Weed and Pest Control Conference*, **41**, 286-87.

Elmer, P.A.G.; Gaunt, R.E. (1986) A survey of fungicide insensitivity in *Monilinia fructicola. Proceedings 39th New Zealand Weed and Pest Control Conference*, **39,** 166-169.

Elmer P.A.G.; Gaunt RE; (1993) Effect of frequency of dicarboximide applications on resistant populations of *Monilinia fructicola* and brown rot in New Zealand orchards.*Crop Protection*, **12**, 83-8.

Elmer P.A.G.; Gaunt R.E. (1994) The biological characteristics of dicarboximide resistant strains of *Monilinia fructicola. Plant Pathology* (In Press).

Milgroom, M.G.; Levin, S.A.; Fry, W.E. (1989) Population genetics theory and fungicide resistance. In: *Plant disease epidemiology: genetics, resistance, and management.* (Ed. by Leonard, K.J. and Fry, W.E.), pp. 340-367. McGraw-Hill Publishing Company.

Penrose, L.J. (1990) Prolonged field persistence of resistance to benomyl in *Monilinia fructicola. Crop Protection*, **9**,190-191.

Prince, R.W.; Moore; M.S.; Wearing, C.H. (1989) The New Zealand Committee on Pesticide Resistance 1989 summary. *42nd Proceedings New Zealand Weed and Pest Control Conference*, **42**, 278-292.

Sanoamuang N.; Gaunt R.E. (1991) Survival of *Monilinia fructicola* resistant to MBC and dicarboximide fungicides on twig cankers and mummified fruits. *Proceeding New Zealand Weed and Pest Control Conference*, **44,** 225-228.

Sanoamuang N.; Gaunt R.E. (1995) Apothecial production and genetic basis of MBC resistance in *Monilinia fructicola* (Wint) Honey. *Mycological Research* (in press).

Sanoamuang N.; Gaunt R.E.; Fautrier A. (1991) Inheritance of MBC resistance in *Monilinia fructicola* (Wint.) Honey. *Proceeding New Zealand weed and pest Conference*, **44,** 229-231.

Willetts H.J.; Harada Y. (1984) A review of apothecial production by *Monilinia* fungi in Japan. *Mycologia,* **76**: 314-325.

Zehr E.I.; Toler J.E.; Luszcz L.A. (1991) Spread and persistence of benomyl-resistant *Monilinia fructicola* in South Carolina peach orchards. *Plant Disease,* **75**, 590-593.

Hindle, R.W., Moore, M.S., Jennings, [illegible] (1989). The New Zealand Committee on Pesticide Resistance 1989 summary. Proceedings of the New Zealand Weed and Pest Control Conference, 42, 228–232.

Sanoamuang N., Gaunt R.E. (19[illegible]) [illegible] MBC and dicarboximide fungicides [illegible] carbendazim and dicarboximide [illegible]. Proceedings of New Zealand Weed and Pest Control Conference [illegible]

Sanoamuang N., Gaunt R.E. ([illegible]) [illegible] and fitness [illegible] MBC resistance in [illegible] fructicola [illegible]. Mycological Research [illegible]

Sanoamuang N., Gaunt R.E., [illegible] resistance in Monilinia fructicola (Wint.) Honey. Proceedings of New Zealand Weed and Pest Control Conference, 44, 275–[illegible]

Wiggins [illegible], Harada Y. ([illegible]) A review of [illegible] in Japan. Mycologia [illegible] 314–325.

Zehr E.I., Toler J.E., Luszcz L.A. (1991) Spread and persistence of benomyl-resistant Monilinia fructicola in South Carolina peach orchards. Plant Disease, 75, 590–593.

MONITORING STUDY OF THE RESISTANCE OF *BOTRYTIS CINEREA* TO BENZIMIDAZOLE AND DICARBOXIMIDE FUNGICIDES IN GRAPES IN HUNGARY

T. DULA, T. KAPTÁS

Plant Health and Soil Conservation Station of County Heves, Eger Pf.218. H-3301 Hungary

I. APONYI

Plant Health and Soil Conservation Station of the Capital, Budapest Pf.127. H-1502 Hungary

GY. JOSEPOVITS

Research Institute for Plant Protection, Hungarian Academy of Sciences, Budapest, Pf.102. H-1525 Hungary

ABSTRACT

According to monitoring studies, which began in 1981, resistance to benzimidazole fungicides has developed in strains of *B. cinerea* in grapes. Negative cross-resistance to phenylcarbamates can be detected. Since 1988, a decrease in sensitivity to dicarboximides has also been observed. A simulation model was developed to predict the development of resistance. The model has been compared to the results of monitoring studies and is considered to be suitable for choosing spray programmes to prevent the build-up of resistance to dicarboximides.

INTRODUCTION

Grapevines are grown in Hungary on an area of 110,000 ha. There are three important regions (Northern Hungary, Transdanubia and the Plain) and these include several wine districts. One dangerous pathogen occurring in the vineyards is *B. cinerea*, which causes severe damage in some years depending on weather conditions. Infection may occur before and during flowering, at closing of the bunch growth stage and at the start of ripening. Treatments are made with multi-site inhibitors (e.g. chlorothalonil, dichlofluanid, folpet and thiram) and dicarboximides (e.g. iprodione, procymidone and vinclozolin). Dicarboximide fungicides are mainly used during the second half of the growing season; earlier application is only justified by extreme rainy weather. In the past, benzimidazole compounds were also used (e.g. benomyl and thiophanate-methyl). They were introduced in 1972 and were widely used from the second half of the decade. A decrease in the field efficacy of benzimidazoles was first observed in 1981. In laboratory studies, it was confirmed that this loss in efficacy was due to a decrease of pathogen sensitivity to benzimidazole fungicide (Kaptás & Dula 1984). This prompted the start of the monitoring studies for resistance, which were mainly performed in the Northern grapevine growing districts of Hungary but which included other parts of the country.

The main objectives of study were: to detect occurrence and frequency of resistance to various fungicides; to determine properties of resistance, cross-resistance, duration of resistance; and to study any correlation between the frequency of treatments and the change of sensitivity level of pathogen to fungicides.

MATERIALS AND METHODS

Ten infected bunches were collected randomly at vintage from each field. *B. cinerea* was isolated onto 2% malt-czapek agar and was tested for carbendazim (as 'Kolfugo 25 FW') and iprodione (as 'Rovral') sensitivity at rates of 0, 1, 5, 10, 50 and 100 mg AI/l. Results are expressed as the concentration of carbendazim and iprodione required to reduce colony diameter by 50% (EC_{50}) when compared to colonies growing on agar without fungicide. Recommendations of literature are studied when using the method (Leroux *et al*, 1982; Leroux & Besselat, 1984). The factor of resistance (FR) was determined based on EC_{50} values compared to a sensitive (reference) isolate.

RESULTS

Resistance of *B. cinerea* to benzimidazole fungicides on grapevine is common in Hungary. Based on various studies, resistance to benzimidazoles is persistent and because of this, they are no longer used. No positive cross-resistance has been observed to other fungicide groups. However, a negative cross-resistance to phenylcarbamates (e.g. diethofencarb) was detected (Kaptás & Dula, 1988; Kaptás *et al*, 1990). Phenylcarbamates are not yet registered against *B. cinerea* in Hungary.

Dicarboximides were first registered in Hungary in 1978 and their became widespread in the 1980's. No general resistance to dicarboximides has developed up to now in Hungary. However, a decrease in sensitivity has been observed since 1988 (Table 1). The frequency of the resistant strains changes with years. The decrease of benzimidazole + dicarboximide sensitivity occurring at the same time is typical (Kaptás *et al*, 1990).

Table 1. Distribution of grapevine *B. cinerea* strains according to Factor of Resistance (FR) in Hungary, between 1987 and 1992.

Years	Distribution of *B. cinerea* strains in % classes of FR[x]					
	1	1-5	5-10	10-15	15-20	20
1987	78	22	0	0	0	0
1988	65	21	0	14	0	0
1989	0	86	7	7	0	0
1990	0	93	0	7	0	0
1991	24	57	5	14	0	0
1992	16	54	6	3	15	6

[x]FR = $\frac{\text{EC50 of field strain}}{\text{EC50 of sensitive (reference) strain}}$

Comparison of the above results with the simulation model

Results of the monitoring studies have been compared with values calculated by a simulation model, which was modified from an earlier general resistant model (Josepovits & Dobrovolsky, 1985; Josepovits, 1989).

Calculation with the simulation programme shows the decrease of resistance level in crops not treated with dicarboximides. More than two treatments a year induce an increase in the frequency of resistance. This increase is the highest if the dicarboximide treatment is made at the risky period of flowering and early ripening. The monitoring data confirm the reliability of the calculation, i.e. the frequency of resistance ranges between 7 and 30% during 5 years, while the average number of the dicarboximide treatment was annually 1 and 2. For the evaluation of the increase in 1992, it should be considered that most resistant isolates are from the same crop varieties of one county. No close correlation between the treatments and the resistance frequency could be made from the monitoring data because of the spread of conidia by wind.

Results obtained with the simulation model show also that the application of the mixture of multi-site fungicides and dicarboximides does not offer a real alternative for a rotation of treatments because of their greatly different action. To control *B. cinerea*, experimental data are available for carbendazim + diethofencarb among fungicides of similar or better action. At the application of this combination, it should be considered that the effect of their components is not added according to the Abbott's formula (Abbott, 1925) and that it offers satisfactory efficiency at a concentration in which each component has adequate activity to the strains which are more sensitive to them. The simulation model suggests that a triple combination of the above products with a dicarboximide could be used positively during the period of the rapid spread of the pathogen when the use of a dicarboximide alone is to be avoided because of the risk of resistance.

With this paper on the resistance of *B. cinerea* to various fungicide, the authors wanted to contribute to the picture known in the Western European countries' grapevine growing districts with their results obtained in Hungary.

REFERENCES

Abbott, W.S. (1925) A method of computing effectiveness of an insecticide. *J. Econ. Entom.* **18**, 265-267.

Josepovits, G. (1989) A model for evaluating factors affecting the development of intensitivity to fungicides. *Crop Protection*, **8**, 106-113.

Josepovits, G.; Dobrovolsky, A. (1985) A novel mathematical approach to the prevention of fungicide resistance. *Pesticide Science*, **16** (1), 17-22.

Kaptás, T.; Dula, B. (1984) Benzimidazol tipusú fungicidekkel szembeni rezisztens *Botrytis cinerea* Pers. törzs kialakulása szölöben. (Resistance to benzimidazole fungicides built-up in strains of *Botrytis cinerea* Pers. in a vineyard). *Növényvédelem*, **20**, 174-182.

Kaptás, T.; Dula, B. (1988) Új tapasztalatok a szölöszürkerothadás (*Botrytis cinerea* Pers.) fungicidekkel szembeni rezisztenciájáról. (New observations on resistance to fungicides in *Botrytis cinerea*). *Növényvédelem*, **24**, 116-119.

Kaptás, T.; Dula, Leiner, I. (1990) A szölöszürkerothadás kórokozója (*Botrytis cinerea* pers.) néhány fungicid készitménnyel szembeni érzékenység csökkenésének ujabb tapasztalatai. (Recent observations on the decrease of sensitivity of *Botrytis cinerea* to some fungicides). *Növényorvoslás a kertészetben*, **11**, 105-114.
Leroux, P.; Besselat, B. (1984) Pourriture grise: La resistance aux fongicides de *Botrytis cinerea*. *Phytoma*, **Juin**, 25-31.
Leroux, P.; Lafon, R.; Gredt, M. (1982) La resistance de *Botrytis cinerea* aux benzimidazoles et aux imides cycliques: situation dans les vignobles alsaciens, bordelais et champenois. *Bull. OEPP*, **12**, 137-143.

BENZIMIDAZOLE RESISTANCE IN *RHYNCHOSPORIUM SECALIS* IN NORTHERN IRELAND AND ITS IMPLICATIONS FOR DISEASE CONTROL

P.J. TAGGART, L.R. COOKE, P.C. MERCER

Department of Applied Plant Science, The Queen's University of Belfast, Newforge Lane, Belfast, Northern Ireland, BT9 5PX.

ABSTRACT

Before 1990, there was little evidence of benzimidazole resistance in the Northern Ireland *Rhynchosporium secalis* population. By 1992, resistance had become common and widespread. In field trials in 1993, carbendazim controlled *R. secalis* well at a site where the majority of *R. secalis* isolates recovered were benzimidazole-sensitive. At another site where most *R. secalis* isolates proved resistant, carbendazim, both alone and with propiconazole, apparently increased the severity of *R. secalis* infection, relative to the untreated control and the propiconazole treatment respectively.

INTRODUCTION

Rhynchosporium secalis, which causes leaf blotch, is the most prevalent and damaging disease of winter barley in N. Ireland (Mercer & Easson, 1987). Fungicides offer the only effective means of control in autumn sown cultivars. Recently a decline in the sensitivity of *R. secalis* to DMI fungicides in the British Isles has increased interest in the benzimidazole fungicide carbendazim for control of leaf blotch (Kendall *et al.*, 1993). Although benzimidazole resistance has not until recently been detected in field populations of *R. secalis* in the British Isles, highly resistant and fit strains of other cereal pathogens, e.g. *Pseudocercosporella herpotrichoides* (King & Griffin, 1985), are widespread. This paper reports the results of surveys of benzimidazole resistance in *R. secalis* in Northern Ireland (1990-1993) and the results of 1993 field trials to investigate the effect of resistance on disease control in the field.

MATERIALS & METHODS

Survey of benzimidazole resistance in *R. secalis*

R. secalis isolates were collected from barley crops (mainly autumn sown) throughout Northern Ireland from 1990 until 1993. The isolates were assayed on yeast malt agar containing different concentrations of technical grade carbendazim, added from stock solutions in dilute hypophosphorous acid before autoclaving. Fungal growth was assessed after ten days at 18°C and sensitivity to carbendazim expressed as the minimum inhibitory concentration (MIC) value (the lowest fungicide concentration which completely prevented fungal growth).

Field trials

Two sites were chosen, both commercial crops of winter barley, cv. Fighter, near Portaferry and Strangford, Co. Down, respectively. A randomised block design, with four blocks and five treatments, was used at both sites. The treatments are summarised in Table 1. Fungicide sprays were applied on two occasions, growth stages 33 and 57 (as defined by Zadoks *et al.*, 1974).

R. secalis isolates were recovered from plots before treatments were applied and tested, *in vitro*, using the method described above, for their sensitivity to carbendazim. Foliar disease assessments were made on several occasions during the growing season, using the key described by James (1971). Grain yields were recorded at the end of the growing season.

TABLE 1. Field trial treatments

Proprietary name	Manufacturer	Active ingredients	Application rate (g/ha)
'Derosal WDG'	AgrEvo	carbendazim	250
'Tilt 250 EC'	Ciba Agriculture	propiconazole	125
'Hispor 45 WP'	Ciba Agriculture	propiconazole + carbendazim	125 + 100
'Legend'	Zeneca	propiconazole + fenpropidin	125 + 562.5

RESULTS

Survey of benzimidazole resistance in *R. secalis*

All isolates tested in 1990 proved benzimidazole-sensitive (MIC values 1 mg/l or less), but since 1991 substantial proportions of benzimidazole-resistant (MIC values >10 mg/l) isolates have been obtained (Figure 1). All isolates with MIC values of >10 mg/l grew well at higher carbendazim concentrations (25 mg/l). Resistant isolates were not restricted to certain regions, but were recovered from all cereal growing areas (Table 2) and in many cases both resistant and sensitive isolates were recovered from the same site (Table 3).

FIGURE 1. Sensitivity of *Rhynchosporium secalis* to carbendazim in N. Ireland, 1990-93

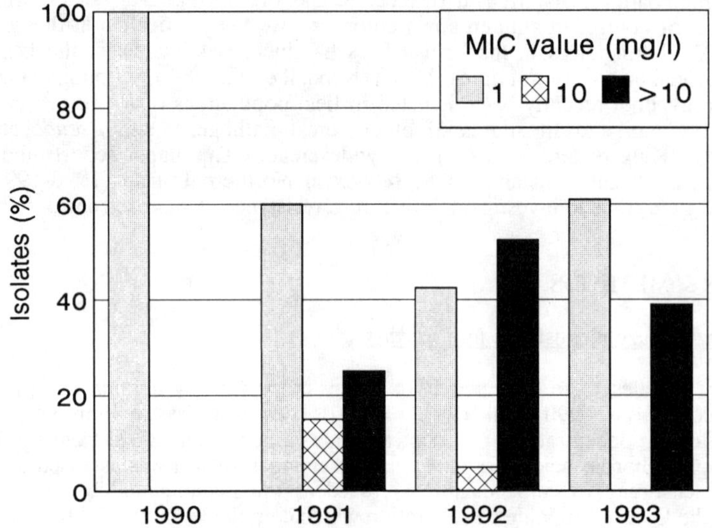

TABLE 2. Geographical distribution of benzimidazole-resistant *Rhynchosporium secalis* isolates collected in 1993

County	Number sensitive	Number resistant	% resistant
Antrim	7	5	42
Armagh	2	9	82
Down	18	4	18
Londonderry	17	10	37
Total	44	28	39

TABLE 3. Distribution of benzimidazole-resistant *Rhynchosporium secalis* isolates within sites in 1992-93

Year	Number of sites with			Total number of sites
	Sensitive only	Resistant only	Resistant+ sensitive	
1992	8	6	8	22
1993	5	2	14	21

Field trials

Only 7.6% of the isolates recovered from the Portaferry site were benzimidazole-resistant (MIC value >10 mg/l), whereas 85.7% of isolates from the Strangford site were resistant. The numbers of isolates recovered and assayed from the two sites were 156 and 119, respectively.

At the "sensitive" Portaferry site all fungicides controlled *R. secalis* (Table 4). There was no indication that propiconazole + carbendazim, or propiconazole + fenpropidin provided better control than propiconazole alone. However, propiconazole controlled *R. secalis* so effectively that any benefit from carbendazim would have been difficult to detect. At the "resistant" Strangford site only the propiconazole and propiconazole + fenpropidin treatments gave adequate disease control. The severity of *R. secalis* infection was greater with carbendazim than with the control and greater with propiconazole + carbendazim than with propiconazole alone. There were no significant differences in yield between treatments at either site.

TABLE 4. Infection of winter barley cv. Fighter by *Rhynchosporium secalis*, 1993

Treatment	*Rhynchosporium secalis* infection (arcsin %) 21 June		Yield (t/ha) at 16% moisture
	Leaf 1	Leaf 2	
Portaferry			
none	24.3	25.3	5.9
carbendazim	7.0	9.6	5.8
propiconazole	5.8	9.6	6.1
propiconazole + carbendazim	8.6	10.1	5.8
propiconazole + fenpropidin	4.0	5.4	6.9
S.E. (12 D.F.)	3.17	2.13	0.32
Significance[a]	**	***	ns
Strangford			
none	31.7	19.6	4.1
carbendazim	29.7	33.7	3.8
propiconazole	8.1	9.2	5.4
propiconazole + carbendazim	15.1	25.4	5.0
propiconazole + fenpropidin	7.2	10.1	5.3
S.E. (12 D.F.)	3.74	2.70	0.61
Significance[a]	***	**	ns

a Significance, ns = P>0.05, ** = P<0.01, *** = P<0.001

DISCUSSION

In Northern Ireland surveys of 1988-89 (Cooke, L.R., unpublished data) as well as in 1990, no isolates of *R. secalis* were confirmed as benzimidazole-resistant. In 1989, one isolate grew on carbendazim at 1 mg/l in an initial test, but died before it could be re-tested. No benzimidazole-resistant isolates of *R. secalis* were obtained in ADAS surveys in England and Wales in 1987 and 1989 (Jones, 1990). However, the 1991-93 surveys reported here clearly demonstrate that benzimidazole resistance is now both common and widespread in the *R. secalis* population of Northern Ireland. During the same period, further surveys carried out by ADAS have confirmed the occurrence of benzimidazole-resistant *R. secalis* in commercial winter barley crops in England and Wales (Locke, 1994).

In the field trials in 1993, carbendazim alone gave good control of *R. secalis* at the "sensitive" site, but no control at the "resistant" site, providing strong evidence that benzimidazole-resistant *R. secalis* is associated with poor disease control by carbendazim. A more surprising finding was that the use of carbendazim, either alone or in combination with propiconazole, apparently increased the severity of *R. secalis* infection at the "resistant" site. Some degree of disease control by carbendazim would have been expected, even at a site where the majority of the *R. secalis* population was benzimidazole-resistant. The reasons for the apparent stimulation by carbendazim of *R. secalis* at the "resistant" site are unclear and will be investigated further.

ACKNOWLEDGEMENTS

We are grateful to the Home-Grown Cereals Authority for funding this work and supporting P.J. Taggart and thank Ian Wilson and Richard Binks who assisted with the practical work in 1990 and 1991, respectively.

REFERENCES

James, W.C. (1971) An illustrated series of assessment keys for plant disease, their preparation and usage. *Canadian Plant Disease Survey,* **51**, 39-65.

Jones, D.R. (1990) Sensitivity of *Rhynchosporium secalis* to DMI fungicides. *Brighton Crop Protection Conference - Pests and Diseases 1990*, **3**, 1135-1140.

Kendall, S.J.; Hollomon, D.W.; Cooke, L.R.; Jones, D.R. (1993). Changes in sensitivity to DMI fungicides in *Rhynchosporium secalis. Crop Protection,* **12**, 357-362.

King, J.E.; Griffin, M.J. (1985) Survey of benomyl resistance in *Pseudocercosporella herpotrichoides* on winter wheat and barley in England and Wales in 1983. *Plant Pathology*, **34**, 272-283.

Mercer, P.; Easson, L. (1987) Fungicides on winter barley. *Agriculture in Northern Ireland,* **1**(6), 8-9.

Phillips, A.N., Locke, T. (1994) Carbendazim resistance to *Rhynchosporium secalis* in England and Wales. *Proceedings BSPP/BCPC Fungicide Resistance Symposium, Reading, 28-30 March 1994*, this volume.

Zadoks, J.C.; Chang, T.T.; Konzak, F.C. (1974) A decimal code for the growth stages of cereals. *Weed Research,* **14**, 415-421.

RESISTANCE TO IPRODIONE IN *ALTERNARIA LINICOLA*

P.C. MERCER, L.R. COOKE

Plant Pathology Research Division, Department of Agriculture for Northern Ireland, Newforge Lane, Belfast BT9 5PX

ABSTRACT

Resistance by *Alternaria linicola* to iprodione was first discovered in linseed in the UK in 1986 and spread rapidly to affect 85% of all seed samples by 1988, leading to the abandonment of the treatment. However, in 1993, resistance could no longer be detected. The maximum incidence of resistance within a sample was estimated at 24% in 1987.

INTRODUCTION

The Department of Agriculture for Northern Ireland (DANI) acts as the Official Seed Tester for seed-borne disease of linseed in the UK. In 1986, when the most commonly used seed-treatment was the fungicide iprodione ('Rovral', Rhône-Poulenc), several treated samples were received with levels of the pathogen *Alternaria linicola*, which exceeded the official 5% limit and consequently failed the test. These samples were further treated with iprodione in case there had been a failure to cover the seed adequately, but on retesting, *A. linicola* was again isolated at levels which resulted in failure. As previous research (Mercer *et al.*, 1985) had shown iprodione to be highly effective in the control of *A. linicola* on linseed, a programme of work was started to investigate possible resistance to the fungicide by *A. linicola*. Results are reported below.

TESTS FOR RESISTANCE

In 1986, seven isolates of *A. linicola*, from seed samples treated with iprodione and where there had been problems with disease control, were compared for sensitivity to iprodione with three isolates from seed samples where there was no obvious control problem. Isolates were plated out on 2% malt agar to which a suitable range of iprodione concentrations in acetone had been added as the agar was cooling (10 ml stock solution/litre agar). At least five concentrations and a control (acetone only added to the agar) were used for each isolate. There were five replicates. ED_{50} values (mg/l) for mycelial growth after nine days' growth in the dark at 20°C were estimated using logarithmic probability paper. Two types of responses were observed:

1. Completely sensitive (the three isolates from samples where there was no disease control problem and two isolates from seed samples where resistance was suspected): ED_{50} values in the range 0.7 - 1.4 mg/l.

2. Resistant (five isolates from the suspect samples): moderate inhibition of mycelial growth (*ca.* 24%) at 0.8 mg/l but inhibition only increased to *ca.* 36% at 500 mg/l. One isolate was tested at 1000 mg/l but inhibition was still only 48% . Repeat tests of resistant isolates, maintained on malt agar without the addition of iprodione, showed no change in resistance over five years.

In 1987, 17 further isolates (from seed samples where resistance was suspected) were tested as in the previous season, except that iprodione concentrations were restricted to 10 and 100 mg/l. One isolate proved to be sensitive and fourteen of the others resistant in a similar way to those in found to be resistant in 1986. However, there was also a third type of response (two isolates) in which growth was increased by iprodione at 10 mg/l (0.4% and 72.4%) and more so at 100 mg/l (28.4% and 84.1%).

In 1993, 25 isolates of *A. linicola*, taken from samples where there was an incidence of the pathogen of over 15%, were tested on agar containing 10 mg/l of iprodione. All isolates were sensitive.

INCIDENCE OF IPRODIONE RESISTANCE

A dramatic increase in the detection of iprodione-resistant *A. linicola* isolated from linseed samples received by DANI occurred between 1986 and 1988 (Figure 1). By 1988, 85% of all seed samples yielded some isolates of *A. linicola* showing resistance to iprodione.

FIGURE 1. Percentage of linseed samples received by DANI from 1986-88 in which iprodione-resistant *A. linicola* was detected.

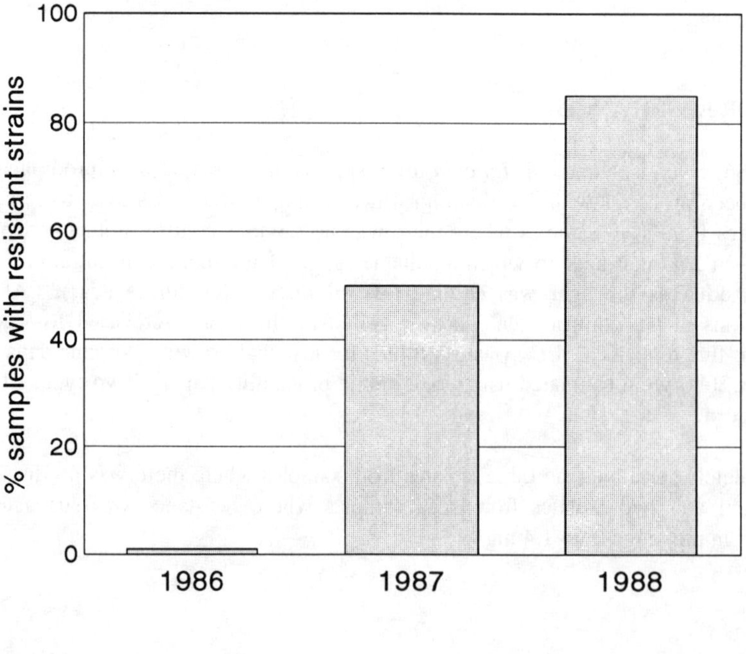

In 1987, estimations were made of iprodione-resistant and -sensitive populations of *A. linicola* in untreated and iprodione-treated seed-samples received by DANI for seed-testing. Samples of untreated and of iprodione-treated seed (37 and 34 respectively) were selected at random. One hundred seeds from each sample were plated out on 2% malt agar and a further 100 on 2% malt agar containing 10 mg/l iprodione, incubated for 7 days at 20°C and 12 h light, 12 h n.u.v. light before being examined for the presence of *Alternaria linicola*. The percentage of seed from which *A. linicola* was isolated when plated out on malt agar was graphed against that obtained when plated out on malt agar containing iprodione.

All samples of seed, apart from one treated sample, yielded isolates of *A. linicola* (Figure 2). Fifty-one percent of samples of untreated seed and 65% of samples of iprodione-treated seed yielded isolates of *A. linicola* resistant to iprodione. Samples of seed treated with iprodione yielded fewer isolates than samples of untreated seed, when grown on untreated malt agar - 12% of seed compared with 46%, indicating that many of the iprodione-sensitive strains had been suppressed. Treatment also reduced the percentage of samples yielding only sensitive isolates - 35% compared with 49%. A possible reason for the existence of **any** sensitive isolates following treatment may lie in poor coverage of the seed in some samples.

FIGURE 2. Percentage isolations of *A. linicola* from seed samples treated or untreated with iprodione and plated out on agar either containing or not containing 10 mg iprodione/l.

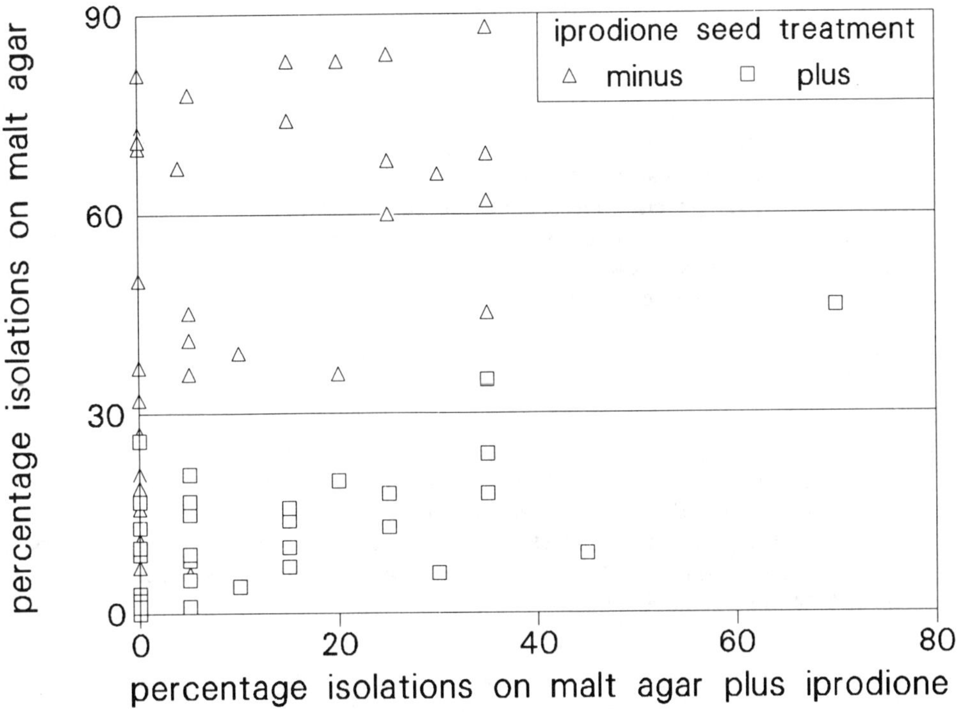

Estimates of the percentage of isolates showing resistance were obtained from counts of isolates from treated or untreated seed plated out on agar containing iprodione. Results from both sets of seed were similar, showing that when resistant strains were detected within a sample they were found in 18.6 - 20% (average 19%) of the seeds. A similar estimate obtained from counts from untreated seed plated out on untreated agar showed that when resistant strains were present within a batch, 59% of seeds yielded isolates of either types, *i.e.* approximately 19% resistant and 40% sensitive. If isolates on individual seeds were always either resistant or sensitive, a maximum ratio of resistant to sensitive isolates of 1:2.1 is obtained. However, if both resistant and sensitive isolates were present in the same seed, the ratio would be reduced as the number of sensitive strains would be underestimated. In any untreated or untreated sample, therefore, the maximum chance of isolating a resistant isolate of *A. linicola* would have been 24% or 31% respectively (51% or 65% x 1/2.1).

INTERACTION OF CULTIVAR AND IPRODIONE-TREATMENT ON INCIDENCE OF *A. LINICOLA*.

In 1989, an survey was made of the incidence of *A. linicola* in a random selection of iprodione-treated and untreated seed of three cultivars, Antares, Atalante and Lidgate. Results showed 78% fewer isolates of *A. linicola* from treated than from untreated samples of Antares but 99% and 96% fewer of Atalante and Lidgate, respectively, indicating a very low level of iprodione-resistance in *A. linicola* from these two cultivars.

DISCUSSION

After iprodione-resistant *A. linicola* was first detected in 1986, its incidence increased rapidly, resulting in the virtual abandonment of iprodione and its replacement by prochloraz (Mercer *et al.*, 1988). Resistance appeared to be higher in cultivar Antares, probably because this was the single most popular cultivar in the early to mid 1980s. Recent testing of isolates detected no iprodione-resistant strains, suggesting that the resistant isolates were less fit than their sensitive counterparts. However, if iprodione treatments were to be reintroduced on a large scale, resistance might well recur.

REFERENCES

Mercer, P.C.; McGimpsey, H.C.; Black, R.; Norrie, S. (1985) The chemical control of *Alternaria linicola* on the seed of linseed. *Tests of Agrochemicals and Cultivars 6. A Supplement to Annals of Applied Biology,* **106**, 56-57.

Mercer, P.C.; McGimpsey, H.C.; Ruddock, A. (1988) The control of seed-borne pathogens of linseed by seed treatments. *Tests of Agrochemicals and Cultivars 9. A Supplement to Annals of Applied Biology,* **112**, 30-31.

CARBENDAZIM RESISTANCE IN *RHYNCHOSPORIUM SECALIS* IN ENGLAND AND WALES

A.N. PHILLIPS

ADAS Wolverhampton, 'Woodthorne', Wergs Road, Wolverhampton, WV6 8TQ

T. LOCKE

ADAS Worcester, Whittington Road, Worcester, WR5 2LQ

ABSTRACT

A survey of winter barley crops in England and Wales in 1993 found that 45.9% contained carbendazim resistant isolates of *Rhynchosporium secalis*. Of the 639 isolates tested 16.6% were resistant to carbendazim at 1.0 µg/ml.

INTRODUCTION

Rhynchosporium secalis is a common disease of barley in England and Wales (Polley *et al*, 1993) and can cause considerable yield losses as a result of leaf infection (James *et al*, 1968). Methyl bendizimidazole carbamate (MBC) fungicides have given useful control of this pathogen since being introduced for use on barley in 1975. Monitoring of the *R. secalis* field population from 1980-1989 failed to detect any strains resistant to MBC, but laboratory mutants possessing resistance can be produced readily (Kendall *et al*, 1994).

In 1991 ADAS confirmed that MBC-resistant isolates of *R. secalis* were present in a crop of winter barley in Dyfed, Wales and in 1992 an *ad hoc* survey was carried out of winter barley crops in England and Wales. This survey of 19 crops found that 47.4% contained MBC-resistant isolates of *R. secalis* and that 14.2% of the isolates examined were resistant at 1 µg/ml. The following year ADAS undertook a stratified survey of winter barley crops to determine the level of resistance in England and Wales.

MATERIALS AND METHODS

Ninety-five winter barley crops were sampled at GS 30-31 and the pathogen was isolated on to Czapek Dox agar amended with bacteriological peptone and streptomycin. A maximum of 10 pure cultures from each crop were then tested for carbendazim sensitivity at 1 µg/ml, alongside standard isolates known to be sensitive and resistant at this level of fungicide incorporation. Cultures were incubated for 14 days in darkness at 17°C and then scored for spore germination and growth. Isolates making no growth on the carbendazim-amended agar were classed as sensitive, whilst those making similar growth on amended and unamended agar were classed as resistant.

TABLE 1. Geographical and cultivar distribution of winter barley crops from which *Rhynchosporium secalis* was isolated

	CULTIVAR						
Location	**Pastoral**	**Pipkin**	**Fighter**	**Puffin**	**Marinka**	**Halcyon**	**Others#**
Cumbria/Northumberland	6	0	2	0	0	1	3
Yorkshire/N.Humberside	1	2	4	2	0	0	1
Norfolk/Lincs./Cambs./S.Humberside	1	1	1	3	0	2	3
Suffolk/Essex	0	7	0	0	0	0	0
Kent/Sussex	2	1	1	3	1	0	1
Shropshire/Staffs./Herefords./Oxon.	0	3	4	0	0	0	1
Cornwall/Devon/Somerset/Avon/ Dorset/Hants./Berks.	2	0	0	0	1	0	2
Wales	2	0	1	3	4	1	1
England and Wales	14	14	13	11	6	4	12

Other cultivars were 2 each crops of Sprite and Kira, one each of Willow, Magi, Bambi, Bronze, Frolic, Maris Otter, Igri and one crop consisting of a mixture of cultivars

TABLE 2. Incidence of carbendazim resistance in *R. secalis* by crop and isolate according to cultivar

Winter Barley cultivar	CROP Number sampled	CROP % with resistant isolates present	ISOLATE Number tested	ISOLATE % resistant to carbendazim
Marinka	6	83.3	53	47.2
Pastoral	14	64.9	114	36.0
Fighter	13	46.2	121	13.4
Halcyon	4	50.0	40	10.0
Puffin	11	36.4	96	9.4
Pipkin	14	14.3	113	2.7
Other cultivars	12	33.3	102	7.8
All cultivars	74	45.9	639	16.6

TABLE 3. Geographic distribution of carbendazim - resistance in *R. secalis*, by crop and isolate

Location	CROP Number sampled	CROP % with resistant isolates present	ISOLATE Number tested	ISOLATE % resistant to carbendazim
Cumbria/Northumberland	12	66.7	103	24.3
Yorkshire/N.Humberside	10	10.0	89	1.1
Norfolk/Lincs./Cambs./ S.Humberside	11	27.3	102	15.7
Suffolk/Essex	7	0	54	0
Kent/Sussex	9	66.7	67	20.9
Shropshire/Staffs./ Herefords./Oxon.	8	25.0	80	5.0
Cornwall/Devon/Somerset/ Avon/Dorset/Hants./Berks.	12	75.0	109	31.2
Wales	5	100	35	34.3
England and Wales	74	45.9	639	16.6

RESULTS

R. secalis isolates were successfully cultured from plants from 74 crops (Table 1) and 639 were screened for carbendazim resistance (average 8.6 isolates per crop). Resistance was detected in 16.6% of isolates, and 45.9% of the 74 crops had such isolates present. The level of resistance varied according to the cultivar from which the isolate was obtained (Table 2) and geographic location of the crop (Table 3).

DISCUSSION

The incidence of carbendazim resistance detected in the 1993 survey was very similar to that found in 1992 and therefore probably represents an accurate picture of the situation in England and Wales. More resistance was found in the South West of England and in Wales - areas of higher rainfall and higher rhynchosporium leaf blotch incidence in most years. Crops in these areas are likely to have more frequent applications of fungicides and hence the selection pressure for resistance would be greater. The situation concerning variations in resistance levels in different cultivars needs to be explored further.

The geographic distribution of the cultivars sampled in the survey was variable, with cv. Pipkin being the only type encountered in Suffolk/Essex. As this cultivar had the lowest incidence of recovery of carbendazim-resistant isolates of *R. secalis*, some caution is needed in interpreting the survey finding that no resistance was confirmed in those counties.

The low incidence of resistance in cv. Pipkin suggests that a genetically-based disassociation may exist between virulence for host resistance factor Rh_4 in that cultivar and sensitivity to carbendazim. As a consequence of these findings advice to farmers on their strategy for the control of *R. secalis* will have to be examined.

ACKNOWLEDGEMENTS

This work was funded by the Home-Grown Cereals Authority as part of a project on fungicide resistance in cereal pathogens. The help of Mr. R. W. Polley, CSL Harpenden is gratefully recorded.

REFERENCES

James, W. C.; Jenkins, J. E. E.; Jemmett, J. L. (1968) The relationship between leaf blotch caused by *Rhynchosporium secalis* and losses in grain yield of spring barley. *Annals of Applied Biology*, **62**, 273-288.

Kendall, S.; Hollomon, D. W.; Ishii, H.; Heaney, S. P. (1994) Characterisations of Benzimidazole-Resistant strains of *Rhynchosporium secalis*. *Pesticide Science* (in press).

Polley, R. W.; Thomas, M. R.; Slough, J. E.; Bradshaw, N. J. (1993) Surveys of diseases of winter barley in England and Wales, 1981-1991. *Annals of Applied Biology*, **12**, 287-307.

DETECTION OF BENZIMIDAZOLE-RESISTANT DRY ROT IN POTATO SEED LOTS USING A MODIFICATION OF THE "BAG TEST."

P. Nolte

University of Idaho, Idaho Falls Research and Extension Center, 1776 Science Center Drive, Idaho Falls, Idaho, USA 83402

ABSTRACT

Benzimidazole resistance in the potato dry rot fungus (*Fusarium sambucinum*) has been encountered with increasing frequency in Idaho and other potato-producing areas of the United States in the last five years. The traditional "bag test" is a simple, on-farm test that can be performed by growers and fieldmen to test for the dry rot potential within a given seed lot. A modification of the bag test referred to as the "modified bag test," has been developed to include the identification of benzimidazole resistance. Information gathered from modified bag test results has been employed as a resistance monitoring tool to estimate the frequency and distribution of benzimidazole resistance in the Idaho potato industry. The test also serves in an educational capacity by providing growers with information on benzimidazole resistance and effective alternative seed piece treatments for control of Fusarium dry rot.

INTRODUCTION

Benzimidazole resistance in the dry rot fungus (*Fusarium sambucinum*) was first confirmed in Idaho seed potatoes in 1990 (Nolte, 1993-a,b) . A seed lot which had been cut, treated with a benzimidazole fungicide, and stored for five weeks before planting displayed virtually 100 percent incidence of dry rot decay with severities that averaged 50 percent or more. In the following two growing seasons, several additional instances were encountered in seed lots cut, treated with benzimidazoles 'Topsin' (thiophanate-methyl) or TBZ (thiabendazole) and stored for several weeks before planting. Because of the threat of benzimidazole resistance, growers who make a practice of cutting and storing their seed before planting have been warned that the practice can be very risky (Nolte, 1993-a).

The resistance problem in cut seed has an even greater significance in view of the fact that thiabendazole is the only fungicide approved for post-harvest application on potatoes in the USA. In many cases, it appears that the resistant strains of *Fusarium sambucinum* can be brought in on the seed and be transmitted to the daughter tubers. This is thought to occur because benzimidazole-treated seed will decay in the rhizosphere providing resistant inoculum to infect daughter tubers during the rigours of harvest and handling (Author, unpublished). If thiabendazole is applied post-harvest, selection pressure again favours the resistant strains. The use patterns of this class of chemicals in Idaho would seem to be a classic case of over-reliance on a single-site action fungicide for control of a fungal disease. In some areas the benzimidazoles have been used for both seed treatment application and post-harvest fungicide application for several years in a row. The development of resistance under this kind of selection pressure seems almost inevitable.

Effective management strategies depend heavily on grower awareness of the problem. The modified bag test is a tool that educates growers about fungicide resistance and provides them with information about the potato seed lots that they are selling or purchasing (Nolte, 1993-b). The kit also contains information on fungicide resistance management including the use of alternate and benzimidazole-combination products for seed decay control as well as Integrated Pest Management techniques for disease control (Brent, 1988; Delp, 1988; Wade, 1988).

MATERIALS AND METHODS

Several chemical companies, including Snake River Chemical, Caldwell, Idaho; Wilbur-Ellis, Twin Falls, Idaho and Plant Health Technologies, Boise, Idaho have produced and made pre-packaged test kits available to Idaho growers who want them. The pre-packaged kits are very convenient and may be one reason for the large number of growers (hundreds of kits were distributed in 1993) who are performing these tests. General instructions for performing the modified bag test as they were provided to Idaho growers are included below.

The Modified Bag Test

The familiar "bag test" was developed to enable growers to determine whether or not there is a potential for dry rot problems in their seed potatoes (Secor, *et al*, 1985, 1992). The modified bag test employs exactly the same methods but has some simple added features for fungicide resistance determination. The traditional bag test will be described first, followed by the modified test.

The Traditional Bag Test

"To do the bag test you will need a knife, a large paper shopping bag, a large plastic trash bag (large lawn bags are ideal), and at least twenty-five 6 to 10 oz seed tubers from the storage or seedlot that you wish to test. Pick the tubers at random but don't include any that have obvious decay problems.

Once you have all these items gathered, the test is simple to perform. First, cut all of the tubers into seed pieces with the knife and place them in the large paper shopping bag. Cutting the tubers into quarters with a longitudinal cut followed by a cross section cut exposes a lot of wounded tissue. Seed pieces made this way are ideal for the test. Next, roll the top down on the paper bag and gently shake the bag for at least 30 seconds. You may wish to invert the bag several times during this process. The shaking will serve to inoculate the freshly-cut surfaces with any Fusarium spores that might be present. Don't worry if the top of the bag unrolls when you are finished.

Then put the paper bag (with the cut seed still inside) into the large plastic trash bag. DO NOT SEAL THE PLASTIC BAG! Instead, just allow the top of the bag to droop over so that the seed in the paper bag is not subjected to direct air currents. The aim here is to trap most of the humidity from the cut seed in the plastic bag while still allowing for some oxygen exchange. Without some humidity, the seed pieces will simply dry out and without oxygen (if the trash bag were sealed) they will probably decay due to soft rot.

Now store the bag test in a place where you can leave it undisturbed for 3 to 4 weeks at a reasonably warm temperature. Room temperature (72 F) is about the upper limit but the temperature should be at least 65 F. A corner of your office or under a bench in a heated shop is an ideal location.

Examine the cut seed after 3 to 4 weeks for decay. If the seed pieces are whole and there is little or no dry rot present, you should have no problems with your seed. If there is a lot of dry rot, however, your seedlot may have a high dry rot potential. This factor will have to be taken into account when planting or selling the seed next season.

The bag test is easy to perform and provides growers with important information about their seed. Growers should perform this simple test on their seedlots every year."

Determining Benzimidazole Resistance

"Now, we'll make one minor modification to the traditional bag test which will turn it into a tool for determining benzimidazole fungicide resistance. Gather the materials for doing a bag test, only this time we'll triple the number of tubers and the number of shopping bags. This means three shopping bags and at least 75 tubers. Divide the 75 tubers into three equal groups. Do the bag test exactly as it is described above for one of the three groups of tubers. Consider this to be the control.

With the other two groups we will do almost the same thing but we'll make one modification: *After the cut seed has been shaken in the paper bag to inoculate it, open the bag and add a seed treatment.* Use a benzimidazole treatment such as TBZ or 'Topsin' in one bag and either a combination or an EBDC such as mancozeb in the other. It would be best if you used a small scale to weigh the cut seed to determine the proper amount of treatment. Here's a simple formula you can perform on a hand calculator to determine the amount of seed treatment you'll need: 16 X (the weight of your seed in lb. /100) = __ oz of seed treat for the test.

If you don't have this equipment, you can still get close enough to the right amount to do a valid test. The following values are very close to the proper amount: 25 tubers X 8 oz (avg.) = 12.5 lb. of cut seed. Most seed treatments go on at the rate of 1 lb. (16 oz) / cwt., so 16 x (12.5 / 100) = 2.0 oz of seed treat. To make things even easier, we have determined that 1 oz of seed treat is just about exactly 3 level tablespoons. If you use an actual tablespoon, take steps to ensure that no one uses it for preparing food afterwards.

After you add the seed treatment, tightly roll the top of the bag down and shake it gently for at least 30 seconds to coat the seed pieces with the treatment. You may wish to turn the bag upside down several times during this procedure to ensure good coverage. Let the roll at the top of the bag relax.

Now place your control and your bags of treated seed in the plastic trash bag and store as described above. At the end of 3 -4 weeks examine both treated and untreated seed for decay. If the benzimidazole treated seed has as much or more decay than the control and the combination or EBDC treatment has little or no decay, you probably have a resistant strain of *Fusarium* in your seedlot. Call your County Agent or Extension Specialist if you find, or think you have found, this problem. Once the problem has been identified, it is relatively easy to control."

RESULTS AND DISCUSSION

Results of all of the hundreds of modified bag tests performed in 1993 were not provided to University of Idaho investigators. Apparently many who performed the test preferred to keep the information for their own use. Results from tests on 68 seed lots from all over the state were reported to or personally evaluated by the Author. Of these, only 15 showed sensitivity to benzimidazole with the other 53 showing resistance (Nolte, 1993-a,b). In other words, 78 % of the growers who performed the modified bag test and reported their results to us had some level of benzimidazole resistance in their seed lots. Modified bag tests have been distributed to growers all over the state of Idaho for the 1994 season. Results will be collected and compared to 1993 data. Hopefully, with increased awareness of the problem and the adoption of resistance management techniques, over the next few years we will see a reduction in the amount of resistance reported. The modified bag test has played and should play a major role in this educational effort. There are additional problems, however. The role of soil-borne *Fusarium* that may carry resistance remains an unknown quantity. If soil-borne fungi provide a significant source of resistant inoculum, then the wisdom of the continued use of unprotected (not in combination with a fungicide of alternative activity) benzimidazoles for control of potato dry rot both in storage and on cut seed should be seriously questioned.

REFERENCES

Brent, K.J., (1988) Monitoring for Fungicide Resistance. In: *Fungicide Resistance in North America.* Charles J. Delp, (Ed), St. Paul, Minnesota, APS Press. pp. 9-11.

Delp, C.J., (1988) Resistance Management Strategies for Benzimidazoles. In: *Fungicide Resistance in North America.* Charles J. Delp, (Ed), St. Paul, Minnesota, APS Press. pp.41-43.

Nolte, P. (1993a) The Affect of Fungicide Resistance on Fusarium Decay in Cut Potato Seed. *American Potato Journal,* **70**(11):pp.832 (Abstract).

Nolte, P. (1993b) Detection of Benzimidazole resistant *Fusarium sambucinum* Using a Modification of the "Bag Test." *Phytopathology,* **83**(8):pp.885 (Abstract).

Secor, G.A., Preston, D.A., Gudmestad, N.C. and Lamey, H.A., (1992) Fusarium Dry Rot *North Dakota State University Agricultural Experiment Station Extension Bulletin. PP-1039,* pp. 4.

Secor, G.A., Gudmestad, N.C. and Preston, D.A., (1985) Disease Control Guidelines for Seed Potato Selection, Handling and Planting. *North Dakota State University Agricultural Experiment Station Extension Bulletin. PP-877, pp.* 6.

Wade, M., (1988) Strategies for Preventing or Delaying the Onset of Resistance to Fungicides and for Monitoring resistance Occurrences. In: *Fungicide Resistance in North America.* Charles J. Delp, (Ed), St. Paul, Minnesota, APS Press. pp.41-43.

RAPID DETECTION OF BENZIMIDAZOLE RESISTANCE IN *RHYNCHOSPORIUM SECALIS* USING ALLELE-SPECIFIC OLIGONUCLEOTIDE PROBES.

IAN WHEELER, SHEILA KENDALL, JENNY BUTTERS AND DEREK HOLLOMON.

Department of Agricultural Sciences, University of Bristol, AFRC Institute of Arable Crops Research, Long Ashton Research Station, Bristol BS18 9AF, UK.

ABSTRACT

Benzimidazole fungicides are important mixture components in strategies to combat fungicide resistance in *Rhynchosporium secalis*. These strategies have recently been threatened by the detection of benzimidazole resistance in field strains of *R. secalis*. To evaluate the significance of this resistance, and monitor its spread, a rapid and accurate method for identifying resistant alleles has been developed. Hybridisation of allele-specific oligonucleotide (ASO) probes to Polymerase Chain Reaction(PCR) amplified β-tubulin gene fragments has successfully detected single DNA base-pair mutations in several Rhynchosporium strains. Combining ASO detection with sequencing of β-tubulin gene fragments only one point mutation, substituting glycine for glutamic acid at codon 198, occurred in benzimidazole-resistant field populations. Two probes, one sensitive and one resistant, were sufficient to monitor current field populations. A second mutation substituting lysine for glutamic acid, which was generated in the laboratory, has not so far been detected in field populations. This technology should improve efforts to evaluate anti-resistance strategies in practice.

INTRODUCTION

Barley leaf blotch, caused by *Rhynchosporium secalis* (Oudem) J. J. Davis, is an important disease in the cool maritime regions of the world. It commonly causes a 2-5% yield reduction but, in severe epidemics losses can reach 40% (Jenkins & Jemmett, 1967). It is a difficult wet weather disease to control, particularly when spraying is delayed and disease thresholds are missed. Resistant cultivars have not been durable because of rapid selection of novel races from the varied gene pool, and fungicides play a major role in control measures. Fungicides that inhibit the C-14 demethylation step in sterol biosynthesis (DMIs) have been very successful in controlling the disease, but efficacy is now being eroded by the development of resistance to some members of this group (Kendall *et al.*, 1993)

Monitoring *R. secalis* populations throughout the U.K. between 1980 and 1989 failed to detect strains resistant to benzimidazoles (benomyl, carbendazim, thiophanate methyl). Consequently, mixtures of benzimidazoles and DMI fungicides were recommended for leaf blotch control in an effort to combat further spread of resistance to DMIs (Hollomon, 1992). In 1990, the first benzimidazole-resistant strains were detected in the U.K. at an incidence of 1% (Hollomon & Kendall, 1993). Since then, U.K. wide monitoring has revealed frequencies of benzimidazole resistance in field populations ranging from 1% in parts of Northern England (Phillips & Locke, 1994), to over 50% in N.Ireland (Taggart *et al.*, 1994).

It is important to monitor for the development and spread of benzimidazole resistance so that strategies can be evaluated, and advice changed if necessary. Unfortunately, current methods to detect resistance based on bioassay are of limited value. Resistant mutants can only be detected when their frequency reaches 1% or higher in the population, and the methodology to achieve this is both labour-intensive and time-consuming. Isolation of pure cultures of *R. secalis* from infected field samples can take up to 6 weeks, with isolation becoming increasingly difficult as the season progresses. Conclusive results obtained by *in vitro* bioassay can take a further 2 weeks, but this fails to distinguish between different benzimidazole-resistant alleles. It has previously been shown that benzimidazole resistance in *R. secalis* is correlated with decreased binding of these fungicides to tubulin-like proteins (Kendall, *et al*, 1994). To improve efficiency of monitoring procedures, we are seeking to develop a more rapid detection technique. In this paper, the successful application of allele-specific oligonucleotide probes to detect benzimidazole resistance caused by a single base change at codon 198 in the β-tubulin gene is reported.

METHODS

Isolation and bioassay of *R. secalis*.

Single spore isolates were obtained from diseased leaves using methods fully described elsewhere (Hollomon, 1984), and assayed in 25-well repli-plates on Czapek dox agar with 0.5% mycological peptone (CDM) according to procedures given in Kendall, *et al.*, (1993). Carbendazim was a gift from BASF, Limburgerhof, Germany, and diethofencarb was supplied by Sumitomo Chemical Company, Takarazuka, Japan. Both fungicides were technical grade.

DNA preparation, amplification and sequencing.

DNA was prepared from freeze-dried cells (10mg) by vortexing with glass beads in 600μl buffer (lysis) and extraction with phenol/chloroform. DNA was precipitated from the aqueous phase with ethanol, resuspended in water and treated with RNAase. This DNA was used directly as template for both PCR amplification using two 24-mer oligonucleotide primers designed to produce a 416 bp fragment encompassing codon 198 of the *R. secalis* β-tubulin gene, and sequenced. PCR(100μl) containing primers, template DNA, nucleotides and Taq-polymerase were carried out in an Autogene temperature cycler (Grant Instruments, Cambridge, UK.) and involved an initial five minute denaturation step followed by 30 cycles of anealing at 65°C for one minute, extension at 72°C for one minute and denaturation at 94°C for one minute. Production of the target DNA was checked by electrophoresis on 3% Nusieve agarose gel. This product was used directly in the ASO probe assay or was sequenced after purification.

Allele-specific oligonucleotide (ASO) probe detection

The PCR generated β-tubulin gene fragment was dot-blotted onto either a nylon membrane for radioactive (^{32}P) detection, or onto a nitro-cellulose membrane where a non-radioactive labelling system was used. Fifteen-mer oligonucleotide probes designed to match the expected sequence surrounding codon 198 were end-labelled with either ^{32}P or biotin, and hybridised to target DNA using standard procedures. Control of washing temperature allowed mismatched probes to be washed away. Hybridised, and correctly aligned, ASO probes were visualised either by radioautography, or by a non-radioactive detection system (Boehringer Mannheim) which uses streptavidin/alkaline phosphatase and nitro-blue tetrazolium salt as the substrate. Figure 1 shows the full procedure used to detect these single DNA base changes.

Results

Sequencing an amplified fragment of the β-tubulin gene of *R. secalis* has linked benzimidazole resistance with a single DNA base change at codon 198. Table 1 shows results for five strains representing the different sequences so far encountered. In all resistant strains so far isolated from field crops, adenine is replaced by guanine causing the substitution of glycine for glutamic acid at amino acid 198. These strains always show negativ cross-resistance between carbendazim and diethofencarb. In a u.v. irradiated resistant mutant (BEN 22), derived from the wild-type strain K1124, replacement of a guanine in codon 198 with adenine causes substitution of lysine for glutamic acid.

ASO probe detection: Effects of washing temperature on stringency.

Stringency of the washing temperature required to remove any mismatched probe was determined by the probe being used. No selectivity was observed at 40°C, and probes bound to DNA fragments amplified from benzimidazole-resistant and -sensitive strains alike. At 48°C, all mismatched sensitive ASO probe was washed away from DNA of resistant strains following three two-minute washings, allowing specific detection of sensitive strains. For an ASO specific for the resistant mutation GAG--GGG, it was necessary to raise the washing temperature to 50°C in order to selectively detect this mutation. This same probe did not detect the resistant mutation GAG--AAG when membranes were washed at 50°C. So far we have not detected this resistant mutation in field strains. This procedure has correctly identified the resistance status of all strains for which we have bioassay and sequence data.

Discussion

Single base-pair mutations in *R. secalis* causing amino acid substitutions and correlated with benzimidazole resistance, were similar to those found in resistant field strains of other plant pathogenic fungi (Koenraadt & Jones, 1992; Martin, et al, 1992; Yarden & Katan, 1993). In both *R. secalis*, *Botrytis cinerea* (Yarden & Katan, 1993), and *Venturia inaequalis* (Koenraadt, *et al*, 1992) substitution of a lysine for glutamic acid results in a loss of the negatively correlated cross-resistance between benzimidazoles and diethofencarb, although this mutation has yet to be found in field populations of *Rhynchosporium*.

ASO probes provide a rapid diagnostic technique which takes less than 24h to complete. At present, the method requires DNA template extracted from pure cultures of *R. secalis*, but we are currently exploring nested primer PCR approaches for direct detection from diseased leaves. This should reduce the time taken to obtain a result from field samples to 24-48h, compared with 6-8 weeks now required for conventional isolation and bioassay. This technology not only provides an accurate detection of resistance but can distinguish between different resistant alleles. Although up to 10 sites within the β-tubulin gene have been identified as conferring benzimidazole resistance (Osmani & Oakley, 1991), it is clear that mutations in field strains of phytopathogens are restricted to codons 198 and 200 (Koenraadt & Jones, 1992; Adachi, *et al*, 1993; Yarden & Katan, 1993). Consequently, the number of different ASO probes needed to monitor field populations is quite limited; only four probes were needed to accurately identify benzimidazole resistance in a large world-wide collection of *V. inaequalis* strains (Koenraadt & Jones, 1992). For obligate or slow growing pathogens rapid detection of resistance using ASO probes should allow more samples to be tested and so contribute to more effective monitoring of the performance of strategies used to combat resistance. Where efforts to assess the risk of resistance to new compounds identifies resistant mutants at an early stage in development, rapid, DNA based, detection offers a more effective way to identify rare mutants in field populations, than conventional monitoring methods.

Acknowledgements

We thank MAFF and H-GCA for funding this work.

References

Adachi, K., Namba, S. & Tsuchizaki, T. (1993) Use of allele-specific polymerase chain reaction for detection of the mutation in beta-tubulin gene from field isolates of benomyl-resistant *Penicillium digitatum. 6th International Congress of Plant Pathology*. **Abstract 3. 1. 39**. 56

Hollomon, D. W. (1984) A laboratory assay to determine the sensitivity of *Rhynchosporium secalis* to the fungicide triadimenol. *Plant Pathology*, **33**, pp 65-70.

Hollomon, D. W. (1992) Monitoring and forecasting fungicide resistance in *Rhynchosporium. Home-Grown Cereals Project Report*, **59**, London, 13pp.

Hollomon, D. W., Kendall, S. J. (1993) Monitoring and detecting fungicide resistance in cereals. *Arable Farming Plus*, **March 1993**, 34-39.

Jenkins, J. E., Jemmett, J. L. (1967) Barley leaf blotch. *NAAS Quarterly Review*, **75**, 127-132.

Kendall, S. J., Hollomon, D. W., Cooke, L. R. & Jones, D. R. (1993) Changes in sensitivity to DMI fungicides in *Rhynchosporium secalis*. *Crop Protection* **12(5)**, 357-362

Kendall, S.J., Hollomon, D.W., Ishii, H. & Heaney, S.P. (1994) Characterisation of benzimidazole-resistant strains of *Rhynchosporium secalis*. *Pesticide Science*, **40**, 175-181.

Koenraadt, H., Somerville, S. C., & Jones, A. L. (1992) Characterization of mutations in the beta-tubulin gene of benomyl-resistant field strains of *Venturia inaequalis* and other plant pathogenic fungi. *Phytopathology*, **82**, 1348-1354.

Koenraadt, H. & Jones, A.L. (1992) The use of allele-specific oligonucleotide probes to characterise resistance to benomyl in field strains of *Venturia inaequalis*. *Phytopathology*, **82**, 1354-1358

Martin, L.A., Fox, R.T.V., Baldwin, B.C. & Connerton, I.F. (1992) Use of polymerase chain reaction for the diagnosis of MBC resistance in *Botrytis cinerea. Proceedings Brighton Crop Protection Conference- Pests and Diseases-1992*, 207-214

Osmani, S.A. & Oakley, B.R (1991) Cell cycle and tubulin mutations in filamentous fungi. In: *More gene manipulations in fungi.* (eds J.W. Bennett & L.L. Lasure), 107-122.

Phillips, A. N., Locke, T. (1994) Carbendazim resistance in *Rhynchosporium secalis* in England & Wales.(**These proceedings**).

Taggart, P. J., Cooke, L. R. & Mercer, P. C. (1994) Benzimidazole resistance in *Rhynchosporium secalis* in Northern Ireland and its impact on disease control. **(These proceedings)**

Yarden, O. & Katan, T. (1993) Mutations leading to substitutions at amino acids 198 and 200 of beta-tubulin that correlate with benomyl-resistance phenotypes of field strains of *Botrytis cinerea. Phytopathology*, 83(12), 1478-1483.

FIGURE 1. Allele-specific oligonucleotide DNA probes

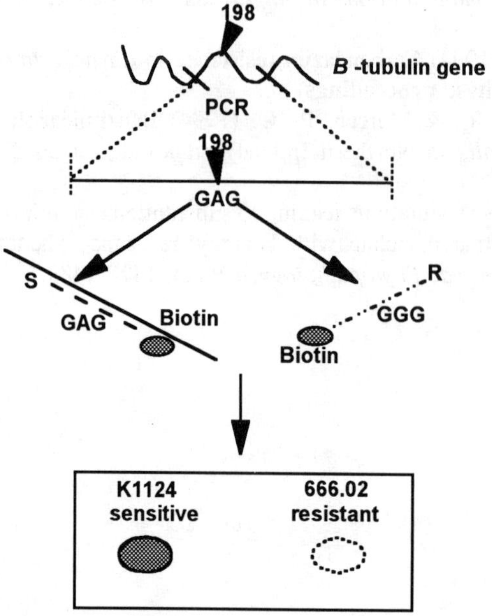

TABLE 1. DNA changes at codon 198 in the *B*-tubulin, and benzimidazole resistance in *Rhynchosporium secalis*..

Isolate	Amino acids						Fungicide sensitivity	
	196 SER	197 ASP	198 GLU	199 THR	200 PHE	201 CYS	Benzimidazoles	N-phenyl-carbamate
K1124	TCT	GAT	*GAG*	ACC	TTC	TGT	*S*	*R*
809.02	TCT	GAT	*GAG*	ACC	TTC	TGT	*S*	*R*
666.02	TCT	GAT	GLY *GGG*	ACC	TTC	TGT	*R*	*S*
769.03.03	TCT	GAT	*GGG*	ACC	TTC	TGT	*R*	*S*
BEN22	TCT	GAT	LYS *AAG*	ACC	TTC	TGT	*R*	*R*

MUTATIONS IN THE BETA-TUBULIN GENE OF BENOMYL-RESISTANT PHENOTYPES OF *BOTRYTIS CINEREA*

O. YARDEN* AND T. KATAN**

*Department of Plant Pathology and Microbiology, Faculty of Agriculture, The Hebrew University of Jerusalem, Rehovot 76100, Israel.

**Department of Plant Pathology, ARO, Volcani Center, Bet Dagan 50250, Israel.

Three phenotypes were identified among benomyl-resistant strains of *Botrytis cinerea* in Israel, when tested for sensitivity to carbendazim (MBC) and diethofencarb (NPC): $Ben^{HR}NPC^{S}$= highly resistant to MBC (EC_{50}>50 µg/ml) and sensitive to 0.5µg/ml NPC; $Ben^{MR}NPC^{R}$= moderately resistant to MBC ($10 \leq EC_{50} < 20$ µg/ml) and resistant to 10 µg/ml NPC; and $Ben^{HR}NPC^{R}$= highly resistant to MBC and resistant to NPC. A 1-kb fragment of the wild-type gene encoding for beta-tubulin (designated *benA*) in *B. cinerea* was cloned and sequenced. The deduced partial amino acid sequence of the *B. cinerea* beta-tubulin showed a high degree of similarity to beta-tubulins of other filamentous fungi. A PCR approach was used to amplify and sequence 992-bp *benA* fragments from strains representing the three phenotypes. In the eight Ben^{R} strains analyzed, three single base-pair mutations were identified and found to correlate with the different phenotypes: codon 198, encoding glutamic acid in the wild type, was changed to an alanine codon in the $Ben^{HR}NPC^{S}$ phenotype, or to a lysine codon in the $Ben^{HR}NPC^{R}$ phenotype; codon 200, encoding phenylalanine was changed to a tyrosine codon in the $Ben^{MR}NPC^{R}$ phenotype. These mutations were similar to those identified in benomyl-resistant field strains of other phytopathogenic fungi.

[illegible]

[illegible]

[illegible] Department of [illegible] [illegible], Japan.

[illegible] fragmented [illegible] cloned and [illegible] sequenced. The [illegible] partial amino acid sequence [illegible] showed high [illegible] [illegible] These mutations were similar [illegible]

RESISTANCE OF BOTRYTIS CINEREA TO DICARBOXIMIDES, BENZIMIDAZOLES AND PHENYLCARBAMATES IN THE CHAMPAGNE VINEYARDS

P. LEROUX

INRA, Unité de Phytopharmacie, 78026 Versailles Cédex, France.

D. MONCOMBLE

CIVC, 5 rue Henri Martin, 51200 Epernay, France.

ABSTRACT

Strains of B. cinerea resistant either to dicarboximides (e.g. iprodione, procymidone, vinclozolin) or to both benzimidazoles (e.g. carbendazim) and phenylcarbamates (e.g. diethofencarb) are commonly found in Champagne vineyards. Limitations in the use of these fungicides are recognized as anti-resistance strategies. However, when the frequencies of resistant strains are too high, they can be temporarily withdrawn. Such advice is given according to the results of the monitoring done every year at vintage.

INTRODUCTION

In french vineyards, B. cinerea remains a parasite feared by vine-growers because of its qualitative and quantitative effects on wine production. The chemical control of this fungus is normaly achieved by three of four treatments applied between the flowering stage and three weeks before vintage. Several families of fungicides are registered in France. The oldest ones are protectants (e.g. chorothalonil. dichlofluanid, folpet, thiram) and have never been affected by resistance because of their multi-site effects. On the other hand, resistance developed towards dicarboximides (e.g. clozolinate, iprodione, procymidone, vinclozolin), benzimidazoles (e.g. benomyl, carbendazim, thiophanate-methyl) and phenylcarbamates (e.g. diethofencarb). The evolution of such phenomena in Champagne vineyards over 10 years and the anti-resistance strategies will be presented in this paper.

CHARACTERISTICS OF THE VARIOUS RESISTANT STRAINS

Dicarboximides are among the most effective fungicides against B. cinerea. However the development of moderately resistant strains (Rd ; Table 1) provided inadequate control of grey mould in Champagne vineyards and also in many other French regions (Leroux and Clerjeau, 1985).

Benzimidazoles which were introduced before dicarboximides, selected highly-resistant strains which were very susceptible to the phenylcarbamate diethofencarb (Rb1 ; Table 1). Since the commercialization of mixtures diethofencarb/carbendazim in 1987, a new phenotype simultaneously-resistant to these two fungicides appeared (Rb2 ; Table 1).

All strains remained sensitive to pyrimethanil, an anilinopyrimidine recently introduced in France (Table 1).

TABLE 1 : Effects of fungicides on the germ-tube elongation of various strains of *B. cinerea*[a] collected in Champagne vineyards in 1992 and 1993

Fungicides	Mean EC50 (mg/l) of phenotypes				
	Sd	Rd	Sb	Rb1	Rb2
iprodione	0.8	5.0	-	-	-
procymidone	0.6	8.0	-	-	-
vinclozolin	0.4	5.0	-	-	-
carbendazim	-	-	0.05	>25	5.0
diethofencarb	-	-	>25	0.05	>25
pyrimethanil	0.07	0.07	0.06	0.07	0.07

a : the tests were conducted according to the method of Leroux and Gredt (1989) on four to eight strains of each type.

TABLE 2 : Relationship between the percentages of *B. cinerea* resistant strains and the efficacies of a programme combining several fungicides (average values obtained in 3 or 4 trials each year)

years	Control plots		Treated plots[a]
	% Rd	% Rb2	% efficacy
1989	25	1	50
1990	31	14	72
1991	38	34	13
1992	20	33	17

a : the programme consisted in three applications of a mixture carbendazim/diethofencarb (500 + 500 g/ha) followed by a mixture vinclozolin/thiram (500 + 3200 g/ha) and a dicarboximide alone (procymidone or vinclozolin at 750 g/ha).

RESISTANCE PHENOMENA IN FIELD TRIALS

From the various trials conducted in Champagne region over the last six years, it has been shown that :

- the selection pressure induced by one application of procymidone or vinclozolin (750 g/ha) towards Rd strains was lower than that of a mixture of carbendazim/diethofencarb (500 + 500 g/ha) towards Rb2 strains (Leroux and Moncomble, 1993).

- at reduced rates (500 g/ha instead of 750 g/ha) procymidone and vinclozolin used in mixture with thiram (3200 g/ha) exerted a selection pressure significantly lower than that of dicarboximides applied alone at their full rate (750 g/ha) (Leroux and Moncomble, 1993).

- the programme based on three applications of a mixture carbendazim/diethofencarb, followed by a mixture vinchlozolin/thiram, and a dicarboximide alone was effective in 1989 and 1990 but failed to control grey mould in 1991 and 1992. It seemed that this was mainly due to the development of Rb2 strains (Table 2).

MONITORING AND RECOMMENDATIONS OF FUNGICIDE USE

A monitoring of resistance of B. cinerea is conducted every year in more than 100 commercial Champagne vineyards at vintage. In each location, 10 to 20 diseased berries are collected and the percentages of the various phenotypes are determined according to the method of Leroux and Clerjeau (1985).

TABLE 3 : Evolution of number of annual fungicide treatments and percentages of B. cinerea resistant strains in Champagne vineyards between 1982 and 1993

years	dicarboximides		carbendazim and diethofencarb		
	Tre.[a]	% Rd	Tre.[a]	% Rb1	% Rb2
1982	4	87	0	85	0
1983	0	72	0	82	0
1984	0	42	0	76	0
1985	0	22	0	94	0
1986	0.5	21	0	97	0
1987	0.8	30	0.8	97	0
1988	0.8	22	1.5	95	2
1989	1.1	30	1.1	77	21
1990	1.6	37	0.9	52	46
1991	1.7	48	0.9	38	62
1992	1.5	56	0.2	54	46
1993	0	34	0	70	29

[a] : average number of annual treatments with dicarboximides (alone or in mixture with thiram) or mixtures carbendazim/diethofencarb.

From the results of Table 3, it appeared that the frequencies of dicarboximide-resistant strains (Rd) as well as those of strains doubly resistant to carbendazim and diethofencarb (Rb2) decreased when the selection pressure ceased (this was not the case with the Rb1 strains). This non- persistant resistance which is probably due to reduced fitness of Rd and Rb2 strains permitted a discontinious use of both dicarboximides and mixtures carbendazim/diethofencarb. The re-employment of these fungicides occured when the average percentages of resistant strains were below 25 % at the preceeding vintage.

This survey confirmed than one annual application of dicarboximides (alone or in mixture with thiram) exerted lower selection pressure than that of mixtures carbendazim/diethofencarb (Table 3).

CONCLUSION

In spite of the limitation of annual treatments against B. cinerea (three instead of four or five in the 1980's) and of combined uses of the various available fungicides (mixtures or alternations), the chemical control of grey mould remains uncertain in Champagne vineyards. As an example, the recommendation for 1994, advises the successive applications of a mixture carbendazim/diethofencarb, of pyrimithanil and of a dicarboximide (alone or in mixture with thiram). This may lead to selection of RB2 strains and possibly Rd ones, which is likely to lead (once more) to the withdrawal of mixtures carbendazim/diethofencarb and dicarboximides in 1995. Other considerations, such as residue levels or negative effects on fermentation are additional constraints for the establishing of grey mould treatment programmes.

REFERENCES

Leroux, P. ; Clerjeau, M. (1985). Resistance of Botrytis cinerea and Plasmopara viticola to fungicides in French vineyards. Crop Protection, **4**, 137-160.
Leroux, P. ; Gredt, M. (1989). Negative cross-resistance of benzimidazole-resistant strains of Botrytis cinerea, Fusarium nivale and Pseudocercosporella herpotrichoides to various pesticides. Netherlands Journal of Plant Pathology, **95 (sup. 1)**, 121-127.
Leroux, P. ; Moncomble, D. (1993). Lutte chimique contre la pourriture grise de la vigne : passé, présent, futur. Phytoma, **450**, 27-30 and Phytoma, **451**, 23-27.

DISTRIBUTION AND INCIDENCE OF BENZIMIDAZOLE-RESISTANT FUSARIUM SAMBUCINUM AND HELMINTHOSPORIUM SOLANI ISOLATED FROM POTATO IN NORTH AMERICA.

G.A. SECOR, D. RODRIGUEZ, J. RODRIGUEZ AND N.C. GUDMESTAD.

Department of Plant Pathology, North Dakota State University, Fargo, ND 58105, USA.

ABSTRACT

Isolates of Fusarium sambucinum and Helminthosporium solani were recovered from dry rot and silver scurf diseased potatoes collected from states and provinces throughout the United States, Mexico, and Canada. Single spore cultures of both fungi were prepared and tested for resistance to thiabendazole (TBZ) by growth reduction (ED50) on TBZ-amended acid PDA (Fusarium) or modified V-8 (H. solani) media. Of 504 Fusarium isolates representing 18 states, 378 or 75% of them, were resistant to TBZ. The resistance ranged from 1 to 64 ppm. The Fusarium isolated was almost exclusively F. sambucinum. Of the H. solani isolates tested representing 7 states, 21/27 or 77% were resistant to TBZ. The resistance ranged from 20 to 50 mg/l. TBZ-resistant isolates were cross-resistant to thiophanate-methyl.

INTRODUCTION

Storage diseases of potato can cause serious problems costing producers and processors large amounts of time and money. Many of these diseases can be controlled by post-harvest application of fungicides as the potatoes are going into storage. The only post harvest chemical approved for potatoes in the US is thiabendazole ('Mertect 340F'). This chemical has been intensively used and has provided excellent primary control of dry rot, caused by Fusarium sambucinum (Fs), for 20 years, and assumed control of silver scurf, caused by Helminthosporium solani (Hs), even though it is not a target disease for this fungicide. Recent outbreaks of both dry rot and silver scurf prompted us to re-examine these diseases, and the role that fungicide application has on disease management, including possible fungicide resistance to TBZ.

METHODS AND MATERIALS

Isolations

Fusarium was isolated from dry rot diseased potato samples of various cultivars collected from throughout the US, Mexico, and Canada. Most samples were sent by colleagues at universities and industry. Fusarium was successfully isolated from samples representing 18 states or provinces (Table 1). Potatoes were surface sterilized (0.5% Na hypochlorite, 15 min) and small pieces of tuber tissue at the margin of the dry rot lesion placed on acidified potato dextrose agar (APDA). After 5-7 days incubation, Fusarium cultures were transferred to PDA medium. Single spore cultures of each isolate were prepared. 504 Fs cultures were collected for testing (Table 1). Carnation leaf agar was used to identify isolates (Nelson et al, 1983).

Hs was isolated from naturally-infected silver scurf tubers recovered from samples collected or sent from 7 states or provinces (Table 2). Surface sterilized tubers, (0.5% Na hypochlorite, 15 min), were incubated in a humid chamber at room temperature. After 5-7 days, conidia were picked from the tuber surface and spread on APDA. After 24-58 hrs, germinated single spores were transferred to modified V-8 agar medium.

Resistance testing

Single spore Fs colonies were tested for TBZ resistance using PDA amended with technical grade TBZ (98.5%), provided by Merck & Co., Rahway, NJ, at 0, 10, 20, 50 and 100 mg/l. Mycelial disks 5 mm in diameter were removed from Fs cultures 5-7 days old, and a single disk was centered in petri plates with or without TBZ. Cultures were incubated at room temperature (approx 25°C) for 7 days under continuous fluorescent light. Each treatment was replicated three times; the experiment was repeated twice.

Single spore Hs colonies were tested for TBZ resistance using modified V-8 (MV8) medium amended by the addition of technical grade TBZ as above. TBZ concentrations ranged from 0 to 200 ppm. Single mycelial disks 5 mm in diameter were removed from the edge of one-month-old cultures and centered in petri dishes with or without TBZ. Cultures were incubated 21 days at 23°C in the dark. Each treatment was replicated four times; the experiment was repeated twice.

Calculation of TBZ resistance

Resistance data was based on the inhibition of radial mycelial growth of the colony after 7 (Fs) or 21 (Hs) days. Two diameter measurements were taken for each colony at perpendicular axes to each other and averaged. The logarithm of the concentration versus the logarithm of the percentage growth reduction was plotted and regression analysis used to determine graphically the concentration in mg/l of the TBZ causing 50% reduction of colony growth compared to the unamended control; this was defined as ED50 (Trivellas, 1988).

RESULTS

Of the Fs isolates recoverd from potato tubers with dry rot, greater than 95% were identified as Fusarium sambucinum. The results of TBZ resistance testing of these isolates can be seen in Table 1. Overall, 378/504 isolates tested, or 75%, were resistant to TBZ. The ED50 ranged from 1-64 mg/l.

The results of resistance testing with Hs isolates can be seen in Table 2. Of the isolates of Hs recovered from silver scurf affected tubers, 21/27, or 77%, were resistant to TBZ. Resistance ranged from 20-50 mg/l.

CONCLUSION

It is evident that resistance to TBZ is geographically widespread in fungi causing the post- harvest diseases dry rot and silver scurf in North America. This data should be used in formulating management strategies for control of these diseases.

REFERENCES

Nelson, P. E.; Toussoun, T. A.; Marasas, W. F. O. (1983) Fusarium species; an illustrated manual for identification. The Pennsylvania State University Press.

Trivellas, A. E. (1988) Benzimadazole resistance monitoring techniques and the use of monitoring studies to guide benomyl marketing. In: Fungicide resistance in North America, C. J. Delp (Ed.). APS Press. pp. 28-30.

TABLE 1. Distribution and frequency of TBZ-resistant Fusarium[1] isolates in North America

Isolate Source	Number Tested	Percent Sensitive	Percent Resistant	ED50[2]
North Dakota	76	18	82	30
Minnesota	30	30	70	21
South Dakota	2	100	0	<10
Idaho	255	16	84	27
Wisconsin	9	78	22	25
Maine	4	100	0	<10
Texas	1	100	0	<10
Washington	7	100	0	<10
Alaska	4	75	25	53
Colorado	15	93	7	21
New York	39	3	97	29
Michigan	17	0	100	33
Nebraska	12	25	75	32
Alabama	4	100	0	<10
Utah	5	0	100	31
Delaware	2	100	0	<10
Wyoming	7	29	71	29
Pr. Ed. Is., Canada	4	100	0	<10
Michoacan, Mexico	12	66	33	13
Total	504	25	75	

[1]Almost exclusively F. sambucinum

[2]Estimated using colony diameter reduction after seven days

TABLE 2. Frequency and distribution of TBZ-resistant <u>Helminthosporium solani</u> isolates from North America

Isolate Source	Number Tested	Percent Sensitive	Percent Resistant	ED50[1]
North Dakota	14	7	93	20-47
Minnesota	2	50	50	50
Wisconsin	2	100	0	--
Alaska	1	100	0	--
Oregon	1	100	0	--
New Brunswick	5	0	100	25-32
Maine	2	0	100	20-32
Total	27	22	78	

[1]Estimated using colony diameter reduction after 21 days

IMPACT OF BENZIMIDAZOLE-RESISTANT FUSARIUM SAMBUCINUM ON DRY ROT DISEASE OF STORED POTATOES.

G. A. SECOR, P. NOLTE, N. C. GUDMESTAD AND J. RODRIGUEZ.

Department of Plant Pathology, North Dakota State University, Fargo, ND 58105, USA.

ABSTRACT

Isolates of Fusarium sambucinum (Fs) with high, medium, and low resistance to thiabendazole (TBZ) or susceptible were evaluated for their impact on dry rot incidence and severity of inoculated potatoes. Potato tubers, cv. Norchip, were mechanically bruised, inoculated with Fs spores, treated with TBZ by dip or low-volume spray, and evaluated after 2,3,4 and 5 months' storage at 10°C. The experiment was conducted twice. In most instances, TBZ resistance resulted in increased incidence and severity of dry rot compared to the control treatments. Method of TBZ application did not appreciably affect the incidence or severity of dry rot in stored potato tubers.

INTRODUCTION

Previous work (Secor et al., these proceedings) has shown widespread resistance of Fusarium sambucinum (Fs) to the fungicide thiabendazole (TBZ). Of 504 isolates of Fs tested, 378, or 75% of them were resistant. The ED50 ranged from 1 to 64 mg/l. TBZ is the only fungicide registered for post-harvest application and has specific activity against Fusarium on a number of crops. Dry rot disease has recently become a serious problem of stored potatoes after twenty years of excellent control which coincides with the widespread use of TBZ. Because fungicide resistance is usually irreversible, it was of interest to determine if the resistance observed has an impact on the ability of TBZ to control dry rot. Similar studies (Hall and Hide, 1992) have shown that resistance to TBZ in Helminthosporium solani does reduce the ability of post-harvest application of TBZ to control silver scurf. The purpose of these studies was to determine the role TBZ resistance in Fs may play in the development of dry rot disease in stored potatoes.

METHODS AND MATERIALS

Potato tubers, cv. Norchip, were used in all studies. In the first year of the study, potato tubers were inoculated with one of eight isolates of Fs that varied in their response to TBZ and could be categorized as susceptible (<10 mg/l) or resistant to low (15-20 mg/l), medium (25-30 mg/l) or high (>40 mg/l) levels of the fungicide. Four isolates of each TBZ response category were recovered from dry rot infected tubers originating from North Dakota and four were from Idaho.

Potato tubers were bruised in a small batch cement mixer for 60 seconds prior to inoculation. Fusarium cultures used for inoculum were grown on PDA at room temperature under continuous fluorescent light. Inoculum was prepared from 5-7 day old single spore cultures of each isolate and diluted to 300,000 propagules/ml. Following inoculation with Fs, the potatoes were treated in one of several ways. The treatments were: untreated (no TBZ), water dip (20 sec), TBZ dip (label rate), TBZ at ultra-low volume spray (label rate). Controls used to monitor natural, background levels of dry rot infection included a) unwashed, unbruised, uninoculated, untreated tubers; b) washed unbruised, uninoculated, untreated tubers; c) unwashed, bruised, uninoculated, untreated tubers and; d) washed, bruised, uninoculated, untreated tubers.

After inoculation and treatment in the above manner, tubers were allowed to air dry. One hundred tubers of each treatment combination (isolate X chemical treatment) were placed into four mesh bags (25 tubers/bag) and placed into a humidified storage (>90% RH) at 10°C. Tubers from each treatment were evaluated for incidence and severity of dry rot after 2, 3, 4 and 5 months of storage. At each sampling date, five tubers/treatment were frozen (-30°C) and sent to Merck Co. (Rahway, NJ) for TBZ residue analysis. Additional tubers representing each isolate were also randomly selected among the treatments at each sampling date. Tuber tissue from the margins of decay was used to isolate the causal Fs. Fusarium isolates obtained in this manner were screened for TBZ resistance and used to make comparisons with isolates originally used in inoculations.

The experiment was repeated with the exception that the two Fs isolates with high levels of TBZ resistance were compared to two isolates of Fs susceptible to TBZ. Dry rot incidence and severity were evaluated 2 and 3 months after inoculation.

RESULTS AND DISCUSSION

Background levels of naturally-occurring Fusarium were insignificantly low at each sampling date on the potato tubers used in the study. Therefore, these treatments were deleted in subsequent analyses. Furthermore, random isolations and TBZ screening demonstrated that the dry rot observed during the course of the study was caused by the isolates used during inoculations.

In the first trial, the incidence of dry rot lesions in potato tubers inoculated with Fs isolates resistant to TBZ was significantly higher than in tubers inoculated with TBZ-susceptible isolates (Figure 1A). The severity of dry rot, as determined by the percentage of tuber tissue rotted, was also greater in tubers inoculated with TBZ-resistant Fs isolates (Figure 1B). However, these differences were only statistically significant at the first and last sampling date.

Similar trends were observed in the second trial when only four isolates were used; however, differences were significant only on the second sampling date. These results indicate that the ability of a Fs isolate to resist TBZ may give it a selective advantage in a mixed population of isolates. This could occur if the fungicide is being used on a farm. Differences in isolate aggressiveness, however, may be responsible for increased severity of disease in addition to fungicide resistance. There is, however, no change in mating type, toxin production, vegetative compatibility or ability to grow *in vitro* as a result of TBZ resistance (Desjardins, et al. 1993).

Few differences were detected in the development of dry rot due to method of TBZ application in the first trial (Figure 2). The incidence and severity of dry rot was not significantly different between tubers treated with TBZ by dipping or by ultra-low volume spray, regardless of sampling date. These results were obtained despite the fact that residue on the periderm of tubers dipped in TBZ (38-42 mg/l) was three times higher than tubers sprayed with TBZ at ultra-low volumes (11-14 mg/l). These results were duplicated in the second trial. This was interpreted to mean that the recent increase of dry rot in stored potatoes is not due to poor coverage of TBZ fungicide applied at ultra-low volumes.

It was observed that the incidence and severity of dry rot in untreated potato tubers was generally not significantly different from TBZ-treated tubers in the first trial (Figure 2). This was unexpected and no explanation can be given for these results. However, in the second trial, the incidence and severity of dry rot in untreated tubers was always significantly higher than in TBZ-treated potato tubers.

It is obvious from these studies that application of TBZ can no longer be relied upon exclusively for the control of Fusarium dry rot in stored potatoes. An integrated approach including wound reduction of potato tubers during harvest and storage conditions that promote wounding healing must be used to effectively manage this disease.

Figure 1. Incidence and severity of dry rot caused by isolates of Fusarium sambucinum differing in response to TBZ.

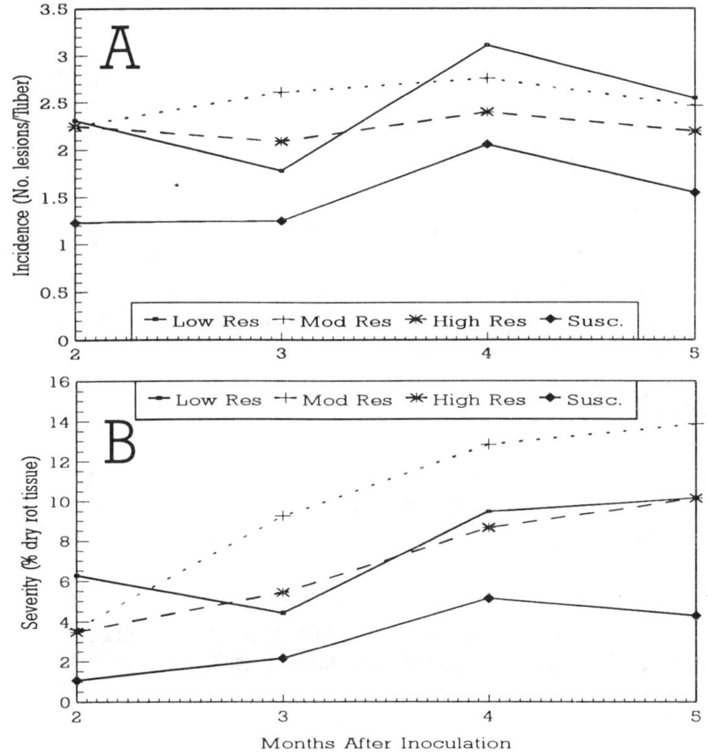

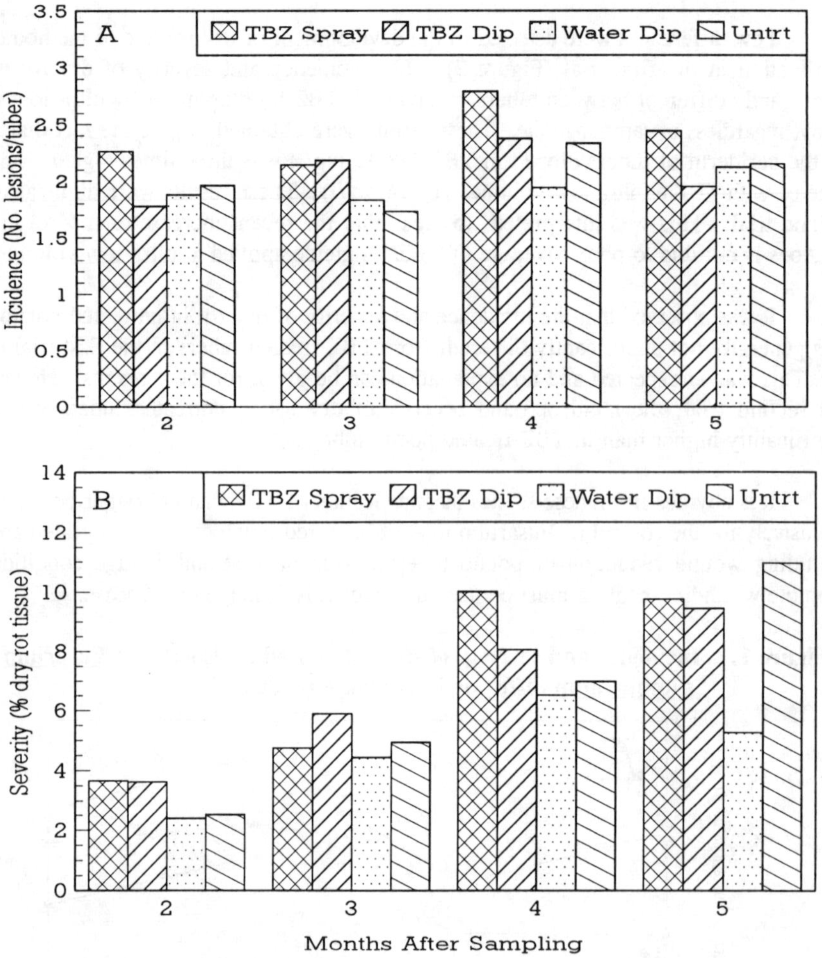

Figure 2. **Effect of post-inoculation treatment; untreated, water dip or TBZ applied by ultra-low volume spray or by dipping, on the incidence and severity of dry rot at four sampling dates.**

REFERENCES

Desjardins, A. E.; Christ-Harned, E. A.; McCormick, S. P. And Secor, G. A. 1993 Population structure and genetic analysis of field resistance to thiabendazole in Gibberella pulicaris from potato tubers. Phytopathology, **83**, 164-170.

Hall, S. M. and Hide, G. A. 1992 Fungicide treatment of seed tubers infected with thiabendazole-resistant Helminthosporium solani and Polyscytalum pustulans for controlling silver scurf and skin spot on stored progeny tubers. Potato Research, **35**, 143-147.

Session 4
Morpholines

Chairman J GILMOUR

Session Organiser D W HOLLOMON

DO MORPHOLINE FUNGICIDES SELECT FOR RESISTANCE?

D.W. HOLLOMON

Department of Agricultural Sciences, University of Bristol, Institute of Arable Crops Research, Long Ashton Research Station, Long Ashton, Bristol BS18 9AF, UK.

ABSTRACT

Morpholine fungicides are considered as low risk for the development of resistance and have been used for more than 25 years without performançe difficulties. Fundamental studies on both the morpholine and piperidine fungicides in this group revealed that they are multisite inhibitors of sterol biosynthesis. Resistance can be generated in the laboratory, but is under polygenic control. Variation in sensitivity in field populations can be detected by a variety of assay techniques, and some selection for reduced sensitivity occurs following treatment. Although seasonal short-term changes in morpholine sensitivity can occur, evidence for a long-term decline in sensitivity is not convincing, and no loss of field performance has been reported.

INTRODUCTION

Morpholines are an important group of systemic fungicides especially useful in controlling powdery mildew diseases on a wide range of crops. Chemically, they are somewhat diverse and include both morpholines (tridemorph, dodemorph, aldimorph, fenpropimorph, Figure 1), and piperidines (piperalin, fenpropidin). The first morpholine introduced was piperalin (Piperonil) in 1960 for control of rose powdery mildew (*Sphaerotheca pannosa*) in protected rose production; it is still in use today without any apparent loss in efficacy (Arnold W., personal comm., 1994). Rose mildew was also the target for dodemorph (Meltatox), but tridemorph (Calixin, Bardew), fenpropimorph (Corbel, Mistral) and fenpropidin (Patrol) have all been used extensively for cereal mildew control. Apart from one isolated report of tridemorph resistance (Walmsley-Woodward *et al.*, 1979) the field performance of morpholines has remained good. Consequently, morpholines are important components, often at reduced dose rates, in mixtures designed to combat the spread of resistance to the sterol 14 demethylase inhibiting (DMI) fungicides (Heaney *et al.*, 1988).

A strong feature of morpholine action against powdery mildews is their rapid "knockdown" action, but, in Northern France especially, there have been suggestions that this action has weakened and more frequent applications, particularly of fenpropimorph, are now needed to achieve a high level of control of wheat powdery mildew (*Erysiphe graminis* f.sp. *tritici*). In addition, there have been reports of shifts in sensitivity to fenpropimorph in France (Andrivon *et al.*, 1987), the Netherlands (De Waard, 1992), and Germany and Switzerland (Lorenz *et al.*, 1992). In the absence of new fungicides with novel modes of action, any development of practical resistance to morpholines would be of great concern in disease control. This paper examines factors of morpholine action which influence the risk of resistance occurring, and reviews evidence available before this meeting, that morpholine resistance is a practical problem.

MODE OF ACTION

Despite several false starts, work on *Botrytis cinerea* and *Ustilago maydis* finally established that tridemorph inhibited sterol biosynthesis at two steps in the pathway, $\Delta^{8\text{-}7}$ isomerase and $\Delta^{14\text{-}15}$ reductase (Table 1). These conclusions were later confirmed using cell-free extracts from yeast (See Köller (1992) for review on morpholine action). Even so, the action of tridemorph against the $\Delta^{14\text{-}15}$ reductase was very weak, whilst fenpropidin hardly affected the isomerase. Much depended on the fungus and in barley powdery mildew (*E. graminis* f.sp. *hordei*) not only were the effects on the accumulation of Δ^{8} and Δ^{14} abnormal sterols small, but other sterol changes were observed (Senior, 1991). Inhibition of $\Delta^{24(28)}$ reductase, Δ^{24} transmethylation, and squalene cyclisation have all been identified as steps inhibited by morpholines (Ziogas, *et al.*, 1991; Schneegurt and Henry, 1992), together with possible feedback inhibition of HMG-CoA reductase. A common feature of these target sites is that all involve high energy carbocationic intermediates, which are mimicked by the morpholines because they are protonated at the ring N at physiological pH. Although the sensitivity of these steps may differ, since morpholines are applied at high rates (3.75 - 5.0 ml product litre^{-1}; 3.75 g a.i. litre^{-1}; 13 mM a.i.) all sites are probably inhibited under field conditions. Because of this multisite action it is unlikely that resistant mutants will arise rapidly. Furthermore, the carbon chain attached to the heterocyclic nitrogen in all morpholines is quite flexible, and can adopt several different configurations, making it difficult for the target site protein to alter to exclude the fungicide.

GENETICS OF RESISTANCE

It is not difficult to generate laboratory mutants resistant to morpholine fungicides, and especially fenpropimorph. In both *U. maydis* (Erg 40, Walsh and Sisler, 1982) and yeast (Parks, personal comm., 1994) these do not involve mutations in the target site genes. The practical significance of these mutations is not clear since they show cross resistance with DMI fungicides, a characteristic generally lacking in DMI resistant field strains. Molecular genetic techniques have not only highlighted the possible uniqueness of tridemorph, but have questioned the involvement of $\Delta^{8\text{-}7}$ isomerase in its action. We have generated targeted mutations in *Neurospora crassa* which are sensitive to tridemorph, but are quite resistant to fenpropimorph and fenpropidin (Hollomon, unpublished observation). Furthermore, an *Ustilago maydis* mutant lacking $\Delta^{8\text{-}7}$ isomerase activity was able to grow almost as well as its wild-type parent (James *et al.*, 1992), although these experiments provided no measure of the effect of $\Delta^{8\text{-}7}$ isomerase mutations on pathogenicity. Nevertheless, genetic studies with *Nectria haematococca* var. *cucurbitae* indicated polygenic control of fenpropimorph resistance which would reduce the risk of resistance occurring in field populations (Demakopoulou *et al.*, 1989).

FIGURE 1. Morpholine and piperidine fungicides used in agriculture.

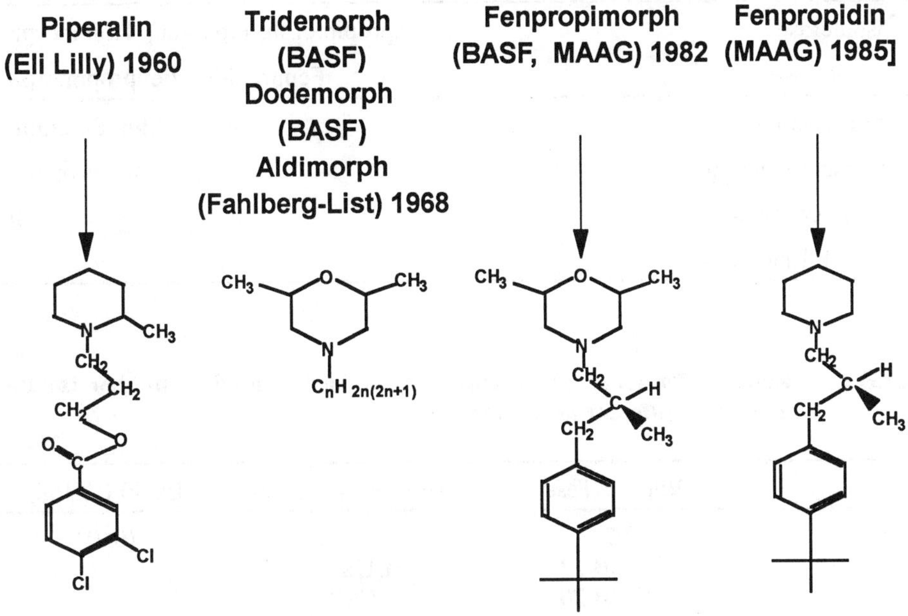

FIGURE 2. Changes in fenpropidin and triadimenol sensitivity in barley powdery mildew 1973-1993.

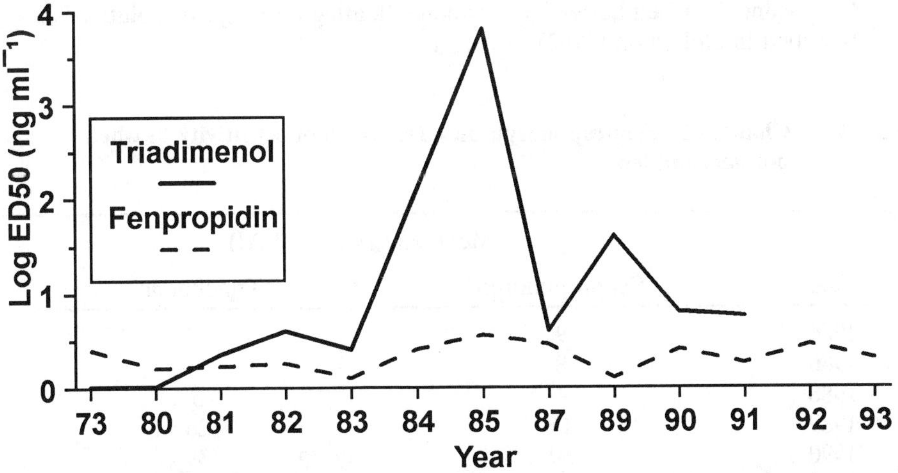

TABLE 1. Inhibition of sterol biosynthesis by morpholines

Δ^{8-7} isomerase	Fenpropimorph, Piperalin, Tridemorph
Δ^{14-15}reductase	Fenpropidin, Fenpropimorph
$\Delta^{24(28)}$ reduction	Fenpropidin, Piperalin
Δ^{24} transmethylation	Fenpropidin
Squalene cyclisation	All
Feed-back Inhibition	?

TABLE 2. Ranking for fenpropidin sensitivity of barley powdery mildew isolates using two different assay methods

	Vapour Test[1]	Detached Leaf Test[2]	ED50 (μg/ml)
Sensitive	L32	1641	(0.01)
	BUSB 22	BUSB 22	
	BUSB 20	CSB 2	
	23D5	BUSB 20	
	CSB 2	23D5	
	DH14	DH14	
Resistant	1641	L32	(0.14)

1 Assay as described by Readshaw and Heaney (1994) except that leaf segments were placed on agar. Fenpropidin was applied to leaf segments.

2 Assay using detached barley leaf segments floating on fungicide solutions as described in Hollomon (1982).

TABLE 3. Changes in Fenpropimorph and Triadimenol sensitivity in wheat powdery mildew.

	Mean ED_{98} (μg ml^{-1} AI)	
Year	Fenpropimorph	Triadimenol
1984	9	22
1986	8	14
1988	4	5
1989	15	26
1990	30	39
1991	9	93

Data from Lorenz and Pommer (1984); Lorenz *et al.*, (1992)

DETECTION OF VARIATION IN SENSITIVITY

Several different assay procedures have been used to measure morpholine sensitivity. Most focussed on cereal powdery mildews, and seedlings have been sprayed, sometimes with upto 14 different dose rates to give meaningful dose/response relationships. Inconsistences in foliar applications arising through different orientations of leaves in relation to the spray nozzle, may be reduced by using root drenches instead. Other monitoring procedures have involved inoculation of treated leaf pieces, placed either on an agar surface or on fungicide containing solutions. All morpholines have some vapour action and this has been exploited for monitoring cereal powdery mildews (Readshaw and Heaney, 1994).

Considerable variation is a component of all assay methods. Brown *et al.* (1991) used Principal Component Analysis (PCA) to overcome variability within seedling assays, but this is generally seen as a preliminary statistical procedure requiring further analysis before conclusions can be drawn about differences between sensitivity classes (Arnold, G.M., personal comm., 1993). Direct comparisons between dose rates used in the field and those applied in the laboratory can be misleading, but dose rates generally used in laboratory and greenhouse assays are well below those recommended for field use.

To allow comparisons between test occasions, standard strains of known sensitivity should be included in all assays. Whilst standard strains of known sensitivity to DMIs have given similar rankings in "ring tests", carried out in different laboratories, the same is apparently not the case for standards used in morpholine tests. In our laboratory, tridemorph resistant isolates identified by Walmsley-Woodward *et al.*, (1979) were ranked no different from other isolates (Hollomon, unpublished data) whilst standard strains CC1, CC139 and CC151 (Brown *et al.*, 1991) were within the wild-type variation encountered in natural populations of barley powdery mildew in Scotland (Zziwa, personal comm., 1993). Choice of the wild-type standard sensitive strain must reflect the mean sensitivity of the whole population, and not just the extreme sensitive end of that variation. Fenpropidin sensitivity determined by the vapour test did not rank strains in the same order as an assay using detached leaf segments (Table 2) floating on fenpropidin solutions. This emphasises that morpholine vapour action may not be identical to that exerted on the pathogen through systemic entry from leaf tissue. Despite these limitations, differences in morpholine sensitivity between isolates clearly exist, and can be detected by several different assay procedures. The range of variation in natural cereal mildew populations may well exceed 50-fold.

TABLE 4. Fungicide efficacy: Spring Barley (1980-1992)

	Mean Disease Control (%)			
	Bayfidan	Tilt	Corbel	Patrol
1980	95*	83	83	-
1984	69	45	79	-
1985	55	20	73	70
1988	21	-	70	66
1990	80	85	66	68
1992	75	82	73	76

* Bayleton

Barley powdery mildew

A collection of *E. graminis* f.sp. *hordei* strains have been maintained in our laboratory since 1973. Data are available, therefore, for fenpropidin sensitivity of strains isolated from the field between then and 1993. Although the numbers tested varied from over 300 strains in 1984 to 10 in 1988, and variation associated with the mean ED_{50} for each year differed, no significant changes in fenpropidin sensitivity were observed throughout this period (Figure 2). By contrast, changes in DMI sensitivity were easily identified in the same strains. Whilst there have been differences each year in the mean sensitivity to fenpropimorph of barley mildew isolates collected in Scotland, no consistent decline in sensitivity was observed over the period 1988-1992 (Zziwa and Burnett, 1994). This contrasts with results of other surveys carried out in Scotland, where significant differences in fenpropimorph sensitivity were detected between 1988 and 1990. But this may simply reflect regional differences arising from clonal selection for reduced sensitivity (Brown *et al.* 1991; Brown and Evans, 1992). Changes in morpholine sensitivity have also been detected within field experiments (Brown and Evans, 1992), but were not correlated with performance changes and may just represent the sensitivity of the surviving small population.

Wheat powdery mildew

A decline in fenpropimorph sensitivity in *E. graminis* f.sp. *tritici* was observed in several parts of Europe between 1988 to 1990 (DeWaard, 1992; Lorenz *et al.*, 1992; Felsenstein, 1991). Changes were not large when compared to the decline in DMI sensitivity that occurred at the same time (Table 3). As with barley powdery mildew, changes in sensitivity appear seasonal. Any decline in fenpropimorph sensitivity observed in Germany in 1990 was reversed the following year (Table 3), when the mean sensitivity of the population returned to the level in 1984. These shifts in fenpropimorph sensitivity were not, it seems, correlated with field performance.

FIELD PERFORMANCE

Yearly variation in field and crop conditions make it difficult to evaluate long term changes in fungicide performance. Because of the vapour activity of morpholines, temperature at the time of spraying can very much influence performance. Comparisons based on the same cultivar and disease levels over several years are seldom available, but the results in Table 4 attempt to combine performance data against barley powdery mildew from a number of ADAS and Long Ashton Spring barley field trials, carried out between 1980 and 1992. Conditions favoured good mildew control in 1980, but since then there has been no dramatic fall in the performance of fenpropimorph or fenpropidin. Evidence for tridemorph is more fragmentary, but again shows no clear change in performance since Calixin was introduced some 25 years ago. This contrasts with the decline in the performance of early triazoles (DMIs) against cereal mildews, which is easily seen alongside the morpholine results (Table 4). In France, it seems that the performance of fenpropimorph against wheat powdery mildew may have weakened in some regions (Maumené, personal comm., 1994), but not that of fenpropidin. With the exception of black Sigatoka disease (*Mycosphaerella fijiensis*) on bananas (Cronshaw *et al.*, 1994), for other diseases where morpholines play a part in disease control strategies, long term performance results are not generally available.

DISCUSSION AND CONCLUSION

Morpholine fungicides have always been considered a low risk for resistance. In part this is based on practical experience, since both piperalin and tridemorph have been used now for more than 25 years without noticeable changes in performance. They have certainly been used in situations where resistance to other groups of fungicides, i.e. benzimidazoles, hydroxypyrimidines and triazoles, has developed rapidly. Work on mode of action has identified morpholines as multisite inhibitors of sterol biosynthesis, and highlighted differences between fenpropidin/fenpropimorph and tridemorph, which are reflected in cross-resistance patterns. Laboratory mutants resistant to morpholines are easy to generate in several fungi, indicating that the biochemistry required to overcome the effects of morpholines is possible. But the exact mechanisms of resistance are unknown, although morpholines are bulky molecules with flexible side chains which may be difficult to accommodate through a change in the target site protein.

It is not surprising that variation in the sensitivity of plant pathogens to morpholines should exist. Despite limitations, several assay procedures have been used to chart the range of this variation, although unfortunately no base-line sensitivity data were collected before morpholines were introduced. Selection of this variation leading to an increase in frequency of field resistant strains is clearly possible, but evidence that changes have caused a decline in field performance is less clear. Indeed, no cases of practical resistance have been reported in the literature for any morpholine fungicide. Field rates may simply be sufficient to control populations with reduced sensitivity. Furthermore, where there have been suggestions that mildew control has not lasted so long after spraying, these reports have centred around fenpropimorph and not fenpropidin, although laboratory studies show cross resistance exists between these two fungicides. It is fortunate that practical resistance to morpholines has not occurred, otherwise control of cereal powdery mildews would be difficult without new fungicides, with novel modes of action to replace them.

ACKNOWLEDGEMENTS

The author wishes to thank the many colleagues who have helped shape the thinking behind this article.

REFERENCES

Andrivon, D.; Limpert, E.; Felsenstein, F.G. (1987) Sensibilité au triadiménol et au fenpropimorphe de populations françaises d'*Erysiphe graminis* DC f.sp. *hordei* Marchal. *Agronomie*, **7**, 443-446.

Brown, J.K.M.; Slater, S.E.; See, K.A. (1991) Sensitivity of *Erysiphe graminis* f.sp. *hordei* to morpholine and piperidine fungicides. *Crop Prot.*, **10**, 445-454.

Brown, J.K.M.; Evans, N. (1992) Selection on responses of barley powdery mildew to morpholine and piperidine fungicides. *Crop Prot.*, **11**, 449-457.

Cronshaw, K.; Lorenz, G.; Mappes, D. (1994) Monitoring results of *Mycosphaerella fijiensis* to tridemorph. *These Proceedings*.

Demakopoulou, M.; Ziogas, B.N.; Georgopoulos, S.G. (1989) Evidence for polygenic control of fenpropimorph resistance in laboratory mutants of *Nectria haematococca* var. *cucurbitae*. *ISPP Chemical Control Newsletter*, **12**, 34.

DeWaard, M.A. (1992) Fungicide resistance strategies in winter wheat in the Netherlands. In: *Resistance 91: Achievements and Developments in combating pesticide resistance*. Eds. I. Denholm; A.L. Devonshire; D.W. Hollomon, Elsevier, London, 48-60.

Felsenstein, F.G. (1991) Virulenz und Fungizidsensitivität des Weizenmehltaus, *Erysiphe graminis* DC. f.sp. *tritici* Marchal in Europa. PhD Thesis, Technische Universität München, Feising-Weihenstephan.

Heaney, S.P.; Martin, T.J.; Smith, J.M. (1988) Practical approaches to managing antiresistance strategies with DMI fungicides. *Proc. Brighton Crop Prot. Conf. - Pests and Diseases - 1988*, 1097-1106.

Hollomon, D.W. (1982) The effects of Tridemorph on Barley Powdery Mildew: its mode of action and cross-sensitivity relationships. *Phytopath*. Z. **105**, 279-287.

James, C.S.; Burden, R.S.; Loeffler, R.S.T.; Hargreaves, J.A. (1992) Isolation and characterisation of polyene-resistant mutants from the maize smut pathogen, *Ustilago maydis*, defective in ergosterol biosynthesis. *J. Gen. Microbiol.*, **138**, 1437-1443.

Köller, W. (1992) Antifungal agents with target sites in sterol function and biosynthesis. In: *Target sites of fungicide action*. Ed. W. Köller. CRC Press, Boca Raton, 119-206.

Lorenz, G.; Pommer, E.-H. (1984) Investigations into the sensitivity of wheat powdery mildew populations towards fenpropimorph (monitoring programme). *Proc. 1984 Brit. Crop Prot. Conf. - Pests and Diseases*, 489-493.

Lorenz, G.; Saur, R.; Schelberger, K.; Forster, B.; Küng, R.; Zobrist, P. (1992) Long term monitoring results of wheat powdery mildew sensitivity towards

fenpropimorph and strategies to avoid the development of resistance. *Proc. Brighton Crop Prot. Conf. - Pests and Diseases - 1992*, 171-176.
Readshaw, A.E.; Heaney, S.P. (1994) Fenpropimorph sensitivity - *Erysiphe graminis* f.sp. *tritici* survey in Northern France. *These Proceedings*.
Schneegurt, M.A.; Henry, M.J. (1992) Effects of Piperalin and Fenpropimorph on sterol biosynthesis in *Ustilago maydis*. *Pestic. Biochem. Physiol.*, **43**, 45-52.
Senior, I.J. (1991) The mode of action and mechanisms of resistance to triazole fungicides in *Erysiphe graminis*. PhD Thesis, University of Bristol, pp.227.
Walmsley-Woodward, D.J.; Laws, F.A.; Whittington, W.J. (1979) The characteristics of isolates of *Erysiphe graminis* f.sp. *hordei* varying in response to tridemorph and ethirimol. *Ann. Appl. Biol.* **92**, 211-219.
Walsh, R.C.; Sisler, H.D. (1982) A mutant of *Ustilago maydis* deficient in C-14 demethylation: characteristics and sensitivity to inhibitors of ergosterol biosynthesis. *Pestic. Biochem. Physiol.*, **18**, 121-131.
Ziogas, B.N., Oesterfelt, G.; Masner, P.; Steel, C.C.; Furler, R. (1991) Fenpropimorph: A three site inhibitor of Ergosterol biosynthesis in *Nectria haematococca* var. *cucurbitae*. *Pestic. Biochem. Physiol.*, **39**, 74-83.
Zziwa, M.C.N.; Burnett, F.J. (1994) The effect of reduced doses on the sensitivity of powdery mildew to fenpropimorph in barley field trials. *These Proceedings*.

THE GENETICS OF THE RESPONSE OF BARLEY MILDEW TO MORPHOLINE AND PIPERIDINE FUNGICIDES

JAMES K.M. BROWN

Cereals Research Department, John Innes Centre, Colney Lane, Norwich, NR4 7UH, England

ABSTRACT

Responses of the barley powdery mildew pathogen, *Erysiphe graminis* f.sp. *hordei*, to three fungicides were studied. Two of these chemicals were morpholines, fenpropimorph and tridemorph, and a third, fenpropidin, was a piperidine. Three classes of *E.g.* f.sp. *hordei* isolate were found. One type was sensitive to all three chemicals, another was resistant to fenpropidin and fenpropimorph, while a third had further resistance to fenpropimorph and also some resistance to tridemorph. The levels of resistance observed were not high enough to have caused serious failures of mildew control. The use of either fenpropidin or fenpropimorph in field trials selected isolates which were resistant to both fungicides, while tridemorph selected against fenpropidin resistance. A single gene controlled responses to these three fungicides, but had no effect on the response to a demethylation inhibitor (DMI) fungicide, triadimenol.

INTRODUCTION

Morpholine and piperidine fungicides have been widely used to control powdery mildew of barley (*Erysiphe graminis* f.sp. *hordei*). The most important of these chemicals are a piperidine, fenpropidin, and two morpholines, fenpropimorph and tridemorph. All three inhibit the synthesis of ergosterol from lanosterol (Baloch *et al.*, 1984).

Reduced sensitivity of *E.g.* f.sp. *hordei* to tridemorph was first reported by Walmsley-Woodward *et al.* (1979), and to fenpropimorph by Wolfe *et al.* (1987). Following continuing reports of diminishing sensitivity (Wolfe *et al.*, 1988), more extensive surveys of the responses of *E.g.* f.sp. *hordei* to these fungicides, and to fenpropidin, were undertaken in 1988 and 1990. This paper reviews this work. Three aspects are considered: (i) the phenotypes of resistance to the three fungicides; (ii) selection for resistance in field trials; (iii) the genetic control of responses to these fungicides.

RESISTANCE AND CROSS-RESISTANCE

In 1988, a national survey of *E.g.* f.sp. *hordei* isolates was undertaken. This was described in detail by Brown *et al.* (1991b). Random samples of airborne spores were collected by a wind-impaction spore trap mounted on a car (Wolfe *et al.*, 1981). In England, samples were obtained from Lincolnshire, Cambridge, Suffolk and Essex. In Scotland, samples were collected along a route which went north from Newcastle to Edinburgh, Perth and Aberdeen, then made a circular tour through Banffshire and Moray, and then returned from Aberdeen to Newcastle.

Single colony isolates were tested for their responses to fenpropidin, fenpropimorph and tridemorph. Isolates from England showed no significant variation in responses to these fungicides, all being similar to isolates collected before morpholines and piperidines came into widespread use.

Isolates from Scotland, however, were polymorphic. In tests of responses to fenpropidin, all isolates fell into one of two distinct groups, and were thus classified as either sensitive or resistant. Responses to fenpropidin were strongly correlated with those to fenpropimorph, in that all isolates which were resistant to fenpropidin had a higher level of resistance to fenpropimorph than fenpropidin-sensitive isolates did. In all, 39 out of 73 Scottish isolates were resistant.

Sensitive isolates, whether from England or Scotland, did not differ significantly in their responses to tridemorph from those of most of the resistant isolates. However, four isolates, all from Moray, showed a three-fold increase in resistance to fenpropimorph over that of the rest of the fenpropidin-resistant class. Furthermore, the sensitivity of these four to tridemorph was halved, compared to that of most other isolates. Three of these four samples were studied further, by virulence tests of ten differential varieties of barley. All of them proved to have the same virulence phenotype. This suggests that this group was genetically homogeneous, and was perhaps a clone.

Three phenotypes of *E.g.* f.sp. *hordei* were therefore observed. One, sampled in both England and Scotland, had baseline sensitivity to fenpropidin, fenpropimorph and tridemorph (phenotype S). Another, found in Scotland, had resistance to fenpropidin and fenpropimorph, but not tridemorph (phenotype R). A third, only found in Moray, in north-east Scotland, had a further increase in fenpropimorph resistance and also had some tridemorph resistance (phenotype RM). However, the resistance of types R and RM were sufficiently low that barley mildew would still have been controlled by either fenpropidin or fenpropimorph.

SELECTION ON RESISTANCE

From the results of the survey in 1988 (Brown *et al.*, 1991b), it was predicted that, in the field, fenpropidin and fenpropimorph would select for resistance to each other, while neither would select for resistance to tridemorph, and that tridemorph would not select for resistance to fenpropidin or fenpropimorph. These predictions were tested in 1990. The experiments were described by Brown & Evans (1992).

Six field trials were carried out, two in England and four in Scotland. In each trial, thirteen different spray treatments were applied to 12m × 3.2m plots in a field of a mildew-susceptible barley cultivar, Golden Promise. These were replicated in two blocks at each site, and the full trial of 26 plots was replicated at the six locations. In each block, one plot was sprayed with water, and four each were sprayed with fenpropidin (Patrol), fenpropimorph (Corbel) or tridemorph (Calixin). For each fungicide, the four treatments were applied at the full recommended rate, or at a half, a quarter or an eighth of that rate. (The full rates are 1.0 l ha^{-1} of Patrol and Corbel and 0.7 l ha^{-1} of Calixin, with a concentration of 750 g l^{-1} of the active compound in each case. Fungicides were applied in 250 l ha^{-1} water.)

Samples of *E.g.* f.sp. *hordei* were obtained from each site except for one of the English ones, which was too lightly infected with mildew for sampling to be worthwhile. Up to 20 single colony isolates were obtained from each plot. Isolates from untreated plots at three sites in Scotland, and at the remaining English site, were put through a preliminary screen for resistance to fenpropimorph. All of the Scottish isolates, and some English isolates, were resistant. Clearly, the frequency of resistance had increased between 1988 and 1990.

Isolates from a site in north-east Scotland, Banff, and from one in eastern England, Levington, in Suffolk, were studied more intensively. The doses used to analyse the 1990 isolates were more closely spaced, and covered a greater range, than those used in 1988. Also, a numerical scale, the minimum inhibitory concentration, was used to analyse responses in 1990, whereas most of the analysis of the 1988 results was done by simply classifying the responses of isolates to each fungicide as resistant or sensitive. More detailed comparisons could therefore be made between responses to the different fungicides. The results are shown in detail in Figure 1.

No polymorphism was observed at Banff, all isolates tested being resistant to both fenpropidin and fenpropimorph. At Levington, as in 1988, there was a strong correlation between responses to fenpropidin and fenpropimorph, such that isolates could be classified as resistant to both, or as sensitive.

The mean resistance to fenpropidin of Banff isolates was the same as that of the resistant isolates from Levington, both being about 25 times more resistant than sensitive Levington isolates.

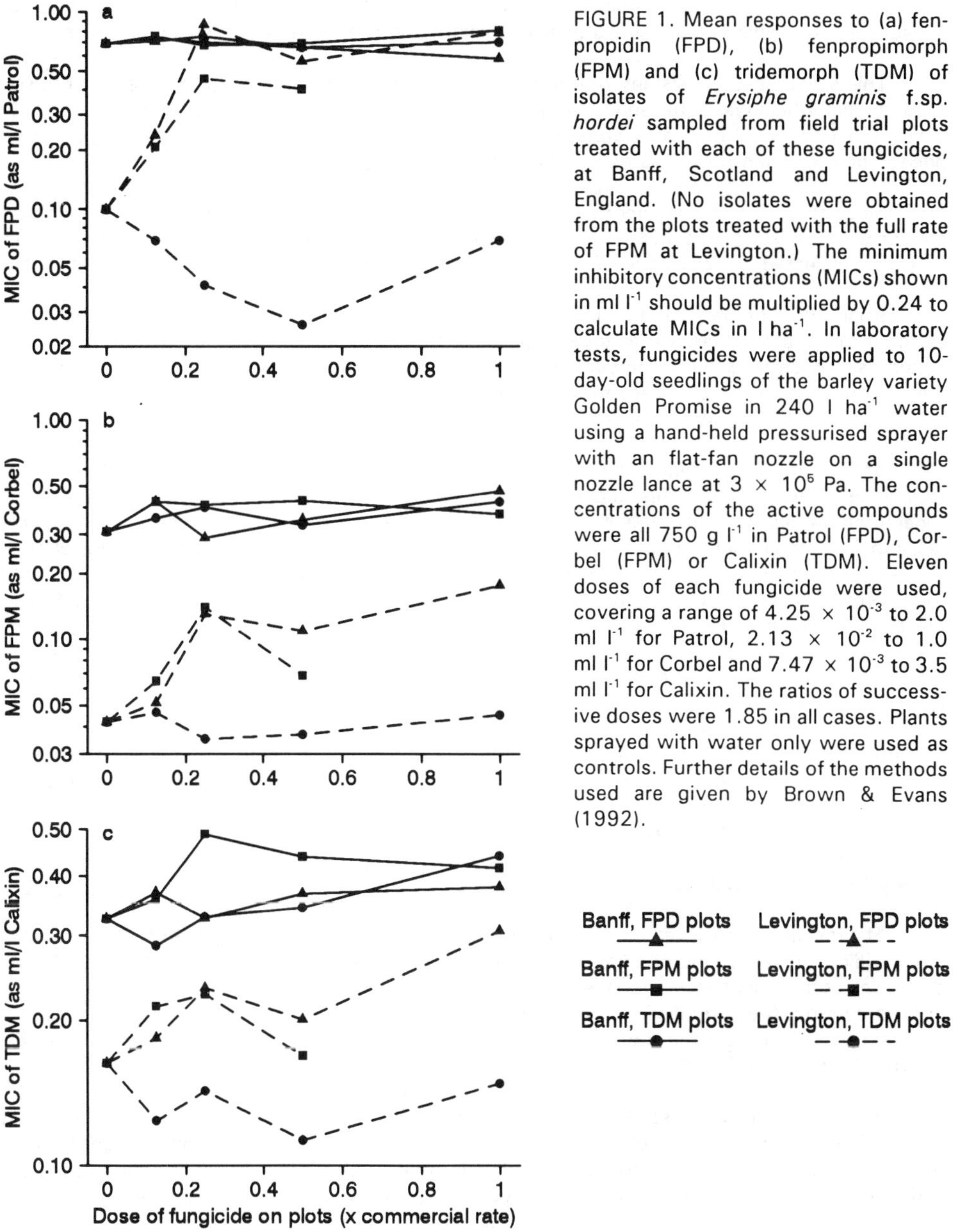

FIGURE 1. Mean responses to (a) fenpropidin (FPD), (b) fenpropimorph (FPM) and (c) tridemorph (TDM) of isolates of *Erysiphe graminis* f.sp. *hordei* sampled from field trial plots treated with each of these fungicides, at Banff, Scotland and Levington, England. (No isolates were obtained from the plots treated with the full rate of FPM at Levington.) The minimum inhibitory concentrations (MICs) shown in ml l^{-1} should be multiplied by 0.24 to calculate MICs in l ha^{-1}. In laboratory tests, fungicides were applied to 10-day-old seedlings of the barley variety Golden Promise in 240 l ha^{-1} water using a hand-held pressurised sprayer with an flat-fan nozzle on a single nozzle lance at 3×10^5 Pa. The concentrations of the active compounds were all 750 g l^{-1} in Patrol (FPD), Corbel (FPM) or Calixin (TDM). Eleven doses of each fungicide were used, covering a range of 4.25×10^{-3} to 2.0 ml l^{-1} for Patrol, 2.13×10^{-2} to 1.0 ml l^{-1} for Corbel and 7.47×10^{-3} to 3.5 ml l^{-1} for Calixin. The ratios of successive doses were 1.85 in all cases. Plants sprayed with water only were used as controls. Further details of the methods used are given by Brown & Evans (1992).

At Levington, isolates which were resistant to fenpropidin were four to five times more resistant to fenpropimorph than sensitive isolates were, while Banff isolates showed a further doubling of fenpropimorph-resistance over the level of the resistant Levington isolates. Furthermore, the Banff isolates were half as sensitive to tridemorph as the resistant Levington isolates. The Banff isolates varied in their responses to fenpropimorph and tridemorph, some being repeatably more resistant than others. All but one of the twelve most resistant Banff isolates had the same set of virulences as the three more fenpropimorph-resistant isolates from Moray sampled in 1988, the odd one differing by only a single virulence. The twelve least resistant isolates from Banff, however, had diverse virulence phenotypes, as did twelve resistant and twelve sensitive isolates from Levington.

This evidence is consistent with most *E.g.* f.sp. *hordei* at Banff being members of the same clone as that which was sampled in Moray in 1988, having phenotype RM, but with a small fraction having phenotype R. It is also consistent with resistant *E.g.* f.sp. *hordei* at Levington having phenotype R and sensitive isolates being of type S.

Since there was no polymorphism for responses to fenpropidin or fenpropimorph at Banff, neither fungicide selected for resistance either to itself or to the other. At Levington, however, both fungicides selected for resistance to each other, since responses to the two were highly correlated.

Application of each of the three fungicides, fenpropidin, fenpropimorph and tridemorph, selected for greater resistance to tridemorph at Banff, although the increase was not large. If the hypothesis concerning the phenotypes of *E.g.* f.sp. *hordei* at Banff, described above, is correct, the pattern of selection for responses to tridemorph is consistent with type RM isolates being selected in plots treated with higher doses of fungicide.

At Levington, treatment with fenpropidin and fenpropimorph also selected weakly for increased resistance to tridemorph. This suggests that type R isolates may have reduced sensitivity to tridemorph. Application of tridemorph, however, had the unexpected effect of selecting *against* resistance to fenpropidin and, to a lesser extent, fenpropimorph and tridemorph. Selection for increased sensitivity was most marked in plots treated with a quarter or a half of the full dose of tridemorph. This suggests that the fitness of type R isolates is lower than that of type S in the presence of tridemorph.

No higher level of resistance was observed in 1990 than in 1988, although resistance had spread to England and was evidently much more common in Scotland. However, the results from the tridemorph-treated plots at Levington suggested an intruiging possibility for crop protection, in that an early spray with tridemorph might select out type R isolates, and thus improve the efficacy of a later spray with fenpropidin or fenpropimorph. Clearly, a proposition such as this, based on a single trial, must be followed up by proper tests in the field.

GENETICS OF RESISTANCE

The results of Brown *et al.* (1991b) and Brown & Evans (1992) suggested that resistance to fenpropidin and fenpropimorph might be controlled by the same gene or genes, while the additional decrease in sensitivity to fenpropimorph and fenpropidin in type RM isolates might be under common genetic control. These predictions were tested by analysing the progeny of genetic crosses. The results are outlined here and by Brown *et al.* (1992a), and will be described in detail elsewhere (J.K.M. Brown, S. Le Boulaire, N. Evans, in preparation).

In a cross of a type R isolate of *E.g.* f.sp. *hordei* with one of type S, resistances to fenpropidin and fenpropimorph co-segregated in all 55 progeny tested. There was no significant correlation of the responses to these two fungicides with that to tridemorph. The segregation ratio of resistance and sensitivity was not simple, however, since there was a large excess of resistant progeny. It is not known whether this was because resistance was controlled by a more complex genetic system than a single gene, or because factors affecting fitness were linked to the resistance gene.

Resistance to fenpropidin and to fenpropimorph also co-segregated in a cross of a type RM isolate with the same type S isolate. Resistant progeny were more resistant to tridemorph than sensitive progeny were, and were also more resistant to fenpropimorph than were resistant progeny of the type R × type S cross. The segregation ratio was not significantly different from 1:1, and was therefore consistent with the responses to all three fungicides being controlled by a single gene. This gene was not linked to one controlling responses to a sterol C14-demethylation inhibitor (DMI) fungicide, triadimenol.

DISCUSSION

Brown *et al.* (1991b) and Brown & Evans (1992) showed that resistance to morpholine-type fungicides had developed in the British population of *E.g.* f.sp. *hordei* between 1986 and 1990. The resistance of R and RM isolates was considerably greater than that of any isolate collected before 1986 (Wolfe *et al.*, 1987).

It has also been shown that a piperidine fungicide, fenpropidin, falls into the same cross-resistance group as a morpholine, fenpropimorph. However, a second morpholine, tridemorph, displays little cross-resistance with these two. The pattern of cross-resistance may be associated with the mode of action of these compounds. Although in yeast, all three inhibit two steps in sterol biosynthesis, tridemorph inhibits $\Delta^8 \rightarrow \Delta^7$ isomerisation better than Δ^{14} reduction, while the reverse is true for fenpropidin and, to a lesser extent, for fenpropimorph (Baloch *et al.*, 1984). It can be hypothesised that type R and RM isolates are both less sensitive to inhibition of Δ^{14} reduction than type S isolates are, while type RM isolates also have reduced sensitivity to inhibition of $\Delta^8 \rightarrow \Delta^7$ isomerisation. Despite the complexity of the cross-resistance relationships of these fungicides, responses of *E.g.* f.sp. *hordei* to these chemicals were under simple genetic control in one cross, as are those to ethirimol and to a DMI, triadimenol (Brown *et al.*, 1992b). Whether the genetics of resistance to fenpropidin and fenpropimorph are more complex in the R × S cross is not yet known.

The increase in resistance to fenpropidin and fenpropimorph up to 1990 was probably caused by heavy usage of these fungicides to control mildew. Virulence tests showed that phenotype R was selected in many different genetic backgrounds (Brown & Evans, 1992). Similarly, resistance to triadimenol and to ethirimol occurred in many different clones of *E.g.* f.sp. *hordei* (Brown *et al.*, 1990). This is probably because many clones infect fields of susceptible barley varieties, and are thus selected for increased resistance by applications of morpholine-type fungicides. No definite explanation can be advanced for the high frequency of one particular clone, with phenotype RM, in north-east Scotland (Brown *et al.*, 1991b; Brown & Evans, 1992). However, clones of *E.g.* f.sp. *hordei* have initiated epidemics of mildew on previously resistant varieties (Brown *et al.*, 1990, 1991a). It is possible that type RM mutants are comparatively rare, and that one such clone was selected by the very high use of morpholine-type fungicides on the mildew-susceptible variety Golden Promise, which was widely grown in north-east Scotland.

The future course of resistance to these fungicides in *E.g.* f.sp. *hordei* cannot be predicted. Although successively higher levels of DMI resistance were selected during the 1980s (Brown & Wolfe, 1991), there is no evidence for significant increases in resistance to morpholine-type fungicides since 1990, even from the relatively low levels seen then (Mitchell & Slater, 1994). This may be because isolates with higher levels of resistance are unfit, or, perhaps more likely, because selection for resistance has been relaxed. Since 1990, many more spring barley varieties with an effective mildew resistance gene, *mlo*, have been grown (Mitchell & Slater, 1993), and the use of all fungicides to control barley mildew, including morpholine-type compounds, appears to have fallen markedly.

ACKNOWLEDGEMENTS

I thank Neal Evans, Stéphanie Le Boulaire, Susan Slater and Karen See for their participation in this work. The first piece of work described here was supported by the Home-Grown Cereals Authority and the second and third by BASF plc. The programme of research on cereals mildews at the JIC is supported by the Ministry of Agriculture, Fisheries and Food.

REFERENCES

Baloch, R.I.; Mercer, E.I.; Wiggins, T.E.; Baldwin, B.C. (1984) Inhibition of ergosterol biosynthesis in *Saccharomyces cerevisiae* and *Ustillago maydis* by tridemorph, fenpropimorph and fenpropidin. *Phytochemistry* ***23***, 2219-2226.

Brown, J.K.M.; Evans, N. (1992) Selection on responses of barley powdery mildew to morpholine and piperidine fungicides. *Crop Protection* ***11***, 449-457.

Brown, J.K.M.; Wolfe, M.S. (1991) Levels of resistance of *Erysiphe graminis* f.sp. *hordei* to the systemic fungicide triadimenol. *Netherlands Journal of Plant Pathology* ***97***, 251-263.

Brown, J.K.M.; O'Dell, M.; Simpson, C.G.; Wolfe, M.S. (1990) The use of DNA polymorphisms to test hypotheses about a population of *Erysiphe graminis* f.sp. *hordei*. *Plant Pathology* ***39***, 391-401.

Brown, J.K.M.; Jessop, A.C.; Rezanoor, H.N. (1991a) Genetic uniformity in barley and its powdery mildew pathogen. *Proceedings of the Royal Society of London, Series B* ***246***, 83-90.

Brown, J.K.M.; Slater, S.E.; See, K.A. (1991b) Sensitivity of *Erysiphe graminis* f.sp. *hordei* to morpholine and piperidine fungicides. *Crop Protection* ***10***, 445-454.

Brown, J.K.M.; Evans, N.; Le Boulaire, S.; (1992a) Responses of barley powdery mildew to morpholine and piperidine fungicides. *AFRC Institute of Plant Science Research: 1991 Annual Report*, 6-7.

Brown, J.K.M.; Jessop, A.C.; Thomas, S.; Rezanoor, H.N. (1992b) Genetic control of the response of *Erysiphe graminis* f.sp. *hordei* to ethirimol and triadimenol. *Plant Pathology* ***41***, 126-135.

Mitchell, A.G.; Slater, S.E. (1993) Mildew of barley. *UK Cereal Pathogen Virulence Survey: 1992 Annual Report*, 26-31.

Mitchell, A.G.; Slater, S.E. (1994) Mildew of barley. *UK Cereal Pathogen Virulence Survey: 1993 Annual Report*, in press.

Walmsley-Woodward, D.J.; Laws, F.A.; Whittington, W.J. (1979) The characteristics of isolates of *Erysiphe graminis* f.sp. *hordei* varying in response to tridemorph and ethirimol. *Annals of Applied Biology*, ***92***, 211-219.

Wolfe, M.S.; Minchin, P.N.; Slater, S.E. (1981) Powdery mildew of barley. *Plant Breeding Institute: 1980 Annual Report*, 88-92.

Wolfe, M.S.; Slater, S.E.; Minchin, P.N. (1987) Mildew of barley. *UK Cereal Pathogen Virulence Survey: 1986 Annual Report*, 26-33.

Wolfe, M.S.; Slater, S.E.; Minchin, P.N. (1987) Mildew of barley. *UK Cereal Pathogen Virulence Survey: 1987 Annual Report*, 22-31.

FENPROPIMORPH SENSITIVITY IN *ERYSIPHE GRAMINIS* F.SP. *TRITICI*; SURVEY OF NORTHERN FRANCE 1991-1993.

A.E. READSHAW, S.P. HEANEY

Plant Pathology Section, ZENECA Agrochemicals, Jealott's Hill Research Station, Bracknell, Berks, RG12 6EY.

ABSTRACT

As part of an annual survey of fungicide sensitivity in cereal powdery mildew, 'single-spore' isolates of *Erysiphe graminis* f.sp. *tritici* were obtained from Northern France in 1991, 1992 and 1993, using a car-mounted Jet Spore Trap. A standard technique was used to measure the sensitivity of all the isolates to the morpholine fungicide fenpropimorph. In 1993, approximately 90% of the isolates were found to be less sensitive to fenpropimorph than wild type sensitive isolates, compared with 55% in 1991. Forty-five *E.graminis* f.sp. *tritici* isolates from various sources were tested against fenpropimorph, tridemorph and fenpropidin, to investigate cross-sensitivity relationships between these three fungicides. While there appeared to be no differences in the responses of the isolates to tridemorph, there was evidence to suggest a correlation between sensitivities to fenpropidin and fenpropimorph.

INTRODUCTION

Annual surveys of fungicide sensitivity in wheat powdery mildew in Northern France were carried out by ICI Agrochemicals (now ZENECA Agrochemicals) between 1991 and 1993. Single spore isolates of *Erysiphe graminis* f.sp. *tritici* were collected using mobile 'Jet Spore Traps'. A standard technique was then used at Jealott's Hill to test samples of the isolates for sensitivity to the morpholine fungicide, fenpropimorph.

Experiments were also carried out on a selection of 45 *E. graminis* f.sp. *tritici* isolates, to observe any correlation between their responses to fenpropimorph, fenpropidin (a piperidine) and tridemorph (a morpholine).

MATERIALS AND METHODS

Collection of 'single spore' isolates

Spores of *E. graminis* f.sp. *tritici* were collected from Northern France in early May (1992 and 1993) or June (1991), using car-mounted Jet Spore Traps (Burkard Manufacturing Co. Ltd., Rickmansworth, UK).

In 1991, 20 'single spore' isolates were obtained from a region between Calais, Reims and Amiens (Figure 1). In 1992, 106 isolates were collected from a similar area. In 1993, the survey area was extended, and 203 isolates were obtained from 9 different *départments* of Northern France, along a route of approximately 800 miles. Isolates from the spore trap were subcultured on untreated wheat prophylls until sufficient inoculum was produced to assay sensitivity to fenpropimorph. Most isolates produced sufficient inoculum within three generations.

Figure 1. Map of Northern France showing spore trap routes in 1991, 1992 and 1993.

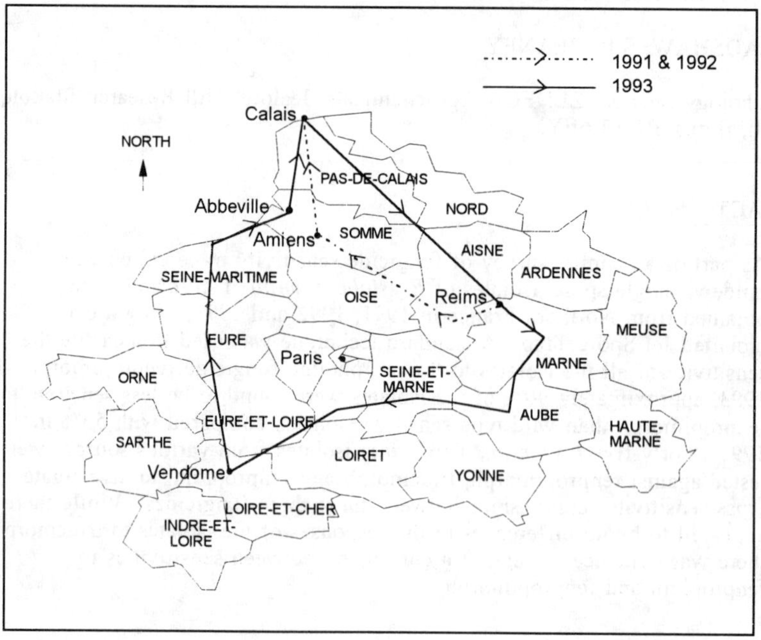

Fenpropimorph sensitivity tests

Test design

The test is designed to exploit the vapour active property of fenpropimorph. Vapour from the fenpropimorph solution in the first row of the test dish diffuses across the dish, inhibiting development of sporulating disease on the leaf pieces at a distance from the vapour source dependent on the fenpropimorph sensitivity of the test isolate. Previous work has shown that the powdery mildew on the leaf pieces in the row of wells furthest from the treated row is unaffected by the fenpropimorph vapour, and therefore acts as an untreated control (figure 2). The test enables the fenpropimorph sensitivities of mildew isolates to be qualitatively compared with one another. Quantitative data, such as estimates of LC50 values or resistance factors are not obtained.

Preparation of test dishes

$2cm^3$ non-sterile de-ionised water was placed in 20 wells of a 25 well plastic test dish. Untreated prophyll pieces (approximately 2.5cm long), were cut from 9 day old wheat seedlings (cv. 'Rapier') grown under constant conditions; (DAY [16h]; 21°C, 60% r.h., 8000 lux: NIGHT; 18°C, 95% r.h.). Prophyll pieces were placed diagonally on the surface of the water in the wells, adaxial surface uppermost. $2cm^3$ aliquots of 5mg/l fenpropimorph solution (made up from 'Mistral'; 75% EC) were added to each well of the remaining row of the dish, immediately prior to inoculation. Leaf pieces, as above, were placed on the surface of this solution.

Figure 2. Fenpropimorph sensitivity test in 25 well plastic dish.

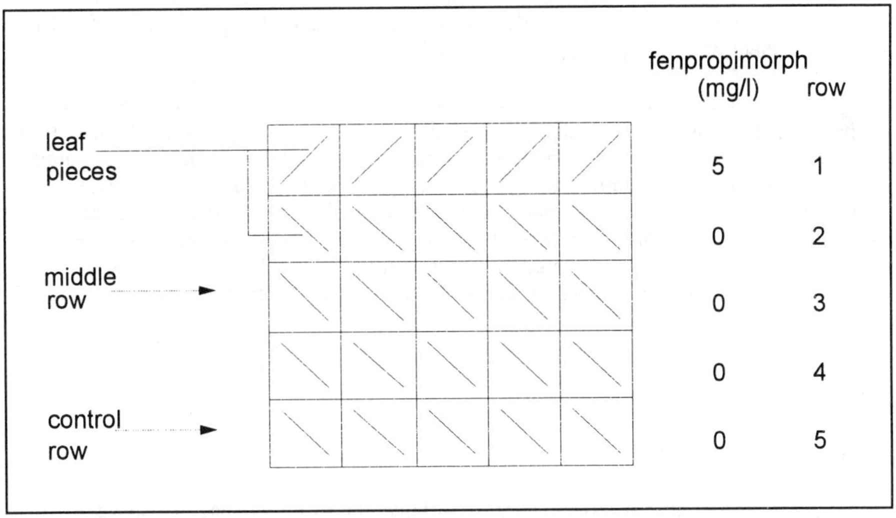

Inoculation of test dishes

Test dishes were inoculated using prophyll pieces (10 per dish) that had themselves been inoculated with spores of the test isolate 7 days previously, and subsequently incubated in a CT (constant temperature) room (DAY [16h]; 21°C, 70% r.h., 8000 lux: NIGHT; 18°C, 95% r.h.). Prepared test dishes were inoculated individually in an extracted fume cupboard, using a small wooden settling tower. Test dishes were sealed with non-porous tape after inoculation. Sealed dishes were immediately placed in the CT room, and incubated for 7 days. Two reference isolates, of known fenpropimorph sensitivity, were included in each batch of isolates tested, to detect any between-test variation.

Assessment of the test and data analysis

After 7 days' incubation, the percentage surface area of each leaf piece covered in sporulating disease was estimated and recorded. An illuminated magnifying glass was used to assist assessment. Percentage values on the middle row (row 3), and on the control row (row 5), were totalled separately. The amount of disease on the middle row was then expressed as a percentage of that on the control row.

This value was used to classify isolates into 3 categories, as follows;-

Category	% of control row covered with sporulating disease
LEAST SENSITIVE	>50
INTERMEDIATE	15-50
SENSITIVE	<15

The results are presented in figure 3.

Cross-sensitivity tests

The sensitivity test technique described above was used to investigate the responses of 45 isolates from various sources to fenpropimorph, fenpropidin and tridemorph. All tests were prepared as described previously, except that either fenpropidin (10mg/l, prepared from Patrol 75% EC), or tridemorph (5mg/l, prepared from Calixin 75% EC) were used instead of fenpropimorph.

Cross-sensitivity tests were inoculated as described previously. The size of the inoculating tower enabled 3 test dishes, treated with either fenpropimorph, fenpropidin or tridemorph to be inoculated simultaneously with a single isolate. Dishes were sealed with non-porous tape after inoculation, and incubated for 7 days in the CT room. The percentage of the surface area of each leaf piece covered with sporulating disease was then estimated by eye and recorded. On the basis of their responses, the isolates were classified as 'least sensitive', 'intermediate' or 'sensitive' to the 3 chemicals. Any correlations between the responses of the 45 isolates to the three chemicals were noted. The results are displayed in figure 4.

RESULTS

Fenpropimorph sensitivity

Figure 3 reveals that the frequency of isolates classified as 'least sensitive' to fenpropimorph increased from approximately 40% in 1991 to approximately 80% in 1993. This is accompanied by a corresponding decrease in the frequency of 'sensitive' and 'intermediate' isolates, suggesting there has been a decrease in the sensitivity of the wheat powdery mildew population in Northern France as a whole. There appeared to be no important regional differences in fenpropimorph sensitivity within the area sampled.

Figure 3. Fenpropimorph sensitivity in Northern France, 1991-1993.

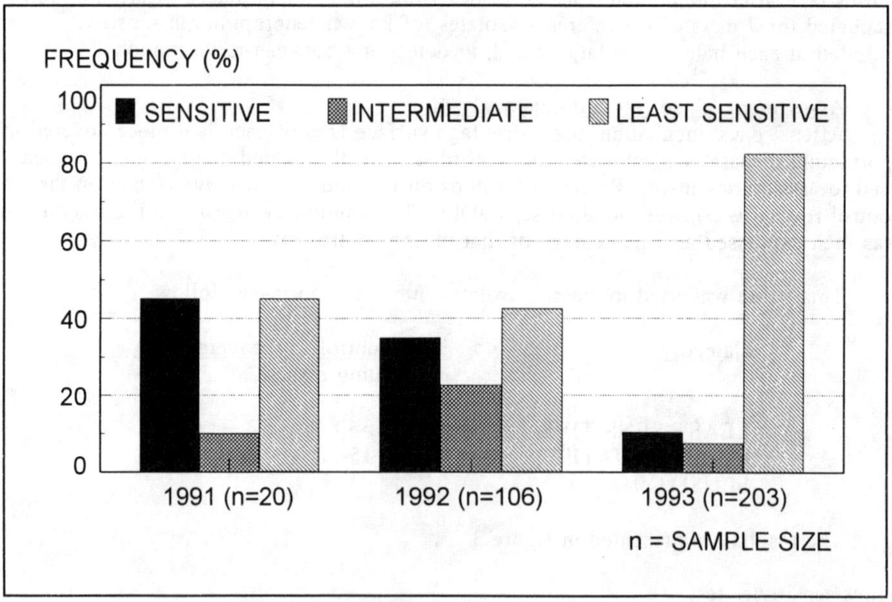

Cross-sensitivity tests

No differences were detected between the responses of any of the 45 isolates to tridemorph. In all tridemorph tests, disease developed to approximately 100% coverage on all leaf pieces on the fourth and fifth rows of the test dish, and none of the 45 isolates produced disease on the third row. Amending tridemorph rates to 25, 10, 3 or 1 mg/l in row 1 of the test dish failed to reveal any differences in dose response between test isolates.

Figure 4. Diagram illustrating responses of 45 isolates to fenpropimorph and fenpropidin.

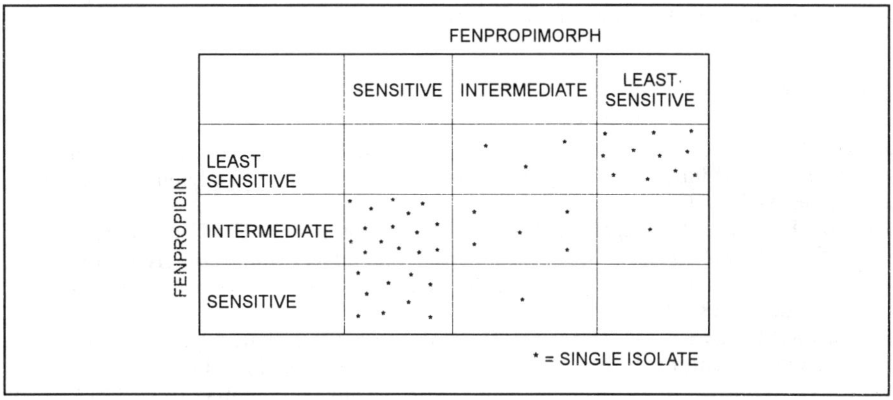

Figure 4 displays the responses of the isolates to fenpropidin and fenpropimorph. It is clear that the distribution of the isolates in this matrix is such that there is a positive correlation between the sensitivities of isolates to these two chemicals. In the absence of such a correlation, the spread of isolates would be random across the grid, with no noticeable clusters in any of the boxes.

CONCLUSIONS

Compared with 1991/1992, the 1993 survey detected a shift towards decreased fenpropimorph sensitivity in the wheat powdery mildew population in all regions of Northern France. The frequency of isolates classified here as 'least sensitive' to fenpropimorph has doubled since 1991. It must be emphasised, however, that no attempt has been made to relate these results to the field performance of fenpropimorph, which remains good (Russell, 1993).

Fenpropimorph acts as a strong inhibitor of sterol Δ^{14} reductase in the sterol biosynthesis pathway. It also has a weaker effect on $\Delta^{8} \rightarrow \Delta^{7}$ isomerase (Baloch *et al*, 1984, Berg *et al*, 1984). As with other sterol biosynthesis inhibitors, the evolution of resistance to morpholines may be a multi-step process, under the control of several genes or alleles. Under continual selection pressure, and in the absence of effective anti-resistance strategies, isolates with increasingly low levels of sensitivity may become more frequent in the population as time progresses. The existence of three different sensitivity classes, as reported here, appears to support this theory. Similar shifts in sensitivity have been reported in previous surveys in Germany and Switzerland (Lorenz *et al*, 1992).

Fenpropidin, (a piperidine), has a similar mode of action to fenpropimorph, and consequently, the possibility of the development of 'cross-resistance' between fenpropimorph and fenpropidin has been considered (Brown & Evans, 1992). This phenomenon occurs when the response of isolates to one fungicide is positively correlated with the response to one or more others. In our survey, there was a marked tendency for isolates with decreased sensitivity to fenpropimorph to also exhibit reduced sensitivity to fenpropidin. This suggests that the sensitivities of *E. graminis* f.sp. *tritici* isolates to these two fungicides are positively correlated. No such relationship was established between tridemorph (another morpholine) and either of the above chemicals (data not shown). All the isolates tested exhibited a similar response to tridemorph, despite over 20 years of continuous usage of this fungicide in France.

REFERENCES

Baloch, R.I., Mercer, E.I., Wiggins, T.E., Baldwin, B.C. (1984) Where do morpholines inhibit sterol biosynthesis? 1984 British Crop Protection Conference - Pests and Diseases **3,** 893-898.

Berg, D.; Kraemer, W.; Regel, E.; Buechel, K-H.; Holmwood, G.; Plempel, M.; Scheinpflug, H. (1984) Mode of action of fungicides: studies on ergosterol biosynthesis inhibitors. 1984 British Crop Protection Conference - Pests and Diseases **3**, 887-892.

Brown, J.K.M.; Evans, N. (1992) Selection on responses of barley powdery mildew to morpholine and piperidine fungicides. Crop Protection **11**, 449-457.

Lorenz, G.; Saur, R.; Schelberger, K.; Forster, B.; Kung, R.; Zobrist, P. (1992) Long term monitoring results of wheat powdery mildew sensitivity towards fenpropimorph and strategies to avoid the development of resistance. 1992 Brighton Crop Protection Conference - Pests and Diseases **1**, 171-176.

Russell, P.E. (1993) Fungicide Resistance Action Committee. Pesticide Outlook **4**, (2) 21-23.

THE EFFECT OF REDUCED DOSES ON THE SENSITIVITY OF POWDERY MILDEW TO FENPROPIMORPH IN BARLEY FIELD TRIALS

M. C. N. ZZIWA AND F. J. BURNETT

Department of Crop Science and Technology, Scottish Agricultural College, West Mains Road, Edinburgh. EH9 3JG

ABSTRACT

Three field experiments were carried out in 1992 and 1993 to study the effect of reduced doses of fenpropimorph alone, or mixed with propiconazole, on the sensitivity of barley powdery mildew *Erysiphe graminis* f.sp. *hordei*. There were no significant differences between the full commercial dose and reduced doses in their effects on the sensitivity of powdery mildew to fenpropimorph. The sensitivity of the experimental isolates fell within the same range of EC_{50} values as that found in previous sensitivity tests with fenpropimorph. It is is concluded that fenpropimorph is still effective in controlling barley powdery mildew.

INTRODUCTION

Considerable efforts have been directed towards breeding barley for resistance to powdery mildew caused by *Erysiphe graminis* f.sp. *hordei*. Although the use of resistant cultivars has reduced infection levels, changes in the virulence spectrum of the pathogen population present problems to the breeder. The disease remains important in barley growing and farmers must continue to rely for control on the application of fungicides. Genetic variation in the mildew population again gives rise to problems, in this case with respect to the level of sensitivity to fungicides. After the development of resistance to some of the early systemic fungicides used for barley mildew control, the demethylation inhibitors (DMI, azoles) and aminopyrimidine (ethirimol) (Fletcher & Wolfe, 1981; Wolfe, 1985; Heaney, Martin & Smith,1988), control of powdery mildew has relied almost exclusively on one class of fungicides. Commonly referred to as the morpholines, the group consists of two morpholines, fenpropimorph and tridemorph, and a piperidine - fenpropidin. These fungicides are marketed commercially as the individual active ingredients, as well as in mixtures with each other or other compounds.

With such heavy reliance for control on one group of fungicides, selection pressure on the mildew population must be significant. Brown and Evans (1992) described isolates that were resistant to reduced doses of tridemorph, fenpropidin and fenpropimorph, and in addition reported cross resistance between fenpropimorph and fenpropidin. They indicated, however, that the levels of resistance they found were unlikely to cause a substantial loss of effectiveness of the chemicals immediately after spraying.

The Scottish Agricultural College at Edinburgh has been monitoring the sensitivity to fenpropimorph of isolates of barley powdery mildew, collected mostly from Eastern Scotland but also from other parts of Britain, since 1988. The results of this survey show that over this period the mean sensitivity of isolates tested to fenpropimorph has changed little, and EC_{50} values for different years fall within the same range as shown in Table 1. There were however significant differences in the mean sensitivity of isolates between seasons, demonstrating that the population is not stable in terms of sensitivity to morpholines.

Table 1. Sensitivity to fenpropimorph in isolates of powdery mildew collected from 1988 to 1992

	Sensitivity to Fenpropimorph mean EC_{50} value in g/l		
Year	Mean	Range	SED
1988	0.057	0.007 - 0.119	$\pm$0.0118
1989	0.021	0.010 - 0.051	$\pm$0.0023
1990	0.033	0.008 - 0.115	$\pm$0.0061
1991	0.082	0.010 - 0.119	$\pm$0.0189
1992	0.029	0.010 - 0.108	$\pm$0.0140

Many farmers use reduced doses of morpholines, usually in mixtures, as standard practice to control mildew infections. Because of the possibility that such practices might influence the selection pressure for insensitive isolates in the mildew population, field experiments were carried out to determine if the use of reduced doses was likely to influence the sensitivity of powdery mildew to fenpropimorph, the most commonly used morpholine.

This paper reports the results of three field experiments carried out in 1992 and 1993 to establish if any shift in sensitivity could be measured following a repeat application of reduced doses as applied in standard practice.

MATERIALS AND METHODS

In the spring of 1992, a large field experiment was laid out at Boghall Farm, at Bush Estate in the Lothian Region of Scotland. The barley cultivar used was Golden Promise. Plot sizes were 24m by 16m. Fertiliser, herbicide and any micro-nutrient treatments were uniform across all plots, and accorded with local practice. Seed for the trial was treated with a single purpose seed treatment only (mercury).

Fungicide treatments consisted of two spray programmes of fenpropimorph alone or in a mixture with propiconazole. The first fungicide application was made when mildew first developed on the plants and the second spray was applied three weeks later. There were eight treatments, shown in Table 2, and three replicates of each laid out in blocks. To facilitate spraying plots within blocks were not completely randomised.

Table 2. Fungicide programmes evaluated in 1992 field experiment

Treatment	First application	Second application
U	nil	nil
A	fenpropimorph 1.0*	fenpropimorph 1.0
B	fenpropimorph 0.5	fenpropimorph 0.5
C	fenpropimorph 0.25	fenpropimorph 0.25
D	fenpropimorph 1.0 +propiconazole 0.5	fenpropimorph 1.0 +propiconazole 0.5
E	fenpropimorph 0.5 +propiconazole 0.25	fenpropimorph 0.5 +propiconazole 0.25
F	fenpropimorph 0.25 +propiconazole 0.125	fenpropimorph 0.25 +propiconazole 0.125
G	nil	fenpropimorph 0.25 +propiconazole 0.125

* dose rates as a proportion of the full commercial dose of the products used: full commercial doses for the products used were as follows:

Active ingredient	Product	g AI / ha
fenpropimorph	Corbel	750
propiconazole	Tilt 250 EC	125

All fungicides were applied using a tractor mounted Allman hydraulic sprayer with standard flat fan nozzles in 270 l / ha of water at a pressure of 2 bars

Infected leaves were sampled from the middle of plots at three times during the season; before spraying and three weeks after both the first and second sprays. Isolates from leaves from each plot were tested for sensitivity to fenpropimorph in the laboratory following the method reported in detail by Robertson *et al.*, (1990). Isolates were cultured on detached leaf segments of Golden Promise and maintained on Davis minimal medium containing 80 mg/l benzimidazole. To determine the sensitivity of isolates in tests, seedlings of Golden Promise were grown to the two leaf stage and then fenpropimorph solutions applied at concentrations of 0.015, 0.029, 0.058, 0.117 and 0.234 g AI / l in a spray cabinet using a Humbrol spray gun for five seconds. Control plants were sprayed with water. Each spray treatment was repeated in the same cabinet for replication. Segments of the treated leaves were then plated on the minimal medium and inoculated with the experimental isolates. The mildew cover after 14 days incubation at 18°C was analysed using a Genstat 5 programme which allowed EC_{50} values to be calculated.

Following the field methodology described, two further experiments were laid out at separate sites at Bush Estate in the spring of 1993 but only fenpropimorph was sprayed at full and at three reduced doses as shown in Table 3. To reduce uncontrolled variation brought about by freely mobile inoculum in untreated plots, there were no unsprayed plots. There were three replicates of each of the four treatments, laid out as before. Plot sizes were 24m by 17m. The seed was treated with guazatine plus imazalil.

Table 3. Fungicide programmes evaluated at two sites in 1993

Treatment	First application	Second application
A	Fenpropimorph 1.0*	Fenpropimorph 1.0
B	Fenpropimorph 0.75	Fenpropimorph 0.75
C	Fenpropimorph 0.5	Fenpropimorph 0.5
D	Fenpropimorph 0.25	Fenpropimorph 0.25

* dose rate as a proportion of the full commercial dose of fenpropimorph as follows:

Active ingredient	Product	g AI / ha
fenpropimorph	Corbel	750

Sampling of infected leaves for tests for sensitivity to fenpropimorph in the laboratory were carried out as in 1992.

RESULTS

The results for the three field experiments are summarised in Tables 4 and 5. Although the sampled isolates varied in their sensitivity to fenpropimorph there were no significant differences between mean EC_{50} values for isolates in relation to the concentration of fungicide to which they had been exposed.

For samples assessed after the second spray application in 1992 (Table 4), the untreated isolates showed an EC_{50} value of 0.056 g/l while those exposed to fenpropimorph alone showed comparable values of 0.032 for the highest dosage rate and 0.025 for the lowest. Where fenpropimorph was combined with propiconazole the equivalent values were 0.023 and 0.086.

Table 4. Sensitivity of isolates from 1992 experiment to fenpropimorph based on mean EC_{50} values in g/l

Sampling time	Fungicide treatments							
	U	A	B	C	D	E	F	G	SED
Before spray	0.092	0.05	0.111	0.001	0.128	0.048	0.244	0.096	-
After 1 spray	0.034	0.035	0.102	0.055	0.118	0.029	0.139	0.077	±0.0470
After 2 sprays	0.056	0.032	0.086	0.025	0.023	0.073	0.136	0.086	±0.0636

In 1993, when fenpropimorph was applied alone at full or reduced rates to all plots at two sites (Table 5), the EC_{50} values at site 1 after the second spray application ranged from 0.120 for the isolates from plots receiving the highest dosage rate to 0.030 for those receiving the lowest. At site 2 the equivalent values were 0.015 and 0.011.

Table 5. Sensitivity of isolates from 1993 experiments to fenpropimorph based on mean EC_{50} values in g/l

Site	Sampling time	Fenpropimorph treatments				
		A	B	C	D	SED
1	Before spray	0.068	0.023	0.111	0.070	
	After 1 spray	0.029	0.181	0.102	0.156	
	After 2 spray	0.120	0.081	0.087	0.030	±0.0695
2	Before spray	0.027	0.102	0.029	0.188	
	After 1 spray	0.059	0.115	0.031	0.012	
	After 2 sprays	0.015	0.038	0.036	0.011	±0.0574

In considering the variation in sensitivity between times of sampling, there were no significant differences between the mean EC_{50} values for isolates sampled before any spray application, after one application of fenpropimorph or after two applications, in both seasons. In 1992, the mean EC_{50} of all the isolates collected before any sprays were applied was 0.096. After the first spray the mean was 0.064 for all treatment plots that had received fenpropimorph alone and 0.095 for those that received fenpropimorph plus propiconazole. After the second spray the comparable values were 0.049 and 0.077. In 1993 the mean EC_{50} for all isolates collected from treatment plots before spray application at site one was 0.068 and at site two was 0.086. After one spray the mean values were 0.117 (site 1) and 0.054 (site 2) and after the second spray 0.080 and 0.025 respectively.

DISCUSSION

There was no evidence from the field experiments conducted in 1992 and 1993 that variation in dose rates of fenpropimorph affected the level of sensitivity of isolates of barley powdery mildew exposed to this fungicide within two growing seasons. There was also no significant difference between the sensitivity of untreated plots in 1992 and those which had received fenpropimorph sprays. The range of sensitivities of the isolates tested in all three trials fell within the range found during routine monitoring from 1988 onwards. Brown and Evans (1992) also reported that they could find no correlation between dose rate and frequency of insensitivity, although they did report an increased frequency of resistance after exposure to fenpropimorph sprays. They concluded that the reductions in sensitivity that they observed were unlikely to result in any reduction in field performance.

Reduced rates of morpholine clearly still provide effective control of mildew in the field situation (Wale et al., 1993). The results of the three trials reported here, where reduced doses of morpholine were applied twice in the season as is common agricultural practice confirm the observations of farmers and advisers that reduced doses have not significantly affected the performance of fenpropimorph against the pathogen. Further trials over several years, however would be necessary to establish the long term effects of reduced dose rates on fenpropimorph sensitivity in fungal populations.

In conclusion there was no evidence that reduced doses of fenpropimorph, applied following normal agricultural practices, are likely to reduce the sensitivities of mildew isolates in treated plots. In keeping with previous observations, fenpropimorph was found to retain its effectiveness as a fungicide for the control of barley powdery mildew in the field.

ACKNOWLEDGEMENTS

We are grateful to Mr. T. Hunter of the Scottish Agricultural Statistics Service for advice. Acknowledgement is made to the Home-Grown Cereals Authority who funded this work.

REFERENCES

Brown , J. K. M. and Evans, N. (1992) Selection on responses of barley powdery mildew to morpholine and piperidine fungicides. *Crop Protection* , **2,** 449-457.

Heaney, S. P.; Martin, T. J. and Smith, J. M. (1988) Practical Approaches to managing anti-resistance strategies with DMI fungicides. *Proceeding Brighton Crop Protection Conference 1988 - Pests and Diseases*, **3**, 1097-1106.

Fletcher, J. T and Wolfe, M. S. (1981) Insensitivity of *Erysiphe graminis* f.sp. *hordei* to triadimefon, triadimenol and other fungicides. *Proceeding British Crop Protection Conference 1981- Pests and Diseases*, **2**, 633-640.

Robertson, S; Gilmour, J; Newman, D. and Lennard, J. H (1990) Sensitivity of powdery mildew isolates to morpholine fungicides. *Proceedings Brighton Crop Protection Conference 1990 - Pests and Diseases*, **3**, 1159-1162.

Wale, S.J.; Oxley, S.J.P. and Bain, R.A. (1993) Effective spring barley control using low-dose fungicide mixtures. *Proceedings Crop Protection in Northern Britain Conference 1993*, 127-132.

Wolfe, M. S. (1985) Dynamics of the response of barley mildew to the use of sterol synthesis inhibitors. *EPPO Bulletin*, **15**, 451-457.

SENSITIVITY TO MORPHOLINE FUNGICIDES IN YELLOW RUST OF WHEAT (*PUCCINIA STRIIFORMIS*)

R. A. BAYLES, P. L. STIGWOOD AND E.G. BARNARD

National Institute of Agricultural Botany, Huntingdon Road, Cambridge, CB3 OLE

ABSTRACT

Isolates of *Puccinia striiformis* collected between 1961 and 1990 were tested for sensitivity to fenpropimorph using wheat seedlings sprayed with low doses of fungicide. Isolates varied widely in their sensitivity, but there was no evidence that sensitivity had declined over the thirty year period. There were indications that sensitivity was related to the geographical origin and specific virulence of isolates. Isolates from the south of the U.K., and those without virulence for the wheat cultivar Hornet, tended to be less sensitive than isolates from the north and those possessing virulence for cv. Hornet. Sensitivity appeared to be unrelated to fungicide applications to the crop prior to sampling. The ranking of a number of isolates for sensitivity to fenpropidin was similar to their ranking for fenpropimorph.

INTRODUCTION

Puccinia striiformis, the causal agent of yellow rust, is a major pathogen of wheat in the U.K. Epidemics of the disease occur on average once every four to five years, depending largely on the susceptibility of cultivars dominating the wheat acreage. Although fungicides of the triazole and morpholine groups have been widely used to control yellow rust, there have been no substantiated reports of loss of disease control in the field.

The main aims of this investigation were to establish a base-line measure of sensitivity of *P.striiformis* to triazoles and morpholines and to determine whether sensitivity has changed over the past thirty years. This paper reports the results for morpholine sensitivity.

Variation in *P.striiformis* for virulence for cultivars has been monitored since 1967 by the U.K. Cereal Pathogen Virulence Survey, which maintains a collection of isolates classified on the basis of origin and virulence. This collection was the main source of isolates for fungicide sensitivity tests.

METHODS

Isolates collected between 1961 and 1990

268 isolates of *P.striiformis* , collected between 1961 and 1990 were tested for sensitivity to fenpropimorph using a seedling test. Isolates were stored as freeze-dried uredospores in sealed glass ampoules. Each test included a standard isolate (83/62) for reference. Ten seedlings of the universally susceptible wheat cultivars Sappo or Vuka were grown in 6.5cm pots. Seedlings were sprayed with fungicide when the first leaf had expanded, 7 to 8 days after sowing, using a field application simulator sprayer, delivering 200 $l.ha^{-1}$ at a pressure of 2 bars. The dose rate of fenpropimorph was 187.5 mg AI/l, corresponding to 1/20 field rate.

24 hours after spraying, seedlings were inoculated in a rotary spore inoculator, using fresh uredospores mixed with acid purified talc. After incubation for 48 hrs at $7^{o}C$ and high relative humidity, seedlings were transferred to a controlled environment growth room with 16 hrs light at $18^{o}C$ and 8 hrs dark at $11^{o}C$.

Four to five days after the appearance of yellow rust pustules on unsprayed seedlings (14 - 15 days after inoculation), the percentage leaf area covered with pustules was assessed on first leaves. An index of infection 'I', comparing the relative infection for the test isolate with the standard isolate, was estimated as follows:

$$I = P_{tf} / P_{to} - P_{sf} / P_{so}$$

where P_t = percentage infection on test isolate
P_s = percentage infection on standard isolate
f = fungicide - treated
o = untreated

Positive values of 'I' indicated higher infection than the standard isolate i.e. lower sensitivity, whilst negative values of 'I' indicated lower infection than the standard i.e. greater sensitivity.

Isolates were classified according to:

1. Year of collection
2. Geographical location, north or south of River Tyne (N or S)
3. Virulence or avirulence for the cultivar Hornet (V or A)
4. Whether or not morpholine fungicide applied to crop prior to sampling (+Mor -M)

Using these classifications, mean 'I' values for contrasting groups of isolates were compared using a t-test, to indicate associations between sensitivity to fenpropimorph and other characteristics.

<u>Comparison of sensitivity to fenpropimorph and fenpropidin</u>

Four isolates were tested for sensitivity to fenpropimorph (187.5 mg AI/l, equivalent to 1/20 field rate) and fenpropidin (375.0 mg AI/l, equivalent to 1/10 field rate). The higher rate of fenpropidin was required to reduce infection to the same degree as fenpropimorph. The isolates comprised two collected from crops in which poor fungicide control had been reported (92/27, from a crop treated triazole and morpholine, and 93/32, from a crop treated with triazoles), one with consistently reduced sensitivity to fenpropimorph (90/20) and the standard isolate (83/62).

RESULTS

Isolates collected between 1961 and 1990

For the 268 isolates tested, values of 'I' ranged from -54 to +41, representing approximately 1% to 65% of the infection level on the nil control (Figure 1).

Figure 1. Infection indices for 268 isolates of *P.striiformis* inoculated onto fenpropimorph - treated seedlings (1/20th field rate)

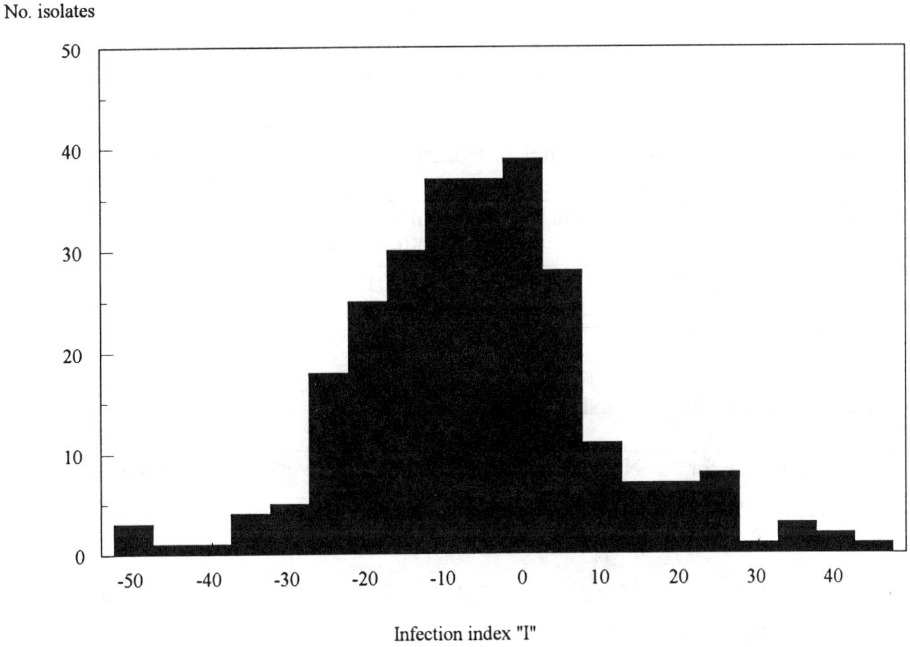

Mean infection indices for contrasting groups of isolates, and the significance of the differences between them, are given in Tables 1a to 1e.

TABLE 1. Mean infection indices for isolates of *P.striiformis* inoculated onto fenpropimorph - treated seedlings (1/20 field rate).

a) Isolates classified by year

Year	No. isolates	mean 'I'	comparison	significance
pre-1989	95	-7.33	pre-'89 v '89	NS
1989	111	-7.89	'89 v '90	NS
1990	62	-9.45	pre-'89 v '90	NS

b) Isolates classified by geographical location

Location	No. isolates	mean 'I'	significance of difference
N	66	-16.25	P = 0.001
S	188	-5.17	

c) Isolates classified by virulence for cultivar Hornet

Virulence for cv Hornet	No. isolates	mean 'I'	significance of difference
V	102	-12.05	P = 0.01
A	157	-5.11	

d) Isolates classified by location and virulence

Location	virulence for cv Hornet	No. isolates	mean 'I'	Comparison	significance
N	V	49	-14.97	N,V v S,V	P = 0.05
	A	4*			
S	V	53	-9.35	S,V v S,A	P = 0.001
	A	132	-3.31		

* number of isolates too small for valid comparisons

e) Isolates classified by morpholine application prior to sampling.

Morpholine application	No. isolates	mean 'I'	significance of difference
+M	38	-6.82	NS
-M	230	-8.33	

Comparison of sensitivity to fenpropimorph and fenpropidin

Results are given in Table 2.

TABLE 2. Percentage leaf area infected with yellow rust (relative to untreated control) for wheat seedlings treated with fenpropimorph or fenpropidin and inoculated with four isolates of *P.striiformis*

Isolate	Fungicide	
	Fenpropimorph (187.5 mg AI/l)	Fenpropidin (375mg AI/l)
83/62	37.5	41.0
90/20	51.5	53.0
92/27	23.5	39.0
93/32	44.5	45.0

DISCUSSION

Isolates of *P.striiformis* collected between 1961 and 1990 varied widely in their sensitivity to a low dose of fenpropimorph, equivalent to 1/20 field rate. However, there was no evidence that the pathogen has become less sensitive in recent years. The mean sensitivity of isolates collected in 1989 and in 1990 did not differ significantly from that of isolates collected between 1961 and 1988.

There was evidence of an association between geographical location and sensitivity, with isolates from the south being, on average, less sensitive than those from the north. One possible explanation for this is that the use of morpholine fungicides to control yellow rust has been more common in the south than in the north. There was also an association between sensitivity and virulence characteristics, such that isolates lacking virulence for the wheat cultivar Hornet were less sensitive than isolates possessing this virulence. This observation is probably related to the north : south difference, since virulence for Hornet first arose in northern populations (Bayles *et al.*, 1989, Bayles and Stigwood, 1991), which were themselves relatively sensitive to morpholines.

Whether or not a morpholine fungicide had previously been applied to a crop appeared to have no effect on the sensitivity of isolates taken from it. However, the number of samples from morpholine treated crops was small compared with those from untreated crops or crops receiving other fungicides, and this result should therefore be interpreted with caution.

There was no evidence from a limited comparison of the reactions of four isolates to fenpropimorph and fenpropidin that sensitivity to the two morpholine fungicides differed. The ranking of isolates was similar for the two chemicals and isolate 90/20 showed a marked reduction in sensitivity to both. The isolate taken from a crop in which a morpholine fungicide had reportedly given poor control, (92/27) proved to be the most sensitive of the isolates, underlining the need to exercise caution in attributing poor disease control to insensitivity.

The most important outcome of this investigation was the establishment of a base line sensitivity with reference to isolates with defined sensitivity characteristics. These isolates can be stored for long periods as freeze-dried spores, with negligible risk of genetic change . The use of reference isolates is considered to be vital, since the precise quantity of fungicide delivered to test seedlings can vary significantly depending on the spraying equipment used.

REFERENCES

Bayles, R.A.; Stigwood, P.L (1991). Yellow rust of wheat. *United Kingdom Cereal Pathogen Virulence Survey Annual Report for 1990*, 15-20.

Bayles, R.A.; Channel, M.H.; Stigwood, P.L. (1989). Yellow rust of wheat. *United Kingdom Cereal Pathogen Virulence Survey Annual Report for 1988*, 11-15.

MONITORING RESULTS OF *MYCOSPHAERELLA FIJIENSIS* TO TRIDEMORPH

K. Cronshaw, G. Lorenz, D. Mappes

BASF AG, Landwirtschaftliche Versuchsstation,
D-67114 Limburgerhof, Federal Republic of Germany

ABSTRACT

Control of black Sigatoka, caused by *Mycosphaerella fijiensis*, presently depends on five fungicide groups, three of which belong to the so called single site inhibitors. Pathogen resistance to benzimidazoles is now general and widespread in most of the banana growing countries of South- and Central-America. The first decline in sensitivity to triazoles has been reported from Belize, Guatemala, Honduras, Costa Rica, Mexico, Panama and Cameroon.

In contrast, monitoring results show that there are no indications of sensitivity changes to tridemorph in any country. The use of tridemorph in fungicide mixtures or alternation programmes is, therefore, an important resistance management strategy to limit further development of resistance to triazoles and benzimidazoles.

INTRODUCTION

TABLE 1.Sequence of products used for the control of Sigatoka and the approximate year of introduction

Year	Product	Application Rate
1934 - 1958	Bordeaux mixture	1000 - 1500 l/ha
1956 -	Mineral oil	7 - 15 l/ha
1958 -	Dithiocarbamates (Mancozeb)	1,5 - 2,5 kg/ha
1970 -	Benzimidazoles	250 - 300 g/ha
1978 -	Chlorothalonil	2 - 3 l/ha
1978 -	Morpholines (tridemorph)	0,6 l/ha
1985 -	Triazoles (propiconazole)	0,4 l/ha

Present day chemical control of *Mycosphaerella fijiensis* depends on five fungicide groups (Table 1), two of which have a multi-site mode of action (mancozeb and chlorothalonil), while benzimidazoles, tridemorph and propiconazole belong to the so called single-site inhibitors. It is widely accepted, that there is little chance of resistance development to the multi-site inhibitors, but they have only protectant activity against fungal infections and, therefore, have to be applied more frequently than the single-site inhibitors, whose limited systemic activity provides a certain degree of curative action.

Benomyl was the first systemic fungicide to be used for Sigatoka control in 1970. Since at that time there was little knowledge of fungicide resistance, benomyl was applied frequently. Reports of Sigatoka resistance to benomyl first occurred in the late 1970's and started the idea of fungicide resistance management in the banana industry.

Tridemorph has now been part of Sigatoka control programmes since 1978 i.e. for up to 16 years in some countries. Based on the experience of benomyl resistance, a periodic monitoring programme was started in 1984 (Cronshaw and Lorenz, 1988) and continued ever since, in order to detect any possible sensitivity changes at an early stage. So far no such changes in sensitivity have been detected.

The triazole fungicide propiconazole was introduced into Sigatoka control programmes in 1985 and is now in its ninth year of use. Monitoring studies were carried out from the beginning and only recently have started to show a reduced sensitivity to propiconazole.

MATERIAL AND METHODS

Materials and methods involved are based on techniques outlined for monitoring the response of *M. fijiensis* to benomyl (Anon. 1983) which were subsequently modified for triazole/morpholine fungicides (Anon. 1988) and are described in some detail by Stover (1992). They are based on the measurement of germ tube growth on water agar amended with various fungicide concentrations.

Generally ascospore-bearing leaf tissue is incubated in plastic bags with moist towelling for 48 hr at room temperature. Leaf tissue for ascospore discharge is removed, cut into pieces of 1 - 2 cm^2, numbered with ink and stapled to filter paper. The filter paper with attached leaf pieces is submerged in tap water for 5 minutes and then placed inside the lid of a Petri dish over 2 % water agar containing a concentration range of tridemorph (0 - 50 ppm a.i.) for 1 hr for ascospore discharge, and then removed. Ascospores are located and identified with the aid of a dissecting microscope. Following ascospore discharge plates are incubated at 26°C for 48 hr. After that time the germ tube length of 50 ascospores per plate is measured. An ascospore is considered to have germinated, if the germ tube is at least 5 μ long. From these data EC_{50}-values can be calculated by using appropriate statistical programmes, or frequency distributions can be drawn using either %

inhibition (based on mean score of the unamended plate) or by graphing the actual germ tube length.

RESULTS

A tridemorph monitoring programme was started in 1984. Since this work was mainly done in the laboratory of Dr. Stover, Honduras, the most consistent data are available from this country. In Table 2 EC_{50}-values are compared from farms where either no tridemorph had ever been applied or where it was part of the commercial spray programmes. Obviously no appreciable differences exist in the sensitivity of ascospores from treated and non-treated areas. When samples are taken at the same date from different locations within the same farm, there is very little variation in sensitivity (Table 2.). On the other hand a certain degree of variation becomes apparent, when samples are taken at the same location but at different sampling dates throughout the year (Table 2), which is probably due to the fact, that ascospores taken during the dry season are presumably less vigourous and consequently more sensitive to fungicides.

TABLE 2.Tridemorph monitoring data from Honduras 1984

	Various farms					
	A (nt)*	B (nt)	C	D	E (nt)	F
ED_{50} (ppm a. i.)	2,04	< 1	1,31	1,52	1,03	1,83

	Various locations within the same tridemorph treated farm			
	A	B	C	D
ED_{50} (ppm a. i.)	1,35	1,37	1,65	1,79

	Various sampling dates from the same non-treated farm				
	Nov.	Dec.	Feb.	Mar.	Aug.
ED_{50} (ppm a. i.)	< 1	< 1	1,4	1,42	3,1

* nt = non-tridemorph treated

TABLE 3.Tridemorph monitoring data from Honduras 1988 - 1992

Santa Catalina Farm (non-tridemorph treated)								
Sample Date	Dec. 88	Apr. 89	Apr. 89	Oct. 89	Feb. 92	Jun. 92	Jun. 92	Oct. 92
EC_{50} (ppm a. i.)	1,98	0,09	1,44	0,11	0,54	0,22	0,23	1,98

Ø = 0,82

Various tridemorph treated farms							
Sample Date	Feb. 88	Feb. 88	Feb. 88	Feb. 88	Oct. 92	Oct. 92	Oct. 92
	Corozal 2	Corozal 16	Santa Rosa 38	Santa Rosa 170	Santa Rosa 40	Corozal 125	Tacamiche
EC_{50} (ppm a. i.)	0,13	0,25	0,05	< 0,01	1,19	1,9	1,09

Ø = 0,66

In 1984 further data were obtained from samples from Panama (Ø EC_{50}-value 1,91 ppm) from Costa Rica (Ø EC_{50}-value 1,48 ppm) and from the Philippines (Ø EC_{50}-value 1,0 ppm) which were comparable to those from Honduras.

Table 3 shows data from Honduras for the years 1988 to 1992. Again values from non-treated locations are compared with those from treated areas. The EC_{50}-values show some variation but the average value from the non-treated control farm (Santa Catalina) is very similar to the average from farms that have received tridemorph applications in commercial spray programmes.

The wide range of variation of EC_{50}-values seems to be a general disadvantage of the germ tube elongation test used for *M. fijiensis*. Sampling date, concentration range chosen, storage conditions and differences between laboratories are some of the factors whose influence on the response of ascospores was investigated by Stover (1992). Besides these, at the last FRAC-meeting of the SBI Banana Working Group the major area for improvement was identified as the method of data analysis. It was suggested to focus rather on specific areas of the sensitivity distribution instead of calculating EC_{50}-values.

In Figure 1, therefore data from Honduras are shown where the percentage of germ tube inhibition is presented as a frequency distribution at a given concentration and again this indicates that there is no difference in sensitivity to tridemorph comparing untreated and treated areas.

FIGURE 1. Frequency distribution of % germ tube inhibition at 0,1 ppm tridemorph

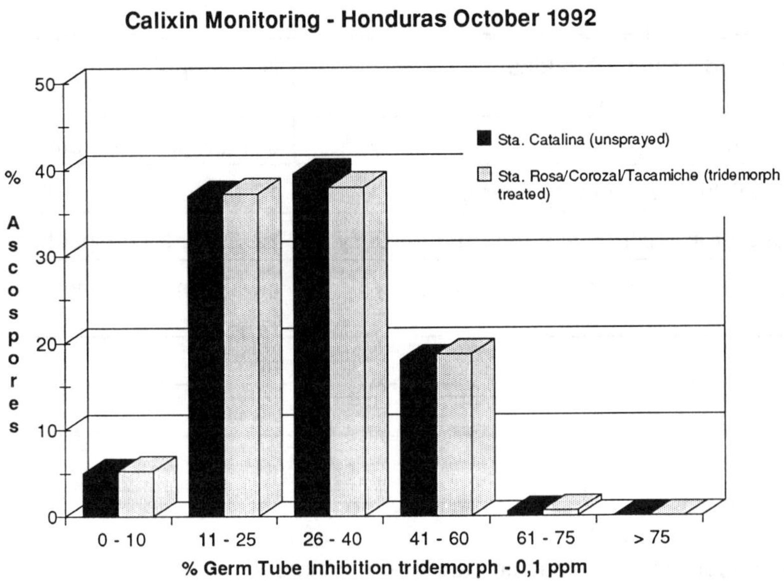

CONCLUSIONS

In spite of the fact that tridemorph has now been a part of the Sigatoka control programmes for up to 16 years in some countries, and was used with a wide range of frequencies, the results of the continuing monitoring programme so far indicate that *M. fijiensis* causing black Sigatoka or black leaf streak, does not show any decreased sensitivity to tridemorph.

The same situation is true for the control of cereal powdery mildews in Europe, where after 24 years of use there are still no reports of reduced sensitivity to tridemorph (Clark 1992). The reasons discussed are possible additional effects on processes not related to sterol biosynthesis i.e. an additional mode of action besides its effects on sterol biosyntheses (James *et al.* 1992). This emphasizes the importance of tridemorph as an alternation or mixture partner to lessen or help prevent further resistance development to benzimidazoles and triazoles. Although no field resistance to tridemorph has occurred, the FRAC recommendations (Table 4) are strongly supported by BASF. The use of tridemorph in a carefully planned resistance management strategy that avoids block treatments and limits the total number of applications per year should be considered as a sensible precaution.

TABLE 4.Recommendations of FRAC, Orlando, January 1993

Fungicides used against Sigatoka	No. of applications / year
triazoles alone	8 (max. 2 consecutive)
triazoles in mixture with other fungicides	10 (max. 2 consecutive)
Calixin alone or in mixture with other fungicides	12 (max. 2 consecutive)
benzimidazoles alone or in mixture with other fungicides	always in alternation
chlorothalonil	no limit
mancozeb	no limit

In addition:

- consecutive applications of triazoles are not recommended
- a 2 - 4 month period free of triazoles is recommended
- in mixtures Calixin 66 % of the normal rate
  triazoles 75 % - 100 % of the normal rate
  benzimidazoles 100 % of the normal rate

REFERENCES

Anon. (1983) Black and yellow Sigatoka: improved identification and management techniques. Du Pont Latin America, 2600 Douglas Road, suite 1000, Coral Gables, Florida.

Anon. (1988) Use of DMI fungicides in bananas. First working group meeting, Miami. ISPP Chemical Control Newsletter **11**, 11 - 13.

Clark, W. S. (1992) Practical aspects of resistance to barley powdery mildew *(Erysiphe graminis)*. Proc. Brighton Crop Prot. Conf., Pests and Diseases Vol **1**, 177 - 192.

Cronshaw, D. K.; Lorenz, G. (1988) Monitoring the sensitvity of black Sigatoka and black leaf streak to tridemorph. Proc. VIII meeting ACORBAT, Santa Marta, Colombia, 149 - 160.

James, C. S.; Hargreaves, I. A., Loeffler, R. S. T.; Burden, R. S. (1992) Effect of morpholine-like fungicides on growth and Sterol composition of a wild-type strain and a sterol mutant of *Ustilago maydis* defective in $\Delta^8 \rightarrow \Delta^7$ isomerase activity. Proc. Brighton Crop Prot. Conf., Pests and Diseases, Vol. **1**, 215 - 220.

Stover, R. H. (1992) Response of the black Sigatoka pathogen *Mycospaerella fijiensis* to Calixin (tridemorph) in vitro. Fruit **47**, 291 - 301.

FIELD PERFORMANCE OF MORPHOLINES

Summary of discussion following the papers presenting recent findings on morpholines.

Chairman: Dr James Gilmour, Scottish Agricultural Colleges, Central Office, West Mains Road, Edinburgh, EH9 3JG

A wide ranging and lively discussion followed presentation of papers in the first part of the session. Inevitably, perhaps, delegates concentrated on morpholine action against cereal mildews. Discussion also focussed on fenpropimorph and fenpropidin; little interest was directed towards tridemorph and other morpholines.

In both England and France advisory workers have noticed some fall in the persistence of Corbel and Mistral, but these perceptions were not generally supported by field experimentation in the Netherlands and elsewhere. The environmental impact, especially of high temperatures at the time of spraying, can be considerable, reducing control levels and generating yearly variation in performance. Where newer triazoles, such as tebuconazole and epoxiconazole have replaced triadimenol as standards, the performance of morpholines may be down graded because of improved control achieved by these standards. Greater mildew susceptibility of current wheat cultivars may increase disease pressures and make adequate control more difficult to achieve. With the notable exception of the very mildew susceptible cultivar Apollo, this was probably not the case in the UK. In comparison with the substantial decline in the performance of early triazoles in different parts of N. Europe, similar changes in morpholine performance had not occurred despite their widespread use.

Different assays of fenpropimorph sensitivity, and comparison with standard isolates, have provided experimental evidence, albeit of relatively small changes in both wheat and barley powdery mildew. Choice of these standard isolates is critical, and their relationship to the natural population has not always been made clear. Equally, so called **ring tests** have not always produced consistent results in different laboratories, in marked contrast to similar tests with isolates used as standards in surveys of azole sensitivity. Nevertheless, there was widespread agreement that significant variation exists in natural populations, and that the range of morpholine sensitivity in wild-type populations is less than in selected ones. This does not necessarily lead to a change in population mean, although this can change from year to year. A consensus view emerged from these discussions of the need to re-examine testing methods to obtain more uniform testing protocols to be used amongst workers.

Shifts in sensitivity as measured by bioassay clearly do not correlate with field performance, although this is often a feature of the early phases of the development of practical resistance. The effect of migration is a very real problem in evaluating selection induced changes in wind-dispersed cereal mildews, especially where the fungicides being examined are in widespread use in surrounding farming regions. With present technologies, small plots are generally inadequate, although just what plot sizes are acceptable is not clear. Shifts in sensitivity might well result from directional selection exerted on the whole population, but may equally well reflect the small population surviving treatment. It should be possible to distinguish between these two possibilities, by examining the overall shape of the population distribution, which should no longer be log normal in the case of the small, surviving population. But morpholine sensitivity assays frequently generate considerable experimental variation making it difficult to detect subtle changes in population distributions.

A consistent feature of laboratory and greenhouse results is the apparent cross resistance between fenpropimorph and fenpropidin, but not with tridemorph. It is perhaps too simplistic to relate this to differences in mode of action, when so many of the biochemical studies have been carried out in yeast, and may not be relevant to plant pathogens. Nevertheless, there was considerable hesitancy at adopting anti-resistance strategies based on mixtures of tridemorph with either fenpropimorph or fenpropidin.

Morpholines are still accepted as low risk fungicides. Attempts to analyse the genetic control of any variation in sensitivity have been inconclusive. Identification of distinct sensitivity classes does not inevitably provide evidence of major gene control, as effects of many genes exerting additive effects within different classes are quite difficult to detect against a background of considerable experimental variation. The picture is not made any clearer by differences in the definition of resistance used by workers, and by references to "less sensitive" isolates, and "low level" resistance. However, in a discussion that pulled together the experiences of many practicising pathologists, there was no convincing evidence of practical resistance to morpholines in any target pathogen.

D.W. Hollomon

PARASEXUAL ANALYSIS OF MORPHOLINE AND TRIAZOLE RESISTANCE IN *PSEUDOCERCOSPORELLA*

M.J. HOCART, J.E. McNAUGHTON

Institute of Ecology and Resource Management, The University of Edinburgh, West Mains Road, Edinburgh EH9 3JG, U.K.

ABSTRACT

Parasexual crosses between the W and R pathotypes of *P. herpotrichoides* and between *P. herpotrichoides* (R-type) and the related *P. anguioides* were used to investigate the inheritance of resistance to the morpholine fungicide fenpropimorph and the triazoles, difenconazole and triadimenol. Both crosses indicated multiple genes involved in fenpropimorph resistance. A gene (or genes) having a major effect on triazole sensitivity was observed in both crosses with additional genes modifying the level resistance expressed. However, the patterns of cross-resistance seen differed between the two crosses suggesting that the 'major' genes detected differ in their effect.

INTRODUCTION

The eyespot pathogen, *Pseudocercosporella herpotrichoides*, exhibits considerable natural variation in sensitivity to the ergosterol biosynthesis inhibiting (EBI) fungicides. In general, the R pathotype of the fungus is less sensitive to DMI fungicides than the W pathotype. These differences in fungicide sensitivity are likely to be important selective factors in the population dynamics of the W and R pathotypes.

Interpathotype crosses in *P. herpotrichoides* are possible by manipulation of the parasexual cycle in this fungus. This system enables genetic analysis of agronomically important characters in this fungus, regardless of the sexual compatibility of isolates (Hocart *et al.*, 1993). Crosses between the W and R pathotypes have been used to investigate the genetic basis of variation in sensitivity to DMI fungicides. Previous work had shown that DMI sensitivity was controlled by several genes in *P. herpotrichoides*. One or more of these genes segregated at or shortly after formation of the diploid fusion product following protoplast fusion, and has a major effect on the level of DMI sensitivity shown (McNaughton & Hocart, 1994). Additional genes, segregating later in the parasexual cycle, modified the level of resistance expressed.

As a part of this analysis of the genetic inheritance of fungicide resistance progeny from parasexual crosses between the W and R pathotypes of *P. herpotrichoides* (WxR) (Hocart *et al.*, 1993) were assessed for their sensitivity to the morpholine fungicide fenpropimorph, and the triazole fungicides triadimenol and difenconazole. This paper compares the interpathotype results with a similar assessment made using interspecific hybrid progeny generated between *P. herpotrichoides* and the weakly pathogenic *P. anguioides* (RxA) (Hocart & McNaughton, 1994).

METHODS

Interpathotype parasexual recombinant progeny, recovered from a cross between the W and R pathotypes of *P. herpotrichoides* (WxR) (Hocart *et al.*, 1993), and interspecific hybrid strains produced by parasexual recombination between an R-type isolate of *P. herpotrichoides* and *P. anguioides* (Hocart & McNaughton, 1994), were used in the analysis. In both crosses the parasexual cycle was initiated by protoplast fusion.

Sensitivity to the morpholine fungicide fenpropimorph, and the triazole fungicides difenconazole and triadimenol, was assessed *in vitro*. Colony diameters measured after ten days incubation at 19ºC were used to calculate relative growth on a range of fungicide concentrations. These data were plotted against fungicide concentration (Log10 μM) and used to determine ED50 values for each strain. Cross-resistance between fungicides was detected by correlation analysis using the Log10 ED50 data.

RESULTS & DISCUSSION

Interpathotype Cross

There was a six-fold difference in the sensitivity to fenpropimorph between the W-type (ED50 = 9.8 μM) and R-type (ED50 = 63 μM)strains used as the parents in the interpathotype cross (Fig. 1a). Fenpropimorph sensitivity of WxR progeny varied from highly sensitive (ED50 = 0.1 μM) to almost as resistant as the R-type parent (ED50 = 47 μM) (Fig. 1a). Some recombinants were super-sensitive to this fungicide. The range of resistance phenotypes obtained strongly suggest that fenpropimorph sensitivity is determined by multiple genes.

The two sub-groups of WxR progeny, sensitive (●) or resistant (○) to the DMI fungicides are clearly seen in Fig. 1d. The 'major' gene(s), having a major effect on DMI sensitivity, confers cross-resistance between the two DMI fungicides used. Additional genes, modifying the level of resistance expressed, were seen to segregate at a later stage in the parasexual cycle (McNaughton & Hocart, 1994).

No cross-resistance was observed to fenpropimorph which could be attributed to the 'major gene(s)' for DMI resistance (Figs. 1b &1c). However, there was some evidence that the 'minor genes' could confer cross-resistance between fenpropimorph and triadimenol, subject to the genetic background of the strains (Fig. 1c).

Interspecific Cross

There was a four-fold difference in the level of sensitivity to fenpropimorph between the *P. anguioides* (ED50 = 6.9 μM) and the R-type *P. herpotrichoides* (ED50 = 28.3 μM) parental strains. Fenpropimorph sensitivity of the interspecific hybrids showed considerable variation, similar to that seen with the WxR cross, suggesting that several genes are involved in resistance expression (Fig. 2a). Both super-sensitive and super-resistant progenies were recovered.

Difenconazole (Fig. 2d) identified a sensitive and resistant subset of progeny of which the difenconazole-sensitive strains were also the most sensitive to fenpropimorph (Fig. 2b). While the triadimenol sensitivity of the hybrids was generally similar to the resistant *P. herpotrichoides* parent (ED50 = 278 μM) (Fig 2c), one strain showed increased triadimenol resistance (ED50 = 976 μM). No cross-resistance was seen between triadimenol and difenconazole (Fig. 2d), in contrast to the interpathotype cross results (cf. Fig. 1d), or between triadimenol and fenpropimorph (Fig. 2c). Evidence was obtained, however, of cross-resistance between fenpropimorph and difenconazole (Fig. 2b).

Clearly the gene(s) having a major effect on sensitivity to difenconazole in this interspecific cross is distinct from that detected in the interpathotype (WxR) cross. In both cases, however, the results demonstrate the involvement of a number of genes conditioning resistance to these sterol-inhibiting fungicides in *Pseudocercosporella*.

The cross-resistance relationships revealed here between the triazole fungicides and the morpholine fungicide fenpropimorph have relevance for the field. The use of these compounds to control other cereal pathogens may result in the selection of particular *Pseudocercosporella* genotypes.

FIGURE LEGENDS

Fig. 1. Fungicide response of *P. herpotrichoides* (WxR) interpathotype progenies

(1a) Fenpropimorph sensitivity of WxR progeny compared with the parental strains (W & R). ED50 values expressed as Log10 μM.

(1b, c, d) Detection of cross-resistance between difenconazole, triadimenol and fenpropimorph through correlation of ED50 (μM) values. Parental strains (W & R) indicated by shaded squares; interpathotype progenies shown as circles and classified as sensitive (●) or resistant (○) to the DMI fungicides.

Fig. 2. Fungicide response of *P anguioides* x *P herpotrichoides* (AxR) interspecific hybrids

(2a) Fenpropimorph sensitivity of hybrids compared with the parental strains (A & R). ED50 values expressed as Log10 μM.

(2b, c, d) Detection of cross-resistance between difenconazole, triadimenol and fenpropimorph through correlation of ED50 (μM) values. Parental strains (A & R) indicated by squares; interspecific hybrids shown as circles.

ACKNOWLEDGEMENTS

The authors thank Ciba Geigy and BASF for providing samples of their fungicides. This work was funded by a grant (No.: P15/550) from the Agricultural and Food Research Council.

REFERENCES

Hocart, M.J.; Lucas, J.A.; Peberdy, J.F. (1993) Parasexual recombination between W and R pathotypes of *Pseudocercosporella herpotrichoides* through protoplast fusion. *Mycological Research*, **97**, 977-983.

Hocart, M.J.; McNaughton, J.E. (1994) Interspecific hybridisation between *Pseudocercosporella herpotrichoides* and *P. anguioides* achieved through protoplast fusion. *Mycological Research*, **98**, 47-56.

McNaughton, J.E.; Hocart, M.J. (1994) Inheritance of EBI fungicide resistance in the eyespot pathogen. (Paper in Preparation).

Fig. 1. Fungicide responses of WxR interpathotype progenies

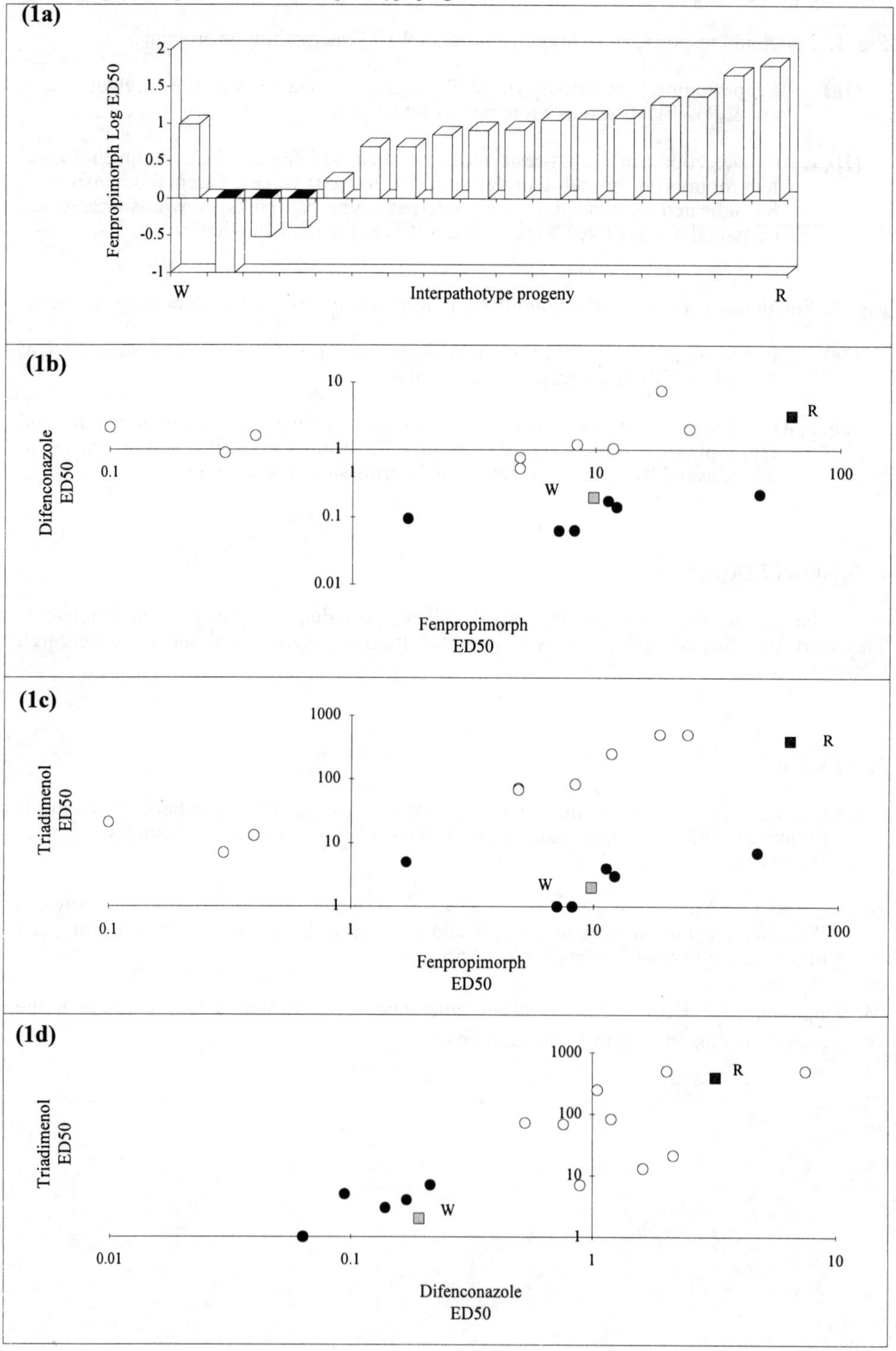

Fig. 2. Fungicide resonses of AxR interspecific hybrids

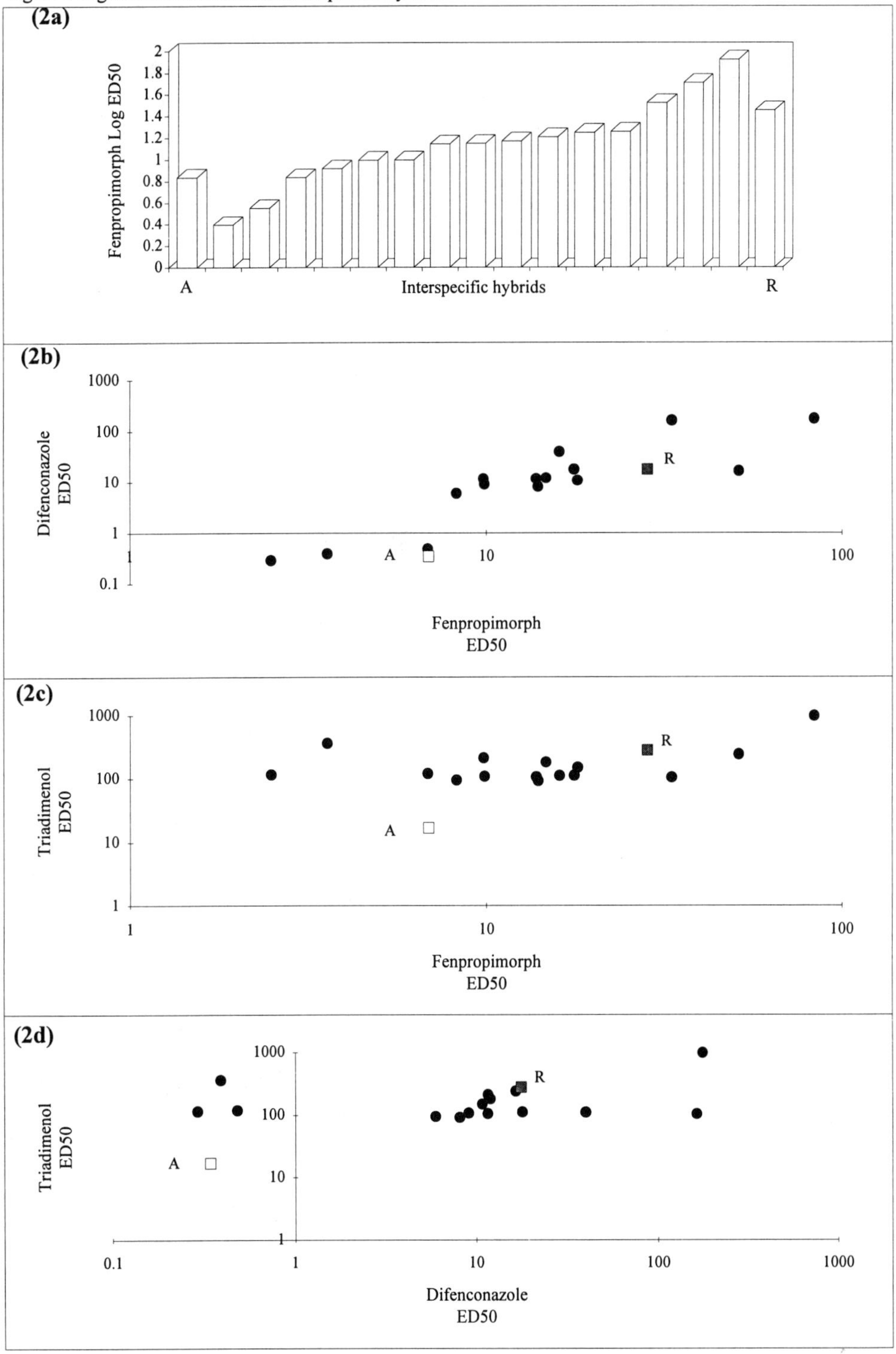

INFLUENCE OF SPLIT APPLICATION OF FENPROPIMORPH MIXTURES ON DISEASE CONTROL AND ON THE SENSITIVITY OF *ERYSIPHE GRAMINIS* F.SP. *TRITICI*

B. FORSTER, D. CHAVAILLAZ, C. STEDEN
Ciba-Geigy Ltd., Plant Protection Division, P.O.B., CH-4002 Basle, Switzerland

W. RADTKE
Pflanzenschutzamt der Landwirtschaftskammer Hannover, Wunstorfer Landstrasse 9, D-30453 Hannover, Germany

M. KÄSBOHRER, A. KÜHL
Ciba-Geigy GmbH, Postfach 11 03 53, D-60323 Frankurt, Germany

ABSTRACT

Mixtures of fenpropimorph with triazoles play an important role in control of *Erysiphe graminis* f.sp. *tritici* and *Erysiphe graminis* f.sp. *hordei*. In Denmark and Northern Germany such mixtures are often used at reduced doses and with more frequent applications than are recommended on product labels. To examine the influence of these split applications on disease control and on the sensitivity to fenpropimorph, field trials were carried out from 1991 to 1993 in Northern Germany. In large plot trials split applications of reduced rates of the fenpropimorph mixture with propiconazole were compared to the recommended two applications of full rates of the mixture and of the solo product fenpropimorph.

Treatments according to the label recommendation provided very good disease control when the timing of applications was adapted to the disease development without producing big changes in fenpropimorph sensitivity at the end of a growing season. Split applications generally resulted in very good powdery mildew control but at the same time showed significant decreases in fenpropimorph sensitivity. Therefore split applications should be avoided.

INTRODUCTION

Mixtures of fenpropimorph with triazoles play an important role in the control of powdery mildew (*Erysiphe graminis* f.sp. *tritici* and *Erysiphe graminis* f.sp. *hordei)* on wheat and barley. Different trials (Bolton and Smith, 1988; Lorenz *et al.*, 1992) have shown the benefit of such combinations for disease control and to combat the development of resistance to either mixture partner. In Denmark and Northern Germany such fenpropimorph mixtures are often used at reduced dose-rates and more frequently than recommended on the label. This may have contributed to a sensitivity shift for fenpropimorph also in this area. Felsenstein (1991) and Lorenz *et al.* (1992) found about a ten-fold decrease of powdery mildew sensitivity to fenpropimorph in this region which occurred already in 1989.

To examine the influence of split applications on the sensitivity to fenpropimorph and at the same time on disease control, specially designed field trials were carried out from 1991 to 1993 in commercial fields in Northern Germany, in an area of high disease pressure for wheat powdery mildew. The influence of split applications of fenpropimorph plus propiconazole was compared to the full rate of fenpropimorph and to the recommended two applications of the mixture.

MATERIALS and METHODS

The field trials were carried out from 1991 to 1993 in Niedersachsen, Germany, on the wheat variety *Kanzler* in different commercial fields. The following treatments were included:

I = check (untreated)
II = fenpropimorph (fpm) at 750 g/ha (recommended rate)
III = fenpropimorph at 300 g/ha in combination with propiconazole (ppz) at 125 g/ha (recommended rate)
IV = fenpropimorph at 150 g/ha in combination with propiconazole at 62.5 g/ha
V = fenpropimorph at 75 g/ha in combination with propiconazole at 31.25 g/ha

TABLE 1. Timing of application related to growth stages (GS) (Zadoks *et al.*, 1974)

Treat-ment	Year	Timing of application (GS)										No. of treatments	% Attack[1] at 1. application
		29	30	31	32	33	34	37	39	49	55		
II	1991			X					X			2	3-4
II	1992			X						X		2	1-3
II	1993				X				X			2	3
III	1991			X					X			2	3-4
III	1992			X						X		2	1-3
III	1993				X				X			2	3
IV	1991			X		X			X	X		4	3-4
IV	1992		X	X				X		X		4	0.5-1
IV	1993			X		X		X	X			4	2
V	1991	X	X	X		X	X		X	X		7	0.5
V	1992		X	X	X			X		X	X	6	0
V	1993			X	X	X		X	X			5	1

[1] percentage powdery mildew attack on green plant tissue

The timing of the first application in the different plots was adapted to the disease development in each year. In treatment V the first application was always carried out when 0-1 % powdery mildew was visible. In 1991 and 1993 the first application in the plot IV was started at the same time as in the full rate plots (II, III) with a disease level of 2-4 %, whereas in 1992 the treatment in this plot was started earlier at 0.5-1 % powdery mildew attack.

The full rate plots (II, III) were treated on average at 24 day intervals; twice per season. The average application interval in the split application plot IV was 10 days and a total of four applications per season were carried out each year. In plot V the spray interval averaged 9 days. In 1991 seven, in 1992 six and in 1993 five applications were carried out in treatment V.

To assess the initial sensitivity to fenpropimorph bulk samples of ten leaves infected with powdery mildew per sample were collected each year before any application was carried out, around growth stage (GS) 29-30 (Zadoks *et al.*, 1974). About 15 days after the last treatment samples were again taken from the different plots. The average number of samples collected in each plot and at the two sampling times was twelve.

Disease rating was carried out by visual assessment of the percentage powdery mildew infection on green plant tissue. In 1991 and 1992 the final rating was carried out at the beginning of July whereas in 1993 this was at the beginning of June.

The average plot size in 1991 was 25000 m^2, 1992 11900 m^2 and in 1993 9900 m^2. However in 1993 the check was integrated as a small untreated part into the middle of a treated area beside the trial plots.

The sensitivity was tested with the test tube method (Sozzi *et al.*, 1991). The concentration range was adapted for fenpropimorph: 0, 0.01, 0.03, 0.1, 0.3 and 1 mg/l AI. EC50-values were determined graphically on a semilogarithmic scale. A sensitive reference isolate (EC50 = 0.03 ppm AI) was included in each test. For a better comparison of results between different tests Q50-values were calculated which represent the ratio of the EC50 value of each tested isolate and the EC50 of the reference isolate in the same test.

RESULTS

Fenpropimorph sensitivity

In table 2 the sensitivity data for the different years are shown as mean Q50-values per treatment. The sensitivity in the different plots did not differ significantly before treatments were carried out. Therefore in the first line the mean Q50-values for all plots are pooled. However the range of Q50-values was broad at the beginning of each tested season: in 1991 Q50-values ranged between 1 and 8, in 1992 between 3 and 9 and in 1993 between 2 and 7.

In 1991 a slight but significant decrease in sensitivity to fenpropimorph was found in all treated plots compared to the check. The most pronounced decrease in sensitivity was detected in the plot with split application IV which differed significantly from the treatment (II) with the solo product fenpropimorph at the full rate. 1992 results show the highest decrease in sensitivity for the split application plot V with 6 treatments. This treatment showed a significant difference to the full rate of the solo product (II) and the mixture (III). Results of 1993 confirm again the highest decrease in sensitivity with the two split applications IV and V. Compared to the previous years, in 1993 the highest Q50-values were found in these plots at the end of the season. Compared to the sensitivity situation early season (reference) the sensitivity decreased by a factor of about two to three. In 1992 and 1993 these changes were more pronounced in treatment V than following fewer applications in treatment IV. In all three years of testing in these plots at the end of the season the very sensitive isolates with Q50-values between 1 and 4 were lost (data not shown).

No significant differences were detected between the application of the solo product and the mixture at the recommended full rate. For both treatments the changes in sensitivity were smaller than for the split applications.

The check-plot in 1993 also shows a higher mean Q50-value at the end of the season. This is probably due to the fact, that the plot size was reduced and that it was not well separated from the trial plots.

TABLE 2. Sensitivity of *Erysiphe graminis* f.sp. *tritici* to fenpropimorph at the beginning and at the end of the growing season in the monitoring trials of the different years.

Treatment	Mean Q50-value for fenpropimorph 1991	1992	1993
reference, early season[1)]	3.56	5.31	4.62
I: check[2)]	4.12 a	5.27 a	6.96 a
II: 2 x fpm 750 g/ha[2)]	6.45 b	6.21 a	6.73 a
III: 2 x fpm 300 g/ha + ppz 125 g/ha[2)]	6.91 bc	5.30 a	6.39 a
IV: 4 x fpm 150 g/ha + ppz 62.5 g/ha[2)]	7.83 c	7.31 ab	10.43 b
V: 5 - 7 x fpm 75 g/ha + ppz 31.25 g/ha[2)]	7.08 bc	8.42 b	12.18 b

Means followed by the same letter in the same column do not differ significantly, LSD P = 0.05.
[1)] before first treatment, all plots pooled [2)] after last treatment

Powdery mildew levels

Powdery mildew levels at the beginning of June and July are shown in table 3 for the different treatments, and for the three trials carried out from 1991 to 1993.

In 1991 the best performance through to the end of the season was obtained with two treatments of the full rate of fpm at 750 g/ha and with the mixture of 300 g/ha fpm plus 125 g/ha propiconazole. The highest powdery mildew level was found with four applications of the mixture of fenpropimorph at 150 g/ha plus propiconazole at 62.5 g/ha. Seven applications of the mixture fenpropimorph at 75 g/ha plus propiconazole at 31.25 g/ha (V) showed very good disease control (about 0.5 % attacked green plant tissue) at the beginning of June. But at the end of the growing season the powdery mildew level increased to 5 % attacked green plant tissue. However generally all treatments performed well.

In 1992 the best performance was obtained with the two split applications (IV, V). Six applications of fenpropimorph at 75 g/ha in combination with propiconazole at 31.25 g/ha (V) even showed better powdery mildew control than four applications of fenpropimorph at 150 g/ha in combination with propiconazole at 62.5 g/ha (IV). However 1992 was an exceptional year due to a severe drought during May with very low disease pressure. The first application in the full rate plots (II, III) was carried out mid May at 1-3 % powdery mildew attack and in the plot IV in the first week of May with an infection level of *Erysiphe graminis* f.sp. *tritici* of 0.5-1 %. The first rainfalls were recorded end of May and the powdery mildew attack in the check plot immediately reached 10 %. The second application of the full rate (II, III) was carried out 1 week later and so the timing of application in these plots was not related to the powdery mildew development.

Performance in 1993 was good in all treated plots, especially in those receiving fenpropimorph alone (II) or split applications (IV, V). However no late disease rating was carried out and therefore no clear conclusions on performance of the different treatments can be drawn.

TABLE 3. Powdery mildew levels on wheat at two dates during three seasons in full dose and split application treatments.

Treatment	% attacked green tissue					
	1991		1992		1993	
	June[1]	July[2]	June	July	June	July
I: check	20	30	19	45	50	nd
II: 2 x fpm 750 g/ha	1	1	3	20	2	nd
III: 2 x fpm 300 g/ha + ppz 125 g/ha	1-2	1-2	7	25	7	nd
IV: 4 x fpm 150 g/ha + ppz 62.5 g/ha	5	7	2	10	2	nd
V: 5 - 7 x fpm 75 g/ha + ppz 31.25 g/ha	1	5	1	5	1	nd

nd: not determined

[1] beginning of June [2] beginning of July

DISCUSSION

As a rule split applications resulted in very good powdery mildew control in all three trial seasons. In these treatments the disease rating at the end of the growing season showed an increase in powdery mildew level compared to the normal applications only in 1991.

With the treatments according to the label recommendation (two applications at the full rate) performance was also very good when the timing of application was adapted to the disease development. The importance of application timing is also shown for the split applications. More frequent treatments generally performed better than fewer treatments. Similar results were obtained by Jørgensen and Nielsen (1992) for *Erysiphe graminis, Puccinia striiformis* and *Septoria* spp. on wheat in Denmark.

The three trials reported here show that with the more frequent application at the reduced dose (split application) bigger shifts towards decreased sensitivity occur than with two applications of the full rate. This is a consequence of a continuous selection pressure resulting in a higher frequency of isolates with reduced sensitivity to fenpropimorph at the end of a season. With two treatments, of either fenpropimorph alone or the mixture with propiconazole, at the recommended rate and timing, changes in sensitivity were smaller and at the same time a good performance was obtained.

These results suggest that the application of the recommended label rates at the right time represent the optimum between the prevention of sensitivity shifts and excellent powdery mildew control. With split applications the margin of error in timing of applications decreases but clearly the risk for sensitivity shifts becomes much greater. Similar results were obtained in monitoring trials carried out in The Netherlands by T. Engels (personal communication, 1994). Thus it appears that the best application would be two treatments of the full rate at the best possible timing. This recommendation was also adapted by the FRAC-DMI working group (1993).

It remains to be determined whether the results shown here may also be extrapolated to other pathosystems.

ACKNOWLEDGEMENTS

The authors would like to thank Mr. H. W. Boedeker for carrying out the field trials and Mrs. O. Schafer for skilled assistance in carrying out the sensitivity tests.

REFERENCES

Bolton, N.J.E.; Smith J.M. (1988). Strategies to combat fungicide resistance in barley powdery mildew. *Proceedings of the 1988 Brighton Crop Protection Conference - Pests and Diseases*, **1**, 367-372.

Felsenstein, F.G. (1991). Virulenz und Fungizidsensitivität des Weizenmehltaus, *Erysiphe graminis* DC. f.sp. *tritici* Marchal, in Europa. PhD Thesis, Technische Universität München, Freising-Weihenstephan.

FRAC-SBI-Arbeitskreis (1993). Sterolbiosynthesehemmer - Resistenzgefährdung und empfohlene Antiresistenz-Strategien. *Gesunde Pflanzen* **5**, 193-196.

Jørgensen, L.N.; Nielsen, B.J. (1992). Reduced dosage of fungicides for controlling wheat diseases in Denmark. *Proceedings of the 1992 Brighton Crop Protection Conference - Pests and Diseases*, **2**, 609-614.

Lorenz, G.; Saur, R.; Schelberger, K.; Forster, B.; Küng, R.; Zobrist, P. (1992). Long term monitoring results of wheat powdery mildew sensitivity towards fenpropimorph and strategies to avoid the development of resistance. *Proceedings of the 1992 Brighton Crop Protection Conference - Pests and Diseases*, **1**, 171-176.

Sozzi, D.; Nuninger-Ney, C.; Staub, T. (1991). A test-tube method for assessment of propiconazole sensitivity in cereal powdery mildew isolates (*Erysiphe graminis*). *Bulletin OEPP/EPPO* **21**, 301-304.

Zadoks, I.E.; Chang, T.T.; Konzak, C.F. (1974). A decimal code for the growth stages of cereals. *Weed Research*, **14**, 415-421.

RECENT EVOLUTION AND CURRENT STATUS OF SENSITIVITY OF ERYSIPHE GRAMINIS F.SP. TRITICI TO FENPROPIMORPH IN DIFFERENT EUROPEAN REGIONS

F.G. FELSENSTEIN

Institute of Agronomy and Plant Breeding, Technical University of Munich, D-85350 Freising-Weihenstephan

ABSTRACT

To study sensitivity and population dynamics of the wheat mildew pathogen towards fenpropimorph on an European scale, a monitoring program has been carried out since 1987. Single isolates out of different regions were analysed. In relation to standard isolates with unselected sensitivity a resistance factor and for samples their mean (MRF) were calculated. The current results show regional differences in sensitivity of the pathogen towards fenpropimorph. In the east and south of Europe populations remained on their former sensitivity level whereas most populations in north-western Europe changed recognizably to a MRF $\geq$ 3 within recent years, due to selection pressure and wind dissemination of the pathogen. Compared with resistance evolution to azoles, sensitivity adaptation towards fenpropimorph appears to progress more slowly. Until 1993 only few populations reached MRFs of about 10. At the same time this sensitivity level seems to be like a first limit for resistance evolution. The latter is discussed to be due to the fact that morpholines are two-site-inhibitors and to the genetical recombination of the wheat mildew pathogen.

INTRODUCTION

Controlling wheat powdery mildew, caused by Erysiphe graminis f.sp. tritici, treatments are up to the present mainly based on two groups of active compounds, namely azoles (DMIs: demethylation inhibitors) and morpholines/morpholine-like-compounds, which have different modes of action inhibiting sterol biosynthesis of the fungus. Whereas during the last decade a partly evident sensitivity loss of the pathogen towards DMIs was monitored mainly for the north-west of Europe (Felsenstein, 1994), and a lot of data are available, there is a lack of information about sensitivity of the wheat powdery mildew towards morpholines on an European scale. This report provides a short survey on the recent evolution and current status of fenpropimorph sensitivity of the wheat mildew pathogen in Europe.

MATERIALS AND METHODS

Random samples of the pathogen were taken out of the air above regions of interest making use of a mobile jet spore trap. Single progenies of collected conidia spores were analysed for their sensitivity towards fenpropimorph. ED50 was calculated for each isolate tested. In relation to the mean ED50 of unselected standard-isolates a resistance factor was determined. To characterize each random sample the mean ED50 (geometric mean) and the mean resistance factor **MRF** is used (more details in Felsenstein, 1991; Felsenstein, 1994).

RESULTS

Current data about fenpropimorph sensitivity of random samples of wheat mildew out of different European regions are presented in Table 1. Region (route) and date of spore sampling, number of isolates tested, mean of ED50 values of single isolates within the random sample, its standard deviations and the total range are listed. For standard isolates the method used at Weihenstephan resulted ED50 values around 0.15 ppm. The mean ED50 of the samples ranged from 0.24 ppm in northern Italy up to 1.98 ppm in Scotland. Thus, at the current status of sensitivity there exists a visible regional differentiation towards fenpropimorph on an European scale. Sensitivity levels close to that of the standard isolates were observed for populations in southern and eastern Europe. Recognizable lower levels were found in general for populations in the north-west of Europe. But despite of a changed sensitivity level isolates with ED50 values close to those of the standards were still obtained in most samples.

A different situation was observed at the beginning of monitoring in 1984 (Lorenz & Pommer, 1984) and at the start of our investigations in 1987. In Table 2 mean resistance factors (MRFs) of the samples out of the wheat mildew populationes listed in Table 1 are presented for a seven year period from 1987 to 1993. In 1987 only two populations, one in Scotland ('Edinburgh-Grantshouse') and one in northern Germany ('Hamburg-Neustadt'), had a markably higher MRF value, if populations with MRF < 3 are considered to possess an unselected or nearly unchanged sensitivity level (Felsenstein, 1991). In the following years more and more populations of north-western Europe attained a level of MRF ≥ 3. As mentioned above, only populations in the south and east of Europe remained on their former

TABLE 1. Sensitivity of wheat powdery mildew from different European regions to fenpropimorph, 1993

region	date	n	**$\bar{x}$ED50***	$\bar{x}$-s - $\bar{x}$+s	min. - max.
GB:					
Edinburgh-Grantsh.	05.07.	10	**1.98**	1.50 - 2.62	1.43 - 3.15
Cambridge-Dover	05.07.	10	**1.74**	1.17 - 2.60	0.61 - 2.44
F:					
Calais-Lille	04.07.	10	**1.13**	0.65 - 1.98	0.24 - 1.63
Paris-Reims	18.06.	10	**0.81**	0.40 - 1.65	0.22 - 1.53
Bourges-Nevers	19.06.	10	**0.97**	0.37 - 2.57	0.18 - 2.37
Auch-Toulouse	26.05.	10	**0.43**	0.26 - 0.71	0.21 - 1.15
DK:					
Nyborg-Kopenhagen	15.06.	20	**1.36**	0.78 - 2.38	0.22 - 2.63
D:					
Hamburg-Neustadt	16.06.	18	**1.46**	0.85 - 2.52	0.23 - 2.63
Hannover-Kassel	12.06.	10	**1.23**	0.89 - 1.71	0.55 - 1.72
Dortmund-Warburg	15.06.	10	**0.75**	0.26 - 2.18	0.16 - 4.60
Magdeburg-Halle	12.06.	10	**0.86**	0.40 - 1.87	0.21 - 1.61
Nürnberg-Freising	05.06.	10	**0.57**	0.21 - 1.57	0.12 - 1.77
Pforzheim-Ulm	04.06.	10	**0.60**	0.30 - 1.22	0.20 - 1.43
CH:					
Baden-Bern	17.07.	10	**0.78**	0.37 - 1.64	0.28 - 1.86
A:					
Marchfeld/by Vienna	11.06.	10	**0.35**	0.19 - 0.62	0.20 - 1.43
I:					
Verona-Venedig	25.05.	10	**0.24**	0.21 - 0.28	0.20 - 0.31
standard isolates:					
"Benno"		5 x	**0.17**	0.15 - 0.20	0.15 - 0.21
"Sappo"		7 x	**0.16**	0.15 - 0.17	0.15 - 0.17
"W72"		4 x	**0.15**	0.12 - 0.19	0.11 - 0.19

* geometric mean of ED50 values of single isolates in the sample; concentration of the active component in mg/l (ppm)

TABLE 2. Mean resistance factors (MRFs) of random samples out of regional wheat mildew populations towards fenpropimorph in Europe, 1987-1993

region	1987	1988	1989	1990	1991	1992	1993
GB:							
Edinburgh-Grantsh.	4.2	1.7	3.1	4.9	4.2	5.6	12.4
Cambridge-Dover	1.8	1.3	2.0	4.0	2.6	7.0	10.9
F:							
Calais-Mons/Lille	1.1	1.8	2.0	2.3	4.8	1.6	7.1
Paris-Reims	1.3	2.6	2.1	1.8	2.2	11.9	5.1
Bourges-Nevers	-	-	-	1.4	1.3	4.0	6.1
Narb./Auch-Toulouse	2.1	-	1.7	-	1.4	1.3	2.7
DK:							
Nyborg-Kopenhagen	2.3	3.5	10.3	7.1	10.5	8.9	8.5
D:							
Hamburg-Neustadt	7.5	4.9	9.0	9.2	10.6	10.5	9.1
Hannover-Kassel	-	-	-	8.3	8.1	6.4	7.7
Dortmund-Warburg	2.1	1.9	11.8	-	10.9	7.3	4.7
Magdeburg-Halle	-	1.4	2.8	3.0	6.8	5.7	5.4
Nürnberg-Freising	2.7	1.5	2.0	1.6	2.9	3.6	3.6
Pforzheim-Ulm	2.3	1.4	1.7	1.6	1.9	5.6	3.8
CH:							
Baden-Bern	-	1.2	1.9	1.5	1.5	2.0	4.9
A:							
Marchfeld/by Vienna	1.4	2.4	2.6	1.7	1.6	1.4	2.2
I:							
Verona-Venedig	-	1.1	1.6	1.3	1.1	1.2	1.5
E:							
Irun-Tudela	0.9	-	1.2	-	0.8	1.2	-

level close to that of the standard isolates.

At the same time it has to be stressed that up to 1993 **no** population could be detected which reached a markably higher MRF than 10. This fact also holds true for the population of the region 'Hamburg-Neustadt' where MRF varied between ca. 5 and 10 during the whole investigation period.

DISCUSSION

Results presented show that the wheat mildew pathogen possesses the ability in sensitivity adaptation towards the morpholine compound fenpropimorph on principle. As MRF values > 3 have been obtained from areas with a relative frequent (e.g. splitting) and extensive use of appropriate fungicides, respectively, it is considered that the loss of sensitivity mainly depends on selection pressure. In the recent past also wind dissemination of the pathogen out of regions with higher MRFs seems to have an increasing influence on the sensitivity of some regional populations. For few populations (e.g. 'Edinburgh-

Grantshouse', 'Dortmund-Warburg') an evident variation in MRF-value from year to year has been obtained. This observation is characteristic for populations which got a wide range within their sensitivity structure.

At the same time results show a relatively moderate population dynamic according to a loss of sensitivity to fenpropimorph during the last decade. In most populations of north-western Europe sensitivity level changed markably at first at the beginning of the 90s. Only population structure did indicate changes at an earlier time (Felsenstein, 1991; Lorenz *et al.*, 1992). The main reason for the moderate resistance evolution and the slightly reduced sensitivities to fenpropimorph might be due to the fact that morpholines are considered to be a two-site-inhibitor (Buchenauer, 1984; Kato, 1986; Kerkenaar, 1987). In addition the polygenic control of resistance evolution towards morpholines combined with a markably genetical recombination of the wheat powdery mildew should be taken into account, as it is discussed for sensitivity of the pathogen to DMIs (Felsenstein, 1994). Because of the yearly redistribution of genes responsible for a loss in fenpropimorph sensitivity, maintenance and multiplication of pathotypes with higher resistance factors are probably hampered.

Nevertheless it is astonishing that despite of a high selection pressure during the last decade sensitivity level of the population in the region 'Hamburg-Neustadt' has **not** changed markably since at least 1987. This observation as well as data from other populations indicate that a MRF of about 10 is a first serious limit towards a furthermore adaptation in fenpropimorph sensitivity of the wheat mildew pathogen. Therefore successful pathogen control seems likely to continue at least in the near future.

ACKNOWLEDGEMENT

The author wishes to thank Maria Goldbrunner, Mariele Felsenstein, Johanna Stieber and Hannelore Meckl for excellent technical assistance.

REFERENCES

Buchenauer, H. (1984) Stand der Fungizidresistenz bei Getreidekrankheiten am Beispiel der Halmbruchkrankheit und des Echten Mehltaus. *Gesunde Pflanzen*, **36**, 161-170.

Felsenstein, F.G. (1991) Virulenz des Weizenmehltaus, Erysiphe graminis DC f.sp. tritici Marchal, in Europa. PhD Thesis. Technical University of Munich, Weihenstephan.

Felsenstein, F.G. (1994) Sensitivity of Erysiphe graminis f.sp. tritici to demethylation inhibiting fungicides in Europe. (In the present volume).

Kato, T. (1986) Sterol biosynthesis in fungi, a target for broad spectrum fungicides. In: *Chemistry of Plant Protection*, G. Haug; H. Hoffmann (Eds), Vol. 1, 1-24.

Kerkenaar, A. (1987) Mechanism of action of morpholine fungicides. In: *Modern Selective Fungicides*, H. Lyr (ed.), Gustav Fischer Verlag, Jena, and Longman Group UK Ltd., London, 159-171.

Lorenz, G.; Pommer, E.-H. (1984) Investigations into the sensitivity of wheat powdery mildew populations towards fenpropimorph (minitoring program). *British Crop Protection Conference - Pests and Diseases*, 489-493.

Lorenz, G.; Saur, R.; Schelberger, K.; Forster, B.; Küng, R.; Zobrist, P. (1992). Long term monitoring results of wheat powdery mildew sensitivity towards fenpropimorph and strategies to avoid the development of resistance. *Brighton Crop Protection Conference - Pests and Diseases*, 171-176.

EFFECT OF DIFFERENT TREATMENT SCHEDULES ON THE SENSITIVITY OF *ERYSIPHE GRAMINIS* F.SP. *TRITICI* TO FENPROPIMORPH

A.J.G. ENGELS and M.A. DE WAARD

Department of Phytopathology, Wageningen Agricultural University, P.O. Box 8025, 6700 EE Wageningen, The Netherlands

ABSTRACT

The effects of Corbel and Tilt Top split applications on disease control, and on sensitivity in *Erysiphe graminis* f.sp. *tritici* to fenpropimorph, were tested in field experiments in 1992 and 1993. In both years, wheat powdery mildew was controlled well in all treated plots. In general, the use of split applications resulted in a better protection against wheat powdery mildew than the use of normal applications. In the 1992 experiment no differences in sensitivity were found between split and normal applications. In the 1993 experiment however, isolates collected from plots treated with split applications had on average the highest EC_{50}- and Q_{50}-values (degree of resistance). This suggests that split application favours reduction in sensitivity to fenpropimorph.

Competitive ability of isolates with a reduced sensitivity to fenpropimorph was studied. Isolates with a reduced sensitivity showed a lower competitive ability than wild type isolates.

INTRODUCTION

One of the main problems in chemical control of plant diseases is resistance development in target organisms. This risk also applies to the fungicide fenpropimorph in the control of *Erysiphe graminis* f.sp. *tritici*, the causal agent of wheat powdery mildew. Some decrease in the sensitivity of this pathogen to fenpropimorph has already been found in a number of West-European countries (De Waard et al., 1992). As far as treatment schedules are concerned, the use of the single product, and the use of a mixture with a companion fungicide both in full rates as well as in split applications of the recommended rates may influence the build-up of strains with decreased sensitivity either positively or negatively.

Resistance development is influenced by the competitive ability of isolates with reduced sensitivity. When competitive ability of less sensitive isolates is lower than those of wild-type isolates, isolates with reduced sensitivity will only develop when a continuous selection pressure of the fungicide is present.

AIM OF THE RESEARCH

I. To study the effect of split application of fenpropimorph, as a single product and as a mixture, on the sensitivity of wheat powdery mildew.
II. To investigate the competitive ability of fenpropimorph-sensitive (FS) and -low resistant (FR) wheat powdery mildew isolates.

MATERIALS AND METHODS

I. Field trials were conducted in the Southern Netherlands in 1992 (pilot experiment) and 1993. Two fungicides, "Corbel" (a.i. fenpropimorph 750 g/l) and "Tilt Top" (a.i. fenpropimorph, 375 g/l and propiconazole 125 g/l), were applied as split (5 x 0.4 l/ha) or normal (2 x 1.0 l/ha) applications. The experimental field of 1993 was situated next to the one used in 1992. Powdery mildew development was assessed in all plots during the growing season. Plots were sampled for plants with mildew symptoms before the start of treatments and after all treatments were applied. Isolates collected were used in foliar spray tests to assess the sensitivity to fenpropimorph. Foliar spray tests were carried out in a spray cabinet in a standardized way. After spraying, the seedlings were inoculated and after 10-12 days of incubation, mildew severity of the seedlings was estimated. EC_{50}- and Q_{50}-values were calculated. The degree of resistance (Q_{50}-value) was calculated by dividing the EC_{50}-value of a field isolate by the average EC_{50}-value of a reference isolate LH (wild type) tested in the same test.

II. To study the competitive ability of FS- and FR- monospore isolates, two FS-isolates (LH and 67) and two FR-isolates (3a and 16c) were mixed on eight day old wheat seedlings (cv Okapi) in 12.5 cm diameter pots. In this way four mixtures, which consisted of four isolates, were made. These four mixtures and the separate FS- and FR-monospore isolates were transferred every week to unsprayed wheat seedlings (cv Okapi). These competition experiments were carried out using inoculum levels that assured 90% infection. The composition of the mixtures was assessed by transfer of 50 colonies to wheat seedlings sprayed with a discriminating concentration (11 $\mu g\ ml^{-1}$) of fenpropimorph. In contrast to FS-colonies, FR-colonies survived this treatment. Tests were done at the start of the experiment (week 0) and 1, 3, 6, 10 and 15 weeks after the start of the experiment.

RESULTS

I. Powdery mildew development

In both years, only small differences in mildew infestation were found between plots at the start of the treatments (DC31). During the course of the season leaves in the untreated plots got more infected than leaves of the other plots. Wheat powdery mildew was controlled well in all treated plots in both years. In general, split application Corbel and Tilt Top resulted in less powdery mildew development than normal application of these fungicides (Figure 1).

Sensitivity to fenpropimorph

The mildew population in the experimental field selected had a decreased sensitivity to fenpropimorph. At the start of the treatments in both years no significant differences in sensitivity were found between isolates from different plots. At the end of the 1992 pilot experiment, the sensitivity of isolates collected from plots with normal and split applications was similar in all treatments (data not shown). At the end of the 1993 experiment, the degree of reduced sensitivity of isolates collected from plots treated with split applications was higher than for isolates from all other plots (Table 1).

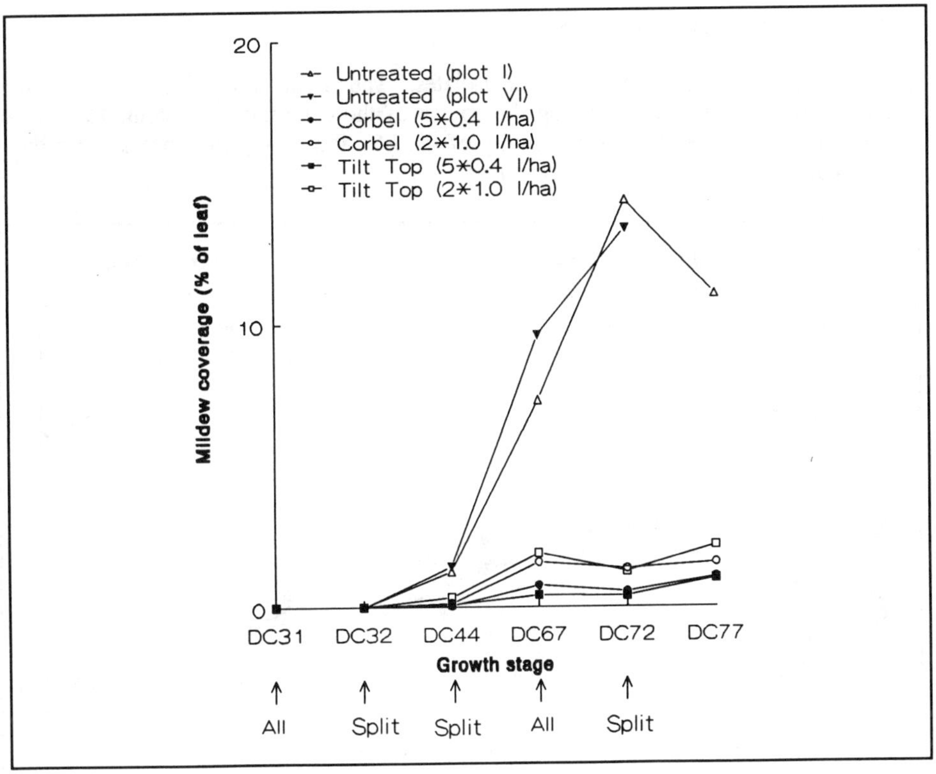

Figure 1. Development of *Erysiphe graminis* f.sp. *tritici* on the second leaf of wheat in six plots during the growing season of the 1993 field trial. Split: fungicide application in split application plots only. All: fungicide application in all plots.

Table 1. Average EC_{50}-values (μg fenpropimorph ml^{-1}) and Q_{50}-values of fenpropimorph for powdery mildew control in foliar spray tests with isolates from six plots in the 1993 field trial, I and VI = Untreated, II = Corbel 5 x 0.4 l/ha, III = Corbel 2 x 1.0 l/ha, IV = Tilt Top 2 x 1.0 l/ha and V = Tilt Top 5 x 0.4 l/ha.

	Treatment						LSD
	I	II	III	IV	V	VI	
EC_{50}	7.9 a	13.8 c	10.0 b	9.8 b	12.6 c	6.6 a	1.7
Q_{50}	6.2 a	11.0 c	8.0 b	7.8 b	10.0 c	5.3 a	1.3

Means followed by the same letter in the same row do not differ significantly, P = 0.05. Total amount (g) of a.i. of fenpropimorph: I and VI: 0, II and III: 1500, IV and V: 750.

II. Competitive ability assessments

Mixtures consisted of 55% FR-isolates. The frequency of FR-isolates in the mixtures gradually decreased during the course of the experiment reaching 15-20% after 15 generations (Figure 2). The sensitivity of the FS- and FR- monospore isolates did not change during the course of the experiment.

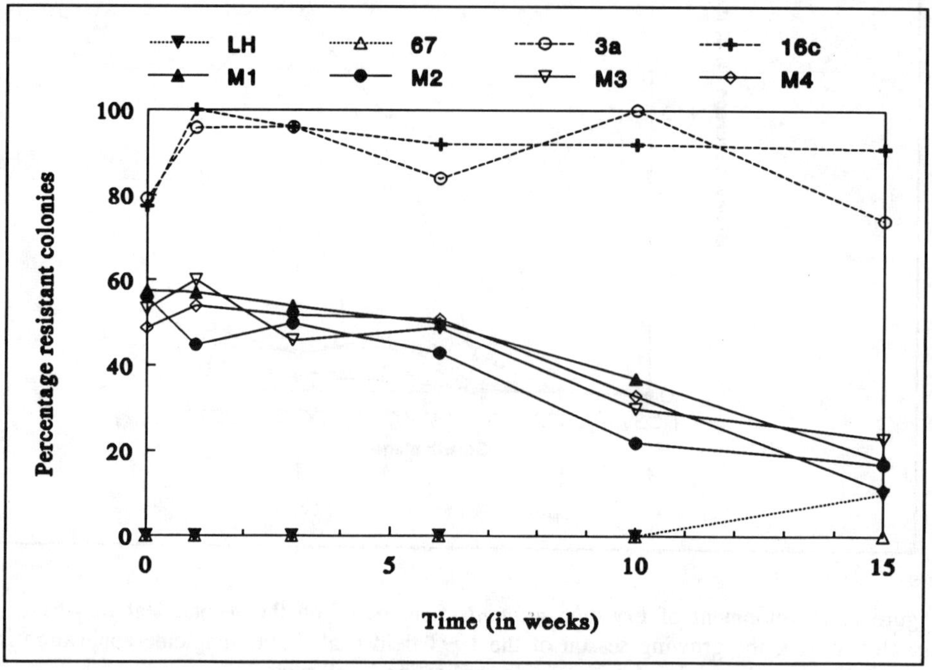

Figure 2. Changes in the composition of mixtures (M1...M4) and monospore isolates (FS: LH and 67; FR: 3a and 16c) at 20°C in the absence of fenpropimorph.

DISCUSSION

I. Powdery mildew development was highest in the untreated plots. In all treated plots mildew was controlled well. In general, split applications performed better in mildew control than normal applications. Similar results were obtained by Jørgensen and Nielsen (1992). Probably, split application results in a better timing of the application in relation to leaf emergence and hence more effectively prevents mildew development. No difference in mildew development between applications of Corbel and Tilt Top were observed. These results were similar in both years.

In both years, isolates collected before the start of the treatments showed no significant difference in sensitivity to fenpropimorph. In the 1992 experiment no differences in sensitivity were found in isolates from plots treated with split and normal applications. In the 1993 experiment, results of foliar spray tests demonstrated that isolates from plots treated with split application had on average the highest EC_{50}- and Q_{50}-values and isolates collected from untreated plots the lowest. This result suggests that in

1993 split applications favour stepwise reduction in sensitivity to fenpropimorph. Forster *et al.* (1994) found similar results in monitoring trials carried out in Northern Germany. A reason for the difference in results of 1992 and 1993 may be that in the 1993 experiment disease pressure was much higher than in the 1992 experiment. Also the fact that 1993 was the second successive year with a high selection pressure may have contributed to the 1993 results.

II. The results show that under crowded conditions the competitive ability of FR-isolates was less than that of FS-isolates. This suggests that in periods without fenpropimorph application and a high mildew infection the frequency of FR-isolates in practice may decrease.

ACKNOWLEDGEMENT

The authors would like to thank Ir. B.C. Mantel for his excellent contribution to the research. This project is fully supported by CIBA-GEIGY AG, Basel and BASF AG, Limburgerhof.

REFERENCES

De Waard, M.A.; Banga M.; Ellis S.W. (1992). Characterization of the sensitivity of *Erysiphe graminis* f.sp. *tritici* to morpholines. *Pesticide Science*, **34**, 365-377.

Forster, B.; Chavaillaz D.; Steden, C.; Radtke, W.; Käsbohrer, M.; Kühl, A. (1994). Influence of split application of fenpropimorph mixtures on disease control and on the sensitivity of wheat powdery mildew. These proceedings.

Jørgensen, L.N.; Nielsen, B.J. (1992). Reduced dosage of fungicides for controlling wheat diseases in Denmark. *Proceedings of the 1992 Brighton Crop Protection Conference - Pests and Diseases*, **2**, 609-614.

Session 5
Strategies for Managing Resistance

Chairman & Session Organiser P E RUSSELL

FUNGICIDE RESISTANCE MANAGEMENT: NEEDS AND SUCCESS FACTORS

P.A. Urech

Ciba-Geigy Ltd., Postfach, 4002 Basle, Switzerland

ABSTRACT

The result of 20 years of fungicide resistance management is encouraging. Modern fungicides continue to add to a high level of disease control despite the fact that many of these are prone to resistance. The strategies implemented did not prevent resistance in all cases but definitely contributed to delay the onset of resistance or to safeguard the high level of field activity. FRAC (Fungicide Resistance Action Committee) played a leading role in this sector.

The analysis of actual operations and organization of resistance management shows weaknesses and suboptimal use of resources which became more scarce in recent years. The squeeze on resources will continue with the ongoing restructuring process and changing priorities in the plant protection industry.

In this paper, ideas are proposed on how to revitalize fungicide resistance management in order to preserve the usefulness of modern fungicides. All of these ideas circle around more intense and constructive cooperation, coordination and communication. Harmonization in methods and information exchange is also an objective. The re-orientation must be done within industry, but should also include the relationship of industry with the public sector including academia. FRAC can and should again take the initiative, with efforts primarily being based in the countries. More regulation is not desired. Cooperation must evolve voluntarily and include all groups involved in plant protection.

INTRODUCTION

Ever since modern fungicides have been developed and marketed, field resistance has been either a reality or a continued, latent threat. As a result, fungicide resistance management became a standard product stewardship procedure in the industry, the main objective being to preserve the usefulness of these products. Resources were made available in many but not all research based companies for these tasks. The plant protection industry, which was so innovative during the last 20 years now faces strong economic head winds. With significant resources bound to safety and environmental challenges, the new situation requires a re-evaluation and a re-orientation in industry, including re-allocation of resources. The aim of this paper is to analyze the achievements and the present environment of fungicide resistance management and to propose new ways whereby this important task could be carried out more efficiently.

ANALYSIS OF PAST ACHIEVEMENTS AND STATUS OF FUNGICIDE RESISTANCE MANAGEMENT

It seems worthwhile to look at the following three sectors in order to clarify the actual situation of fungicide resistance management:

- Achievements
- Operational and organizational aspects
- Economic environment

20 years of fungicide resistance management: A success story?

In the 1970's widespread resistance to fungicides coupled with severe yield losses occurred for the first time. This was the moment when awareness in industry evolved that fungicide resistance management was to be an extraordinary and important task. Individuals with clear vision in industry, public services and academia were the driving forces. As a result of this, fungicide resistance management became a scientific and operational discipline with clear objectives, approaches and tasks allocated (Urech, 1984, 1990) as shown in Table 1.

TABLE 1: Job assignment in fungicide resistance management (according to Urech, 1984)

Active group	Action: Determination of inherent resistance risk	Risk evaluation and design of resistance strategy	Coordination between manufacturers	Implementation: communication, education	Monitoring	Adaption, new design of strategy
Industry	•••	•••	•••	•••	•••	•••
Academia Official Services	•			•••	••	•
Farmers				•••	•	
	Development stage		Marketing stage			

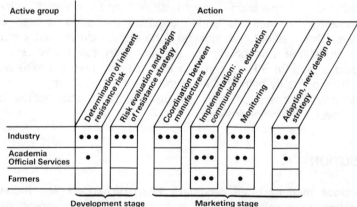

FRAC (Fungicide Resistance Action Committee), which was founded in 1981, was a breakthrough in industry cooperation and helped to shape the fungicide resistance landscape with big impact and authority not only by support of conceptual efforts including strategies, but also by training and education (Highwood, 1989). The results of these efforts confirm that the resources were well invested and that fungicide resistance management practices can be regarded as a success. In particular:

• Phenylamides remain a very valuable tool to control oomycetes despite resistance in foliar pathogens since 1980.

• Dicarboximides are still widely used despite the discovery of less sensitive or resistant strains.

• Even benzimidazoles have not disappeared from the market despite the fact that widespread resistance occurred more than 20 years ago.

• DMI fungicides are growing in importance despite the occurrence of less sensitive strains in some target pathogens.

Public and industrial research supported and accompanied practical management efforts. Although not all the phenomena observed in practice have been clarified (major gaps are present in our knowledge of the epidemology of resistant or less sensitive strains, in the resistance mechanisms and their genetic background) these efforts were of vital importance for risk assessments, modelling, cross resistance studies, synergism studies, etc. As a whole, for us there is no doubt that our joint efforts in fungicide resistance management (and by this we mean not only industry, but at the same time also official services and farmers) have preserved the usefulness of the modern fungicides. Some probably would have disappeared, others would have decreased in use. Winners are all social groups engaged in disease control, including farmers. Fungicide resistance was not prevented but mastered.

The way we operate in fungicide resistance management

Chemical plant protection is a highly regulated industry. Registration requirements and the FAO code of conduct for instance impose stringent regulations and guidelines on manufacturers and users of chemicals. With the exception of the still to be completed EU registration requirements, regulations have not yet covered fungicide resistance, and it is hoped that this complex matter will not undergo more regulation in future. However, with the exception of product related resistance management strategies, little agreed standards, procedures and approaches have been implemented so far. This makes life difficult for all those who are actively engaged in this field.

A few examples should highlight this point and consequently stress the need for action:

• Monitoring of field efficacy and sensitivity is often done using various methods and by only a few companies or institutions. All others take profit from this knowledge free of charge.

• Industry has little or no cooperative monitoring or resistance research programmes implemented.

• Advertisements using resistance arguments often damage the image of the industry because they are too obviously aimed at gaining commercial advantage.

• Resistance strategies are too often unilaterally proposed and promoted, be it from the official or industrial side.

• Resistance definition often remains a matter of view point, rather than an agreed state of the art, e.g. for triazoles.

• In many countries, developing an own view on fungicide resistance with consequently own studies is more important than cooperative investigation of resistance situations.

• In some countries, officials use legal means in a doubtful manner to regulate the use of fungicides which are prone to resistance, e.g. indexing in Switzerland.

• The registration of mixtures is not or only reluctantly allowed, e.g. India.

The changing economic climate in the plant protection industry

The heavy research and support efforts (search for new products, new formulations and packaging, product stewardship, safety, IPM and others) could be easily financed in many companies when economic growth and healthy productivity were the rule in the industry. Over the last 3 years, however, the economic situation has undergone a significant change and external as well as internal factors have contributed to a gloomy economic outlook in the industry:

• Shrinking fungicide markets in Western Europe due to CAP and in Eastern Europe due to economic difficulties.

• Increased competition leading to lower prices.

• Higher cost of doing business, including discovery, design and registration costs.

• Generic producers entering the market.

The plant protection industry is therefore going through a restructuring process worldwide. As a consequence of decreasing profits, resources for all activities become smaller and managers are asked to focus on the most promising projects. In such a climate, technical programmes aiming at managing fungicide resistance have to compete with many other high priority research and maintenance projects. There is, and will be, less money available for fungicide resistance management. Whether the same is true in official services is open, but we could imagine that the shrinking agronomic sector in Western Europe certainly will have an influence on fungicide resistance research and monitoring programmes. That is where we are today: Successful in fungicide resistance management in the past. But what will the future bring us?

FUNGICIDE RESISTANCE MANAGEMENT IN THE LIGHT OF THE CHANGING ECONOMIC ENVIRONMENT

There is no doubt that effective fungicide resistance management must remain an essential part of product stewardship. It must remain a clear and agreed objective for the industry as well as for all official bodies involved in plant protection. The analysis in the previous section shows us that:

• Success stories in fungicide resistance management prove that the conceptual working procedures are sound.

• Resources available for resistance management will shrink.

• Productivity improvements on the operational and organizational level are absolutely possible. This now gives us the basis to propose ideas for a re-orientation in fungicide resistance management.

The framework could be the following: Continued successful fungicide resistance management will only happen if a closer cooperation between industry members and the industry and public services will be realized. Cooperative and integrated work programmes must be considered combined with a free flow of information and better understanding of research methods and resistance strategies.

These general ideas can be clarified by describing the needs and success factors.

Needs

Needs are defined as those factors which are absolutely necessary for highly effective, but also very efficient fungicide resistance management.

The following is a list primarily seen by an industry member.

1. Resources: Sufficient resources must remain available within industry and public services/academia. All industry members selling products which are prone to resistance must contribute. There is no free ride. Funding of projects or doing own scientific work are possible contributions. Best use of resources can be achieved by focusing, priority setting and high quality work.

2. Cooperation: Project related, national or transnational cooperation must be intensified. "Who does what" agreements must lead to a wise use of available resources. Cooperation must improve among industry and between industry and the public sector. Basic research and field monitoring are the main targets.

3. Strategies: Enforceable resistance strategies based on scientific risk assessments for all fungicides at risk are obligatory. There can be no unilateral ruling in resistance strategies either from industry or from the public sector. There is no veto right and no command. Mutual agreement must be the goal in decision making on strategies. Implementation in practice needs understanding and support in all social groups involved in plant protection. Strategies must be seen as changeable if new data justify this. Communication with one voice increases credibility towards the farmers who are to follow the strategies.

4. Field monitoring: This is a product stewardship task and a must. The originators are in the drivers seat, but should also include licencees. Monitoring methods should become more harmonized in order to save resources, facilitate and improve the quality of interpretation of results. See also under point 1 and 2.

5. Residual fungicides: Must remain in the market in order to be included in the design of resistance strategies.

6. Resistance definition: There must be agreement on how to define field resistance for product groups or for individual products. In a much more responsible way, experts have to clarify what is and what is not applicable. There is an obvious and urgent need to do this for the DMI's.

Success factors

They determine whether and to which degree we will be successful in fungicide resistance management.

1. Information exchange: Honest and open exchange of e.g. monitoring data within project groups, and in a consolidated form to the outside is vital if efficiency in resistance management is to be realized.

2. Role of FRAC: It must continue to support efficiently all efforts in fungicide resistance management. Within industry, it must more than ever encourage cooperation. Outside of industry, it must become more open and willing to cooperate.

3. Generic producers: Must stick to FRAC guidelines and accept their share of costs, e.g. monitoring. Officials could and should become more demanding in this respect.

4. Research: Additional research programmes are needed to better understand some of the open questions, e.g. epidemiology of resistant or less sensitive strains, survival of such strains, genetic and biochemical background of resistance.

5. Farmer support: So far, resistance was not so often an issue to them because new products kept emerging to solve occuring problems. Better explanation at users level as to why resistance management efforts are needed and take place is necessary.

6. EU resistance regulations: Have to be finalized soon. We do not want more regulation but support what is written in the directive. All companies must stick to it.

7. Advertisements: Advertising fungicides using the resistance argument must be done in an objective way and must have the aim to inform users correctly and not to discredit competitors. A lot of credibility is at stake for industry.

8. Marketing and resistance management: Everybody must understand that resistance management is only one element in the use strategy of a product.

Marketing has a vital interest in the performance of a product and its longevity. What marketing people do not want is prohibitive regulation and use.

CONCLUSIONS

Modern fungicides represent a controversial, fragile technology. While their technical potential could present a revolution in disease control e.g. curative use and season long use, resistance, aspects pose severe restrictions on such an unlimited wide application. During the 1980's many new products were developed to the market, but most were DMI's and not many new modes of action were discovered. Some older products disappeared from the market.

Fungicide resistance management is therefore important not only to industry but for all social groups involved in plant protection. As far as industry is concerned one has to accept the fact that R & D resources must be directed to many projects and that some of them have also high priority. Among them are: research for new technologies to control plant diseases, biologicals, chemicals with novel mode of action, diagnostics etc.

The ideas presented in this paper should therefore contribute to improve three aspects of fungicide resistance management:

- Enforcement that fungicide resistance management is important.
- More efficient utilization of available resources, doing work more professionally and with better organization.
- More cooperative understanding within all social groups involved.

How to reach these goals?

FRAC groups must become more active and take the initiative, primarily within industry. Industry should define projects for research and monitoring which would then be carried out as joint programmes or farmed out to public services with corresponding funding.

FRAC groups should communicate more actively with the public sector, being from Steering Committee or being by the working groups. Decisions have not been made, but the need for such action seems obvious. At the country level, industry, advisory services and academia could engage in more cooperative programmes for monitoring, with clear objectives for this work, good definition of methods and more intense data exchange. The dialogue on resistance strategies must lead to agreeable positions, in a partnership like attitude and understanding. There is no place for unilateral actions, because resistance strategies must be supported by all influencers. Clearly, the final use strategy of a fungicide including resistance aspects is within the responsibility of the industry, however, the more agreement that can be reached about the best way to manage fungicide resistance, the better for the products. Fungicide resistance management has to stay: The way we do it has to change and become more professional.

REFERENCES

Highwood D.P., (1989) Fungicide Resistance Action Committee, *Pesticide Outlook,* **Vol. 1**, Issue 3, 1989

Urech P.A., (1984) Management of Fungicide Resistance in Practice, *Bulletin OEPP/EPPO*, Bulletin **15**, pages 571 - 575, 1985

Urech P.A., (1990) Management of Fungicide Resistance - A shared Responsibility, *Proceedings, 3rd International Conference on Plant Protection in the Tropics,* Malaysia, **Vol. III**, pages 105 - 109, 1990

STRATEGIES FOR MANAGEMENT OF FUNGICIDE RESISTANCE: CEREALS

D R JONES,

ADAS Bridgets Research Centre, Martyr Worthy, Winchester, Hants SO21 1AP, UK

ABSTRACT

Fungicides were used on over 95% of winter wheat and over 90% of winter barley crops in England and Wales in 1993. Over 90% of all applications included a DMI fungicide, and over 50% included a morpholine. Most farmers applied between one and three fungicides to winter wheat (mean 2.10) and one or two to winter barley (mean 1.35). A large proportion of applications include a mixture of two or more fungicides. Current ADAS and Fungicide Resistance Action Committee (FRAC) guidelines for reducing the risk of fungicide resistance are discussed. It is concluded that, in relation to fungicide mixture or alternation, current farming practice corresponds reasonably with the guidelines, although there will need to be a greater emphasis on the use of mixtures if threats such as morpholine resistance in *Erysiphe graminis* or DMI resistance in *Septoria tritici* become widespread. The area of greatest concern is the use of DMI seed treatments, used on almost 25% of winter wheat and winter barley crops in 1993. In many cases, alternative treatments from different fungicide groups would be appropriate.

INTRODUCTION

Fungicide resistance has been recognised for many years as an important consideration in the effective use of fungicides in UK cereal crops. A list of resistance problems is given in Table 1 (Fletcher & Locke, 1993). The current status of resistance for many of these pathogen/fungicide combinations, and the risk of resistance for others, is covered elsewhere in this volume and will not be detailed here. The objectives of this paper are to review how and why fungicides are used on cereals, to what extent this corresponds with current guidelines for avoidance or management of resistance, and whether these strategies are appropriate.

FUNGICIDES ON CEREALS

The main pathogens of wheat and barley for which fungicides are available, and the most effective fungicide groups for their control, are listed in Table 2. This list is not exhaustive, and takes no account of differences in efficacy between members of a group, but indicates which fungicides are most likely to be used for control of a particular pathogen. The current dependence on DMI (demethylation inhibitor) fungicides and on morpholines is shown clearly. These two groups constitute the SBI (sterol biosynthesis inhibitor) fungicides. Protectant fungicides are used, mostly in mixture with a DMI fungicide, for control of *Septoria* spp., but for all the other pathogens except *Rhynchosporium secalis* there is virtually complete dependence on DMI and morpholine fungicides.

TABLE 1. Fungicide resistance on cereals in the UK.

Crop	Pathogen	Disease	Fungicide or group name	Frequency
Wheat	*Pseudocercosporella herpotrichoides*	Eyespot	MBC	common
	Septoria tritici	Leaf spot	MBC	common
	Erysiphe graminis	Powdery mildew	DMI (some)	common
	Fusarium nivale	Brown foot rot	MBC	common
	Fusarium culmorum	Ear blight/foot rot	MBC	infrequent
Barley	*Pseudocercosporella herpotrichoides*	Eyespot	MBC	common
	Erysiphe graminis	Powdery mildew	DMI (some)	common
			ethirimol	common
	Rhynchosporium secalis	Leaf blotch	DMI (some)	common
			MBC	common
	Ustilago nuda	Loose smut	carboxin	infrequent
	Pyrenophora graminea	Leaf stripe	organomercury	infrequent
Oats	*Pyrenophora avenae*	Leaf spot	organomercury	common

TABLE 2. Summary of fungicide efficacy against pathogens of wheat and barley in the UK.

Pathogen	Disease	Most effective fungicide group(s)
Septoria tritici	Leaf spot	DMI ± chlorothalonil
Septoria nodorum	Glume blotch	DMI, MBC + protectant
Erysiphe graminis	Powdery mildew	Morpholine
Puccinia spp.	Rusts	DMI, morpholine
Pyrenophora teres	Net blotch	DMI
Rhynchosporium secalis	Leaf blotch	DMI + morpholine or MBC
Pseudocercosporella herpotrichoides	Eyespot	DMI

Data on foliar fungicide use on winter cereals in England and Wales are given in Table 3, and data on seed treatments are in Table 4 (R.W. Polley, pers. comm.). There are no comparable data for spring cereals, which generally receive fewer fungicide applications than winter cereals. On winter cereals, the most commonly used fungicides were the DMIs, followed by morpholines, both at GS30-33 and GS37-69 (Tottman, 1987) (Table 3). In most cases, morpholines were applied in mixture with another fungicide, and only 12.4% of winter wheat and 2.8% of winter barley crops received a morpholine fungicide alone. A substantial number of crops received an MBC fungicide but, in every instance except one, the MBC was in a mixture with another fungicide. The other fungicides used were mostly protectants, such as chlorothalonil or dithiocarbamates. The proportions of winter wheat and winter barley crops which were treated with fungicides were 95.7% and 91.0% respectively, and the mean number of fungicide applications per crop was 2.10 for winter wheat (unchanged from the two previous years) and 1.35 for winter barley (compared with 1.43 in 1992 and 1.51 in 1991).

TABLE 3. Foliar fungicide usage on winter wheat and winter barley in England and Wales, 1993.

Fungicide group	Percentage of treated crops which received fungicides from each group			
	Winter wheat		Winter barley	
	GS30-33	GS37-69	GS30-33	GS37-69
DMI	91.0	93.4	87.9	93.6
Morpholine	48.7	63.4	68.2	63.8
MBC	14.4	28.8	43.6	32.6
Total % crops treated	75.0	86.4	74.6	39.0

Data from MAFF winter wheat and winter barley disease surveys (R.W. Polley, pers. comm.). Total number of crops in surveys: winter wheat 370; winter barley, 354.

TABLE 4. Fungicide seed treatments on winter wheat and winter barley in England and Wales, 1991 and 1993.

Fungicide	Percentage of crops treated			
	Winter wheat		Winter barley	
	1991	1993	1991	1993
Organomercury	79.7	0	72.6	0
Carboxin + thiabendazole [+ imazalil][a]	0	26.4	0	29.7
Guazatine [+ imazalil][a]	0	17.7	0	9.3
Triadimenol + fuberidazole	17.5	22.9	6.6	10.2
Flutriafol + ethirimol + thiabendazole	0	0	15.4	15.6
Treated but product unknown	0	28.1	0	31.5
Untreated	2.8	4.9	5.4	3.6

Data from MAFF winter wheat and winter barley disease surveys (R.W. Polley, pers. comm.). Data from 354 and 345 wheat and 354 and 332 barley crops in 1991 and 1993 respectively.
[a]Imazalil included in barley treatments only.

The seed treatment data show the effect of the withdrawal in 1992 of organomercury seed treatments, which had previously been the standard treatment for over 50 years (Table 4). The 1991 data are for the final year in which organomercury was widely available, and 1993 was the first year in which it was unavailable. On both wheat and barley, the main trend was for treatments based on carboxin or guazatine to be used in 1993 in place of organomercury. There was a small increase in the use of the DMI fungicide triadimenol, but no change in the use of the other DMI, flutriafol, on barley. There was a substantial number of farmers in 1993 who had used treated seed, but did not know which treatment had been applied.

HOW DOES THE FARMER USE FUNGICIDES?

Although most cereal farmers have a general appreciation of when and why fungicides should be applied, few farmers feel confident to make all their fungicide decisions without assistance. In addition, fungicide selection is only one of many agronomic decisions that each farmer must make. Most farmers, therefore, rely substantially on a consultant, who may be concerned to avoid risk of serious disease or other mishaps, and therefore may be inclined to recommend insurance treatments. Farmers themselves are generally risk-averse, and may favour this approach. The large proportion of farmers in 1993 who did not know which seed treatment had been applied (Table 4) is an extreme illustration of the extent to which farmers delegate the decision-making!

Few farmers have the time available to monitor crops frequently for disease, and few consultants are able to visit crops sufficiently often to be confident of early detection of all diseases. In consequence, most farmers/consultants have a basic strategy, which is planned before any fungicides are applied. This strategy is then adjusted according to the perception of disease risk, based either on knowledge of diseases in the crop or the vicinity, or on weather conditions in relation to disease. On winter wheat, many farmers apply a fungicide in the period GS30-33, usually at GS31 or GS32 (Table 3). In some instances eyespot is the main target disease, and on some cultivars there is a clear risk of yellow rust, mildew or *Septoria tritici*, but in many cases there is not a clearly identified disease threat and the spray is intended as an insurance against foliar diseases. Almost all crops receive one or two fungicides in the period from GS37-GS69, often with *S. tritici* as the main target, but also with the aim of preventing other foliar diseases from becoming severe. On winter barley, over 50% of crops received just one fungicide in 1993, in most cases at GS31 or GS32. Further sprays are applied as routine for insurance by some farmers, but only on an "as necessary" basis by the majority. Fungicide treatment of other winter cereals (oats, rye) and of spring cereals is more usually in response to specific disease risks rather than to a pre-determined plan.

Although many farmers and consultants have a structured approach to fungicide timing, they are increasingly flexible in relation to application rates. Reduced rates of fungicides have been used by some farmers for many years, but there is increasing interest in reduced rates as more data become available on their efficacy. On winter wheat, Paveley & Lockley (1993) showed that a single application of propiconazole plus fenpropimorph at one quarter of the recommended rates was as effective against yellow rust as a full rate treatment, provided the sprays were applied in the week that yellow rust was first detected in the crop. When the same treatments were applied one week earlier or later, the low rate treatment was less effective than full rate. Similarly on spring barley, Wale (1993) found that one quarter rate, or even one eighth rate, of propiconazole plus tridemorph was effective against mildew provided that it was timed accurately. If treatment was delayed until mildew was well established, a higher rate was required. It is widely understood among farmers and consultants that low rates of application can be very effective, if timed correctly. It is also appreciated that the appropriate rate of application depends not only on disease severity, but also on crop development, host resistance and other factors. Due to economic pressures, reduced rates are here to stay. It is important that they are used appropriately.

STRATEGIES FOR AVOIDANCE OR MANAGEMENT OF RESISTANCE

In the short term, there are unlikely to be any new fungicides with different modes of action available to the UK cereal grower, so strategies will need to concentrate on maximising the effectiveness of DMI and morpholine fungicides. Current ADAS guidelines for reducing the risk to cereal fungicides are in Table 5 (Jones, 1993), and the recommendations of the Fungicide Resistance Action Committee (FRAC) SBI Working Group are in Table 6.

TABLE 5. ADAS recommendations for reducing the risk of fungicide resistance on cereals.

Reduce disease severity by non-chemical means, like good husbandry, growing resistant varieties, and by varietal diversification.
Ensure that fungicides are applied at the optimum times, to give maximum effectiveness.
Avoid the frequent application to a crop either of the same fungicide or of different fungicides which act in the same way. Whenever possible, a fungicide with a different mode of action should be used for the next application.
Make full use of multi-site fungicides. They are less prone to resistance problems.
Make full use of any appropriate formulated fungicide mixtures or label-recommended tank mixtures. To minimise the risk of resistance, both fungicides in the mixture should be effective against the disease or diseases concerned.
Avoid applying fungicide sprays at times when they are not cost-effective or when they are likely to be only marginally cost-effective - for instance, on cereals in the autumn or after flowering.
On winter cereals, minimise the use of seed treatments containing active ingredients which act in the same way as fungicides likely to be used later to control leaf disease.

APPLICATION OF RESISTANCE MANAGEMENT STRATEGIES TO CEREALS

Several of the strategies listed in the ADAS and FRAC guidelines (Tables 5 and 6) relate to fungicide mixtures or alternation of fungicide groups, and avoiding repeated use of one group alone. Where mixture and alternation strategies have been compared, mixture strategies have proved more effective (Staub, 1991). In practice, there are few problems in fitting these strategies to current farming practice. Most winter wheat and barley crops receive only one or two foliar fungicides; in the 1993 surveys 32.5% of wheat and 6.2% of barley crops received more than two foliar fungicides (R.W. Polley, pers. comm.). A large proportion of these treatments include a mixture of two fungicides from different groups, with mixtures generally selected to increase efficacy, allow lower application rates or broaden the spectrum of activity, rather than specifically as an anti-resistance strategy.

TABLE 6. FRAC general recommendations for reducing the risk of resistance to SBI fungicides.

Repeated application of SBI fungicides alone should not be used on the same crop in one season against a high risk pathogen in areas of high disease pressure for that particular pathogen.
For crop/pathogen situations where repeated spray applications (e.g. orchard crops/powdery mildew) are made during the season, alternation (block sprays or in sequence) or mixtures with a non cross-resistant fungicide are recommended.
Where alternation or use of mixtures is not feasible because of lack of an effective or compatible non cross-resistant partner fungicide, then input of SBIs should be reserved for critical parts of the season or crop growth stage.
Where DMI or morpholine performance is declining generally and sensitivity testing has confirmed the presence of less sensitive forms, SBIs should only be used in mixture or alternation with non cross-resistant partner fungicides. For control of cereal powdery mildews, mixtures or alternation of a DMI with a morpholine fungicide represent the best currently available non cross-resistant combination. Consideration should also be given to restricting DMI use to critical parts of the season or crop growth stage.
Users must adhere to the label recommendations.
Fungicide input is only one part of crop management. Fungicide use does not replace the need for resistant crop varieties, good agronomic practice, or plant hygiene/sanitation.

Farmers would normally use a morpholine rather than a DMI where there is a risk of severe mildew, but since most morpholine applications are made in mixture with a DMI, farmers are, perhaps without realising, taking an anti-resistance measure for mildew, and also rusts. If morpholine sensitivity in the mildew fungi declines such that there is resistance in practice, then greater emphasis will need to be placed on using morpholines in mixture with a DMI, and more attention paid to the differences between DMI fungicides in mildew activity. Since DMI resistance was found in *Rhynchosporium secalis* (Kendall *et al.*, 1993), farmers have readily accepted that a DMI plus morpholine or DMI plus MBC mixture should be used. With MBC resistance also found recently to be widespread in *R. secalis* (Phillips & Locke, 1994), there may need to be increased reliance on DMI/morpholine mixtures. For control of *Septoria tritici*, farmers often use a DMI plus chlorothalonil mixture, but a significant proportion of crops receive a DMI alone. To date, there is little evidence of a significant reduction in sensitivity of *S. tritici* to DMI fungicides, but if such a reduction is detected, the importance of the mixture with a protectant fungicide will be increased, and may need to become a standard recommendation. At present, farmers do not have an alternative to DMIs for eyespot control, but the recent introduction of cyprodinil in France provides farmers there with an alternative mode of action, which may become available in the UK in the future.

The fact that farmers often use application rates lower than label recommendations has been discussed above. In the past, it was believed that the use of low rates could enhance the risk of resistance, but it is now considered that the duration of exposure to fungicide is more important than application rate in determining resistance risk (Heitefuss, 1989). Although general observance of label recommendations must be encouraged, farmers are allowed to apply any rate they wish below the stated application rate, and will continue to do so.

The recommendations on preventing disease by non-chemical means should be encouraged, but options are limited for the cereal grower. The main methods by which the grower can manipulate disease are rotation, cultivar selection, sowing date and nitrogen application rate. However, for those diseases which can be controlled by fungicides, it is often more profitable for the farmer to accept the cost of fungicides rather than introduce other changes which may reduce the yield or quality, and hence profitability of the crop or of the farming system. Cultivars with good resistance to disease have only found wide acceptability if they can produce the same gross margins as more susceptible cultivars. In practice, the only disease which farmers routinely manipulate by crop rotation and delayed sowing is take-all (*Gaeumannomyces graminis*), for which there is no reliable commercially available fungicide treatment. If the cost:benefit ratio for fungicides changes markedly (which may happen as grain prices fall during the next few years), or if a resistance problem develops which makes control of a particular disease difficult to achieve, then farmers will re-examine non-chemical methods of disease control.

The final consideration is that because cereal growers are very dependent on DMI and morpholine fungicides in the spring and summer, there is a strong case for not using them at other times, in order to avoid unnecessary risk of resistance development. This is consistent with both ADAS and FRAC guidelines. Foliar treatment of winter barley for mildew control is sometimes thought necessary in the autumn or winter, although the proportion of crops where it is worthwhile is small (M.J. Hims, pers. comm.). Treatment of wheat before GS30 is very rarely required.

The main area of concern is the use of DMI fungicides in seed treatments. There are instances where a DMI seed treatment is clearly justified, for example in controlling bunt (*Tilletia caries*) of wheat or loose smut (*Ustilago nuda*) of wheat or barley in the early stages of seed multiplication, but there are alternative fungicides effective against bunt for commercial crops, and loose smut in both wheat and barley should not require control on commercial seed stocks if the diseases were controlled earlier in multiplication. There are other circumstances where a DMI seed treatment may be justified, for example on wheat cultivars which are very susceptible to yellow rust, but these represented only a small proportion of the total area of wheat in 1993 (less than 5%). Also, triadimenol plus fuberidazole seed treatment can give useful early suppression of take-all, although it cannot be relied upon, so may be justified for early sowings in fields known to have a high risk of take-all. The resistance risk from use of a DMI seed treatment is probably greatest for

Septoria tritici, which infects most winter wheat crops during the autumn, when some of the active ingredient from the seed treatment will still be in the leaf tissues. There is also a risk with the rusts and, possibly, other diseases. On winter barley, the use of flutriafol plus ethirimol plus thiabendazole is preferable to triadimenol plus fuberidazole in relation to DMI resistance in mildew, but ADAS experiments in the 1980s showed that neither treatment was justified in most situations (M.J. Hims, pers. comm.).

It is encouraging that, following the withdrawal of organomercury seed treatments, there has been only a small increase in the proportion of crops receiving a DMI seed treatment. In view of the relatively good correspondence between current practice with foliar fungicides and what would be recommended to combat resistance (even though strategies are rarely designed specifically for resistance management), it would be unfortunate if these beneficial effects were negated by injudicious use of DMI seed treatments.

REFERENCES

Fletcher, J.T.; Locke, T. (1993) Fungicide resistance review. *BSPP Newsletter* **22**, 25-32.

Heitefuss, R. (1989) *Crop and plant protection: the practical foundations.* John Wiley & Sons, Chichester.

Jones, D.R. (1993) How to resist resistance. *Farmers Weekly Fungicides Supplement*, 12 February 1993.

Kendall, S.J.; Hollomon, D.W.; Cooke, L.R.; Jones, D.R. (1993) Changes in sensitivity to DMI fungicides in *Rhynchosporium secalis. Crop Protection* **12**, 357-362.

Paveley, N.D.; Lockley, K.D. (1993) Appropriate fungicide doses for winter wheat - balancing inputs against the risk of disease-induced yield loss. *Proceedings of the 1993 Cereals R & D Conferences, Robinson College, Cambridge*. Home-Grown Cereals Authority, London.

Phillips, A.N., Locke, T. (1994) Carbendazim resistance in *Rhynchosporium secalis* in England and Wales. *Fungicide Resistance*, BCPC Publications.

Staub, T. (1991) Fungicide resistance: practical experience with anti-resistance strategies and the role of integrated use. *Annual Review of Phytopathology* **29**, 421-442.

Tottman, D.R. (1987) The decimal code for the growth stages of cereals. *Annals of Applied Biology* **110**, 441-454.

Wale, S. (1993) Reducing fungicide use on spring barley with confidence. *Proceedings of the 1993 HGCA Cereals R & D Conference, Robinson College, Cambridge*. Home-Grown Cereals Authority, London.

STRATEGIES FOR MANAGING RESISTANCE TO FUNGICIDES IN PRACTICE IN VITICULTURE

M. CLERJEAU

INRA, Institut de la Vigne de Bordeaux, Station de Pathologie Végétale, BP 81, 33883 Villenave d'Ornon Cedex, France

ABSTRACT

Strategies for managing resistance concern *Botrytis cinerea* (benzimidazoles, dicarboximides, phenylcarbamates), *Plasmopara viticola* (phenylamides) and *Uncinula necator* (DMI). Present measures, proposed and in practice, for the management of resistant pathogen populations are presented here and the difficulties encountered are stressed. In all circumstances, it is preferable to organize treatment programs which limit the use of fungicides at risk. It has been difficult to have this idea accepted even though the alternative products are more effective. Mixtures are generally used to guarantee efficacy (phenylamides – downy mildew) and are not always the best solution (DMI – powdery mildew).

INTRODUCTION

During the last 20 years, viticulture has been increasingly confronted with the problem of resistance to fungicides. There are three diseases of principal economic concern, grey mold, downy mildew, and powdery mildew. Each year, these diseases require a considerable number of preventive treatments as a form of insurance (7 to 9 for powdery mildew, 2 to 4 for grey mold, on average each year in France). These diseases are caused by airborne pathogens which possess a remarkable sporulation capacity rendering them extremely destructive to the grape harvest and its quality. For this reason, the selective pressure exerted by the fungicide treatment on the pathogen population is considerable. This pressure is even greater in regions where the risk of epidemics is higher (i.e. *Plasmopara viticola* in the Atlantic vineyards, and *Botrytis cinerea* in Champagne) because growers in these regions favor fungicides whose most effective active ingredients generally have a unisite mode of action. This has lead to the rapid development of resistant strains 2 years, after benzimidazole was put on the market, then the dicarboximides and the phenylcarbamates (diethofencarb) for *Botrytis* or the phenylamides for *Plasmopara*. On the other hand, several years (5 to 7) of intensive use of DMIs have been necessary to see the first signs of the development of resistant strains of powdery mildew (*Uncinula necator*) in the vineyard. These effects are attributed to whether the nature of the resistance is disruptive or progressive (monogenic for the former, polygenic for the latter) (Clerjeau, 1991).

In response to the problem of the decrease in efficacy of these fungicide treatments, it has been necessary to instruct growers as to how to minimize risk. This has required close cooperation between research, offical and professional organizations and agrochemical firms (monitoring of pathogen populations, experiments, information operations and communication) leading to a code for "living with resistance".

Anti-resistance strategies : probably an ambitious expression

Even though there are good intentions on the part of private companies and professional and offical services to cooperate in the face of the problem of resistance, it must be admitted that the actual putting into practice of anti-resistance measures has lacked total consensus. There has sometimes been disagreement as to the evaluation of risk and measures to be taken. In this context, where problems of individual interests are far from minor, actions taken by the partners are not always coordinated or the objectives are not necessarily a veritable anti-resistance strategy.

To date, in France, the diverse problems of fungicide resistance in the vineyard have led to certain lines of action which can be summarized as : 1) Detection of the first resistant strain by researchers. 2) Establishment of monitoring operations thoughout the country with the aid of private firms. 3) Recommendations from the official services in order to reduce the number of fungicide treatments. 4) Multiyear, multisite experimentation to evaluate the use of mixtures of fungicides, the effect of fungicide rotation programs, and the effect of the applied dose. 5) Adjustment of recommendations in light of acquired information. In as far as possible, the cooperation of private companies is sought in order to harmonize information given to the growers.

Although, until relatively recently, anti-resistance measure were rarely of a preventive nature, today, new fungicides coming on to the market have been developed with relative prudence. This is the case notable for pyrimethanil against *B. cinerea* in spite of the fact that the risks have not yet been assessed.

Principal recommended anti-resistance measures

The principal recommendations given in France will be summarized here.

Botrytis cinerea

Because of a lack of effective methods to predict the risk of disease, measures against grey mold have traditionally relied on a series of four preventive treatments as a form of insurance : A) at the end of flowering ; B) at bunch closing ; C) at veraison ; D) 3 weeks before harvest. Treatment D is rarely justified.

Few conventional fungicides (thiram, dichlofluanid, chlorothalonil) are as effective as growers would wish. This explains the success of benzimidazoles in the 1970's, dicarboximides in the 1980's and the combination of carbendazim and diethofencarb at the end of the 1980's which rapidly led to the selection of pathogen strains resistant to these products. The development of strains such as Rb2, doubly resistant to carbendazim and diethofencarb, showed that the idea of negative cross-resistance as a management strategy was rather short lived for these particular compounds.

TABLE 1. Anti-*Botrytis* strategies recommended in France.

A	B	C	D
1. carbendazim + diethofencarb	pyrimethanil	dicarboximide (alone or with thiram)	–
2. carbendazim + diethofencarb	dicarboximide + thiram	dicarboximide (alone or combined)	–
3. carbendazim + diethofencarb	multisite	dicarboximide + thiram	dicarboximide (if need be)
4. pyrimethanil	dicarboximide + thiram	dicarboximide (alone or combined)	–
5. pyrimethanil	multisite	dicarboximide + thirame	dicarboximide (if need be)

The analysis of 10 years of monitoring done by Leroux & Moncomble (1993) showed that the only way to maintain the number of resistant strains below a critical level is to limit the number of treatments. This number can be under no circumstances greater than one treatment for each type of product. The only exception is for dicarboximides where two treatments are allowed in vineyards where Rd strains have not been detected or are not very frequent. Under these conditions, dicarboximides should be used in mixture with thiram at its full concentration (3.2 Kg ai/ha) (Leroux & Moncomble, 1993 ; Bugaret & Lafon, 1990). Nevertheless, in certain vineyards where the number of Rd and Rb2 strains is high (i.e. Champagne), the struggle is at an impass. One must fall back on strict preventive cultivation techniques and the use of conventional fungicides. Recently, the fungicide, pyrimethanil, has been registered for use especially in these vineyard, but its use should be strictly limited.

Several strategies for alternating fungicides have been recommended for other vineyards. These are presented in Table 1.

It is recommended to improve the efficacy of these programs by using an anti-downy mildew fungicide (7 to 8 treatments per year on average) with a secondary action against *Botrytis* (i.e. folpel, fosetyl-Al, methirame-Zn). The program should be chosen taking into account rather the nature of the pathogen population present in the plot (related to past treatment history) than the nature of the pathogen population present on average in the region. *Botrytis* of grapevine, contrary to popular ideas, has, in fact, a short dissemination range.

Plasmopara viticola

Resistance of downy mildew to phenylamides developed rapidly in France in the beginning of the 1980's, and then later in other European countries (Leroux & Clerjeau, 1985). It is not certain that these resistant strains appeared later in countries where these fungicides were used mixed with conventional products than in those where they were used alone. Monitoring activities conducted over several years in France have shown a clear relationship between the frequency of resistant strains isolated in diverse vineyard regions and the frequency of use of these products and the pathogen pressure (Moreau et al., 1984). However, it seems that the mixture of pathogen populations due to their wide dissemination has lead to population structures in the plots which are generally independent of their treatment history. It is in this context that in 1982, the following recommendations were established which are still valid today :

- Temporarily stop treatment with phenylamides in the regions most affected by resistance (i.e. the Cognac region between 1982 and 1985).
- No curative treatments : inactivity of conventional fungicides when combined with phenylamides.
- No treatments in the nursery : these products do not protect well during active vegetative growth. Newly formed organs are not protected when treatments are spaced at 14 days.
- A maximum of three treatments based on phenylamides per year, alternating with fungicides from other families.

The recommendations have been, on the whole, well followed because of the availability of effective alternative products whether penetrating (cymoxanil and dimetomorphe) or systemic (fosetyl-Al). In that the market share of phenylamides among the anti-mildew products is rather low (less than 10%), management of the resistant populations appears to be satisfactory. Even though resistant populations are frequent at certain sites, alternation programs provide growers with an adequate level of protection. This is also due to a synergistic effect between certain phenylamides and the other fungicides with which they are mixed in the commercial formulations. This synergistic effect has been observed for oxadixyl (Grabski & Gisi, 1987), and ofurace (Bugaret et al., 1989). In addition, benalaxyl has been shown to have an original, preventive action on contact with susceptible or resistant strains of downy mildew (Bugaret & Clerjeau, 1991).

It should be noted that, in 1993, insufficient protection was observed in high risk situations because of repeated rainfall during the period of optimal mildew sensitivity. In this situation, the non systemic associated fungicides were not able to provide sufficient protection when applied every 14 days.

Uncinula necator

Although resistance to DMI's has been observed for several years now in most European countries, first in Portugal, then in France (Steva et al., 1989), it is probably the principal current problem in vineyard protection. Many difficulties have been encountered in developing and putting into practice the anti-resistance measures that are summarized here (Anonymous, 1992) :

- The maximum number of treatments with DMI's : two or three per year. This means an alternation with less effective conventional fungicides (sulfur, dinocap, dichlofluanid). Continuous use of 2 or 3 DMI's is considered to be preferable to alternation of DMI's with conventional products when rotation poses practical problems such as a break in the rythm of treatment (10 or 14 days).

- Use of products at their maximum registered dose and with good spraying quality to avoid attack by strains with a weak resistance factor.

- No recommendations encouraging the use of mixtures of DMI with protective fungicides rather than a DMI alone. Due to under-dosage of active ingredients in the mixtures, a more rapid selection of resistant strains has been observed (Steva, 1992) ; in addition, an interaction between DMI (triadimenol) and sulfur has proved to be antagonistic (Steva, 1992).

Several factors play a role in slowing down a significant reduction in the use of DMI treatments :

- The absence of alternative products which offer good control at the same price. As a result, growers only look into their use after the first signs of declining efficacy of DMI's.

- The absence of systematic cross-resistance between the DMI's (Steva and Clerjeau, 1990) : the pathogen populations of powdery mildew preferentially acquire resistance to DMI's used in the plot. As a result, growers alternate different DMI's without reducing the total number of applications. This practice must, of course, be condemned because it leads to the development of strains of the pathogen with multiple resistance.

- The progressive nature of resistance (appearance of strains with higher and higher factors of resistance) : due to a reduction in efficacy observed firstly for weakly resistant strains which have developed under inadequate control conditions (effects of under-dosage of fungicides), growers are encouraged to improve their treatment practices rather than change their program of treatment.

CONCLUSION

If an effective anti-resistance strategy is to consist above all of a good level of prevention, it is necessary to recognize that, in practice, measures are not actually taken until monitoring activities indicate widespread dissemination of resistant pathogen strains and cases of reduced treatment efficacy. The reason for reluctance to adopt alternation programs is that no-risk products are rarely as effective as fungicides at risk. Indeed, recommendation of these programs is often seen as a means to restrict usage of these at risk fungicides. Nevertheless, it is very clear that restriction is the only effective way to prevent development of resistance. Unfortunately, it usually takes several years of experience in the field to evaluate the maximum number of treatments acceptable and this number is generally low (1 to 2 for *Botrytis cinerea* ; 2 to 3 for anti-downy mildew treatments).

And what about the use of mixtures of fungicides to deal with the problem of resistance? Examples found in viticulture do not provide universal answers. For *Botrytis*, the improvement of protection using mixtures is small but appears to be real. For *Plasmopara*, the advantages are more evident. On the other hand, for powdery mildew, the under-dosage of DMI's in the mixtures has contributed to an accentuation of the selection of weakly resistant strains.

Over the last ten years, studies on the development of resistant pathogen populations in relation to the selective pressure of treatment has provided significant progress in our knowledge of the conditions of dissemination of grapevine pathogens (dissemination limited to the parcel for *B. cinerea* and *U. necator*; widespread development for *P. viticola*). It can be said, however, that only real progress in the epidemiology of these fungi allowing better management will give us the upper hand on the problem of resistance.

REFERENCES

Anonymous (1992). Oïdium de la Vigne : Résistance aux IBS et recommandation d'emploi des fongicides. *Phytoma*, **435**, 56.

Bugaret, Y. ; Clerjeau, M. ; Lafon, R. (1989). Résistance du Mildiou aux phénylamides, une donnée nouvelle : la synergie ofurace-folpel. *Phytoma*, **412**, 34–40.

Bugaret, Y. ; Lafon, R. (1990). Iprodione + thirame : une association efficace. *Viti*, **148**, 109–112.

Bugaret, Y. ; Clerjeau, M. (1991). Le bénalaxyl : de surprenantes possibilités. *Phytoma*, **427**, 29–34.

Clerjeau, M. (1991). Questions d'actualité sur la résistance des champignons parasites aux fongicides. *Troisième conférence internationale sur les maladies des plantes, Bordeaux 3–5 Déc. 91, ANPP*, **4 (1)**, 97–104.

Grabski, C ; Gisi, U. (1987). Quantification of synergistic interactions of fungicides against *Plasmopara* and *Phytophthora*. *Crop Protection*, **6**, 64–71.

Leroux, P. ; Clerjeau, M. (1985). Resistance of *Botrytis cinerea* Pers. and *Plasmopara viticola* to fungicides in the french vineyards. *Crop Protection*, **4 (2)**, 137–160.

Leroux, P. ; Moncomble, D. (1993). Lutte chimique contre la Pourriture grise, passé, présent, futur (2ème partie). *Phytoma – La Défense des Végétaux*, **451**, 23–27.

Moreau., C. ; Clerjeau, M. ; Malato, G. (1984). La résistance du Mildiou aux anilides lors de la campagne 1983. *Phytoma*, **357**, 25–29.

Steva, H. ; Cartolaro, P. ; Clerjeau, M. ; Lafon, R. (1989). Premier cas de résistance de l'Oïdium à un traitement fongicide. *Viti*, **137**, 124–125.

Steva, H. ; Clerjeau, M. (1990). Cross resistance to sterol biosynthesis inhibitor fungicides in strains of *Uncinula necator* isolated in France and Portugal. *Medelingen van de Faculteit Landbouwetenschappen Rijksuniversiteit Gent*, **55**, 983–988.

Steva, H. (1992). Résistance de l'Oïdium de la Vigne (*Uncinula necator*) aux fongicides inhibiteurs de la biosynthèse des stérols. *Thèse de doctorat Université Bordeaux II*, 248 p.

protection using [illegible] is [illegible] to [illegible] real. For [illegible] of the advantages are [illegible] [illegible]. On the [illegible] [illegible] the [illegible] of [illegible] has [illegible] contributed to [illegible] of the [illegible] [illegible].

[illegible] the [illegible] studies on the development of [illegible] populations [illegible] to the [illegible] pressure of treatment [illegible] [illegible] in our knowledge of the [illegible] conditions of [illegible] of [illegible] pathogens [illegible] [illegible] and [illegible] [illegible] [illegible]. [illegible] however, will only [illegible] [illegible] in the [illegible] [illegible] [illegible] [illegible] the [illegible] the [illegible] of resistance.

[illegible]

Anonymous (19[illegible]) [illegible] la Vigne: Résistance aux [illegible] et [illegible] [illegible] [illegible] [illegible]

[illegible] (19[illegible]) [illegible] de [illegible] aux phenylamides dans [illegible] [illegible] [illegible] *Phytoma*, [illegible]

[illegible] (19[illegible]) [illegible] [illegible] [illegible] [illegible] [illegible]

[illegible] (19[illegible]) [illegible] [illegible] [illegible] *Phytop*[illegible]

[illegible] (19[illegible]) [illegible] [illegible] de [illegible] [illegible] [illegible] [illegible] [illegible] [illegible] [illegible] [illegible] [illegible] *Phytopathologica*, 4 (1), [illegible]

[illegible] (19[illegible]) [illegible] of [illegible] [illegible] of [illegible] [illegible] [illegible] *Phytopathology*, [illegible]

[illegible] (19[illegible]) [illegible] [illegible] [illegible] [illegible] [illegible] [illegible] *Crop Protection*, [illegible]

[illegible] (19[illegible]) [illegible] [illegible] contre la [illegible] [illegible] [illegible] [illegible] [illegible] [illegible]

[illegible] (19[illegible]) [illegible] de [illegible] [illegible] [illegible] [illegible] campagne 1985. [illegible] 25.

[illegible] (19[illegible]) [illegible] [illegible] [illegible] [illegible] [illegible] [illegible]

[illegible] (19[illegible]) Cross resistance to [illegible] [illegible] [illegible] [illegible] [illegible] [illegible] [illegible] [illegible] [illegible] [illegible] [illegible]

[illegible] (19[illegible]) Résistance de [illegible] [illegible] [illegible] [illegible] [illegible] [illegible] [illegible]

FINANCIAL IMPLICATIONS OF FUNGICIDE RESISTANCE

J-L PASQUEREAU

TRANSAGRA - Advice and Development Bourges, France

ABSTRACT

TRANSAGRA Advice and Development Service works with 2000 farmers who till 174000 ha, 70000 ha of which are cereals.

The use of fungicides on cereals represents a third of the total operating costs. Even though new problems of fungicide resistance may appear in the country, farmers still have to maintain good protection for their crops. The use of diagnostic and forecasting models, new varieties and active materials will all help the farmer to make the correct decisions.

FARMING IN CHER

The Department of Cher is located in the center of France, 240 km south of Paris. It is a large Department of open fields with crops of cereals and oil producing plants : rape seed and sunflower. Maize is grown with spray irrigation.

Climate and soils

In summer, sometimes from April, the climate is dry and warm. The soils (rendzinas) are very thin with limestone. Crops need rain or irrigation, but there is not enough water. (The annual rainfall is 650-750 mm). Seven natural areas constitute the Cher : Champagne Berrichonne covers half of the area.

Production and yields

Cher has 480000 ha of arable surface including 20000 ha of winter wheat and barley and 90000 ha of oil plants. 25000 ha of maize are irrigated. 4000 ha of set-aside are grown with rape-seed and flax as industrial crops. Cattle decreases : Charolais for meat, goats for cheese (Crottin de Chavignol) and poultry (turkeys, chickens) for CODIVOL.

The average yields are around 6.5 - 8.5 t/ha of wheat, 10t - 12t of maize, 3 - 4 t/ha of rape-seed and 2 - 2.5 t/ha of sunflower.

Vineyards cover 3000 ha of Sancerre, Menetou and Quincy, while apple orchards occupy 2000 ha.

Men and machines

4000 farmers cultivate an average 120 ha per man, working with a high degree of mechanization. The individual farm area has increased over the past two years and 150 ha per man will soon be reached.

TRANSAGRA : SURFACE AND ROTATION OF CROPS

Transagra is a cooperative which produces 650000 t of grain per annum and sells seeds, pesticides and fertilizers. The areas grown for harvest 1993 are shown in Table 1 :

TABLE 1. Transagra crop areas 1993

	Hectares		Hectares
Wheat	60 000	Rape seed	23 000
Winter barley	9 000	Sunflower	27 000
Winter oat	900	Peas	5 500
Hard wheat	700	Corn (Maize)	15 000
Spring barley	2 700	Set aside	30 500
		with rape seed	3 000
TOTAL CEREAL	73 300	Flax	1 000
		TOTAL	101 000

Rotation

On loamy soils, a two year rotation is used, rape seed-wheat. On rendzina soils rotation was traditionally on a three yearly cycle of rape seed or sunflower, wheat, barley or second wheat. Now the second wheats are only sown on 5000 ha.

CEREAL DISEASES AND LOSS OF YIELDS

Septoria is the main disease every year, brown rust is also present but less severe.

Eyespot covers a third of the area, mainly in loamy, wet and clay soils. Powdery mildew increases each year, even on the cultivar "SOISSONS".

Field tests

In the past 12 years, we have conducted and harvested 37 fungicide tests on wheat in the different areas, using random block trials with four replicates and plots 3 x 12m.

The 37 trials included 15 for eyespot located in clay and loamy soils, often with flints. The fields are drained but wet in winter. Losses from eyespot were between 0.2 t to 1.5 t/ha. Overall gains due to control of foot and leaf diseases were estimated at:

$<$ 0.6 t/ha : 11 trials
$>$ 0.6 t/ha : 26 trials

The mean yield gain was 1.43 t/ha

Cartography of eyespot in Cher

The DuPont eyespot kit was used by our laboratory to assess eyespot in 200 fields. For two years a third of the fields produced results > 10 UA. In 1993, SPV Orleans tested 900 stems from 21 fields not treated with fungicide. Samples were taken from the fields between 10-15 June and showed all fields to be > 30 UA.

Analysis of the populations showed that the W type was dominant in these 21 fields:

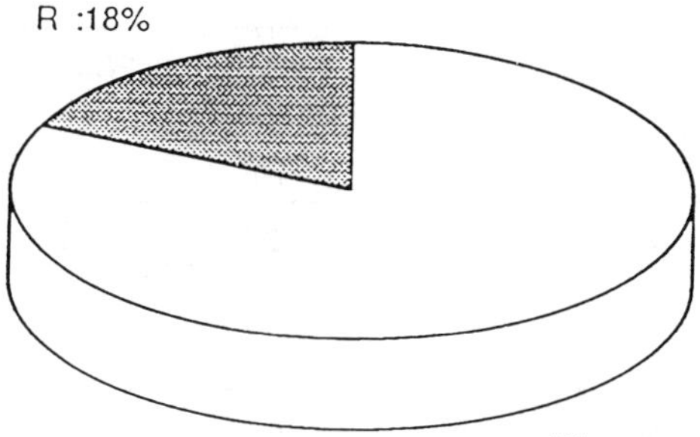

Distribution of W/R types in the population

Isolates from these fields were tested for sensitivity to fungicides. The tests with prochloraz (0,5 and 2 mg/l) and triadimenol (30 mg/l) showed that 70% of isolates were resistant to triadimenol.

At 5 locations, 15% of isolates showed resistance to prochloraz, but only in the laboratory, not in the field.

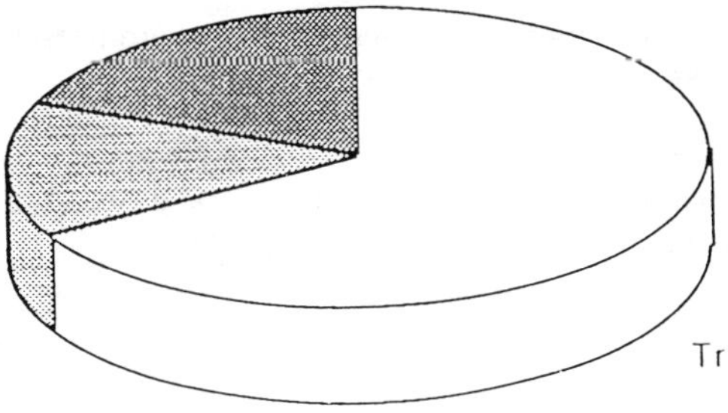

An assessment of the effect of disease on cereals

Winter wheat variety SIDERAL was sown on October 20 1992 at Laverdines, and followed wheat after rape seed.

At the end of March 1993, the "DUPONT" eyespot kit was used to show 125 UA without eyespot visible on plants. A 4 replicate trial was established with treatments as shown in Table 2, applied at GS 31.

TABLE 2. Control of eyespot at Laverdines

Treatment	g AI/ha	Frequency %	Necrosis section %	Efficiency %	Lodging %
Check 0 treatment		100	100	-	100
1 - Prochloraz	450	60	32	68	25
+ Fenbuconazole	67				
+ Fenpropimorph	280				
+ Fenpropidine	94				
2 - Prochloraz	450	76	43	57	50
+ Fenpropimorph	280				
+ Fenpropidine	94				
3 - Fluzilazole	250	92	77	23	100
+ Fenpropimorph	280				
+ Fenpropidine	94				
+ BMC	125				
4 - Epoxyconazole	125	100	94	6	100

The field of the farmer was only treated with Flusilazol + BMC on April 8 at GS 31. This resulted in 100% lodging and 5% efficiency.

As SIDERAL is very sensitive to powdery-mildew, on May 17th, at ear emergence stage, propiconazole + fenpropidine was sprayed over all plots.

The trial was harvested in July and yield data are presented in Table 3. Yield losses due to disease were estimated as:

- for eyespot : 1, 5 t/ha
- for foliar diseases : powdery mildew and Septoria = 1, 6 t/ha
- the best treatment : 1 -> + 3, 52 t/ha

TABLE 3. Laverdines yields

Treatment	g AI/ha	Yield (t/ha)	Gain (t/ha)	Test	Weight of 1000 grains
Check 0 treatment		4, 97	-	c	27, 3 g
1 - Prochloraz + Fenbuconazole + Fenpropimorph + Fenpropidine	450 67 280 94	8, 49	+ 3, 52	a	36, 8
2 - Prochloraz + Fenpropimorph + Fenpropidine	450 280 94	7, 72	+ 2, 75	ab	38, 1
3 - Fluzilazole + Fenpropimorph + Fenpropidine + BMC	250 280 94 125	7, 12	+ 2, 15	bc	34, 2
4 - Epoxyconazole	125	6, 85	+ 1, 88	c	35, 0

CV = 6, 1% ETR : 0, 426 t/ha

A second trial was conducted north of Cher at Aubigny on a loamy wet soil with many stones (flints).

For two years, the area had shown many eyespot R types, but in 1993 only W types were found. Analysis of the trial yield data showed a different picture to Laverdines :

- for eyespot : + 0, 34 t/ha
- for powdery-mildew : + 0, 28 t/ha
- for Septoria : + 1, 49 t/ha

There were no differences between treatments

FINANCIAL IMPLICATIONS OF FUNGICIDE RESISTANCE

With the new Common Agricultural Policy, farmers are seeking to decrease their inputs. Table 4 shows the average operating costs to produce wheat.

Table 4. Average operating costs to produce wheat in Cher

Input	Cost F	
	Minimum	Maximum
Seeds : 350 grains/m2	270	370
P.K. : 50 - 50	100	300
N : 160 - 210	350	460
Herbicides black grass + wild oat + broadleaves	100	300
Fungicides : Eyespot Powdery mildew Septoria - Brown rust	400	600
Insecticides (leafhopper)	50	100
TOTAL	1270 f/ha	2130 f/ha

The cost of fungicides is thus around a third of the operating-costs. Fungicide resistance for eyespot and powdery mildew will have the effect of increasing these costs.

For eyespot control, BMC has been ineffective for 10 years and triazoles (flusilazol, bromuconazole, epoxyconazole) have presented resistance for two years Prochloraz remains effective at present but for how long?

To fight powdery mildew, we have to use morpholines on susceptible cultivars, sometimes at the first node stage in the north area.

For a farmer, the cost of resistance is around 60 F/ha. For the total cereal crops of our farmers, the cost is around 4 000 000 F.

In the future, if problems develop with prochloraz, we can use cyprodinil, but today, it's expensive at 300 F/ha and does not have activity on Septoria and rusts. Farmers will not buy this new fungicide in 1994.

DISCUSSION

What advice can we give to the farmers today? We have to preserve the crop potential from diseases and to make each operating cost profitable.

Resistance is a problem, but we can sow different varieties with different host resistance. Last year, we studied mixtures with SOISSONS + SIDERAL and hybrid MESNIL. There was no difference for the cost of fungicides. For 1994, three hybrids and SOISSONS with two different fungicide protections will be tested.

Today, farmers can choose between several wheat varieties to decrease the cost of fungicides as shown in Table 5.

TABLE 5. The susceptibility of wheat varieties to foliar pathogens

Cultivar	Powdery mildew	Septoria	Brown rust	Yellow rust
EUREKA	+	+	-	-
SOISSONS	(-)	+	+ + +	+ +
ORQUAL	+	+	+ +	+ + +
GASCOGNE	-	+	+ +	+
GASPARD	+ +	+	+ +	-
SIDERAL	+ + +	+	+ +	+ +
THESEE	+ +	+ +	+ + +	+ + +
TEXEL	+ +	+ +	+ +	+ + +
TREMIE	+ +	+ +	+ +	+ +
QUALITAL	+ + +	+	+ + +	+ + +

Our advice to farmers is to examine each field weekly in order to detect disease bulid up.

We will use the eyespot diagnostic kit, for the second year : an experimental Septoria kit and the use of forecasting models for Septoria and brown rust. Among the different fungicides, we prefer mixtures of triazoles + morpholines for foliar control and prochloraz 450 g ai/ha for eyespot.

In the future, I think we shall have to use some cyprodinil and maybe ethirimol to preserve some efficiency.

Plant breeders can also improve their varieties, if farmers use more "certified seeds" But it's not the trend today!

Department of Cher - Natural areas

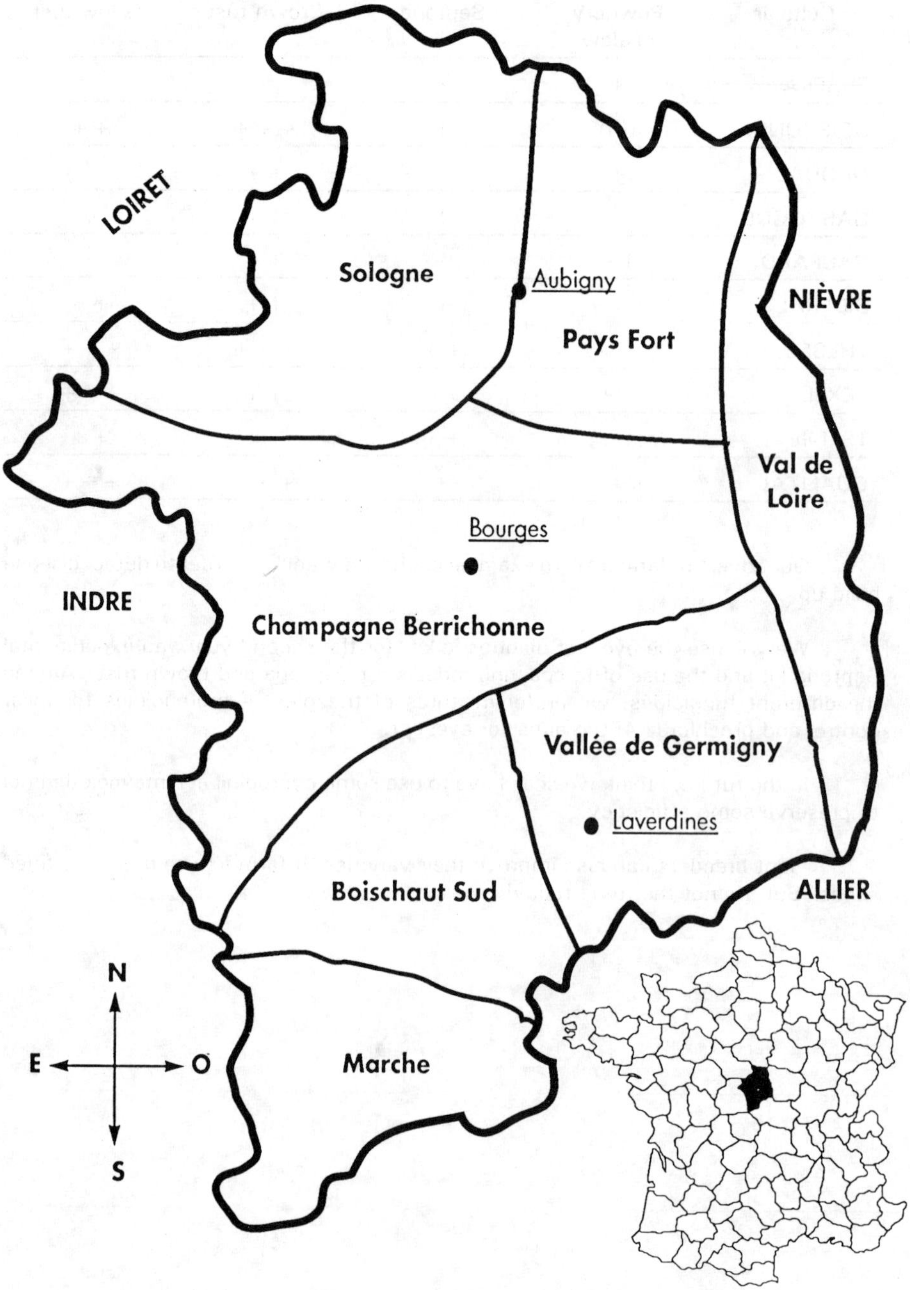

INCIDENCE AND CONTROL OF ORGANO-MERCURY RESISTANT PYRENOPHORA GRAMINEA IN SCOTTISH SPRING BARLEY 1988-1992

V COCKERELL, W J RENNIE AND M JACKS

Official Seed Testing Station for Scotland, Scottish Agricultural Science Agency, East Craigs, Edinburgh, EH12 8NJ

ABSTRACT

Organo-mercury seed treatments effectively controlled barley leaf stripe in Scotland until the mid-80s when the causal fungus P. graminea was reported to be resistant to organo-mercury. By 1990 leaf stripe was widespread in Scottish spring barley. 3454 Scottish spring barley seed samples were tested for P. graminea infection between 1988 and 1992. The percentage of samples infected with P. graminea increased significantly from 1988 to 1990. 26% of certified seed had more than 4% infection in 1990 compared to only 3% in 1988. A voluntary Code of Practice requiring all seed intended for sale to be tested for P. graminea infection was introduced in 1991. Seed with more than 4% infection was to be treated with a seed treatment effective against the disease or discarded. No sample of certified seed failed to meet the 4% standard in 1992. Resistance testing confirmed that all seed infection was caused by organo-mercury resistant strains of P. graminea.

INTRODUCTION

Organo-mercury seed treatments gave good control of barley leaf stripe (Pyrenophora graminea) from their introduction in the 1920s. As a result, the disease was considered to be of little practical importance in the Scottish spring barley crop and was only occasionally seen where untreated or inadequately treated farm-saved seed was sown (Richardson, 1986).

Resistance to organomercury in P. graminea was first reported in the West of Scotland during 1984 where infection was noticed in several fields of Mazurka spring barley grown from one certified seed stock and treated with a commercial organo-mercury fungicide containing phenyl-mercury acetate (PMA) (M R M Clarke, Pers. Comm.). Work by Jones et al. (1989) confirmed the wider occurrence of strains of P. graminea resistant to organo-mercury but concluded that there was no justification to change from the routine use of organo-mercury for leaf stripe control unless seed was being used for further multiplication.

During the summer of 1990 leaf stripe was widespread in the Scottish spring barley crop. Crops grown from both certified and farm-saved seed were apparently equally infected. A voluntary Code of Practice to contain the disease was introduced in January 1991. The Code required all seed intended for sale to be tested for P. graminea infection. Seed with more than 4% infection was required to be treated with a seed treatment effective against the disease or discarded. The

Code was aimed principally at certified seed, but a high proportion of farm-saved spring barley was also tested.

This paper reports the occurrence of seed-borne P. graminea in Scottish spring barley seed from 1988 to 1992 and preliminary work to assess fungicide resistance.

MATERIALS AND METHODS

Seed Testing

The incidence of P. graminea was determined in spring barley seed harvested in each of the 5 years 1988 to 1992. For the years 1988 and 1989, samples were taken at random from seed submitted to the Official Seed Testing Station for Scotland (OSTS) by merchants, for advisory germination testing prior to the processing of seed lots intended for certification, and by growers, who required advisory germination tests on farm-saved seed. Seed harvested between 1990 and 1992 was subject to the Code of Practice and tests were made to meet the requirements of the Code.

Working samples were drawn using a seed sample divider and either 100 seeds (1988 and 1989) or 200 seeds (1990 to 1992) were tested for P. graminea using a standard agar plate procedure (Rennie and Tomlin, 1984).

Seed testing results are reported separately for farm-saved seed and for seed intended for certification.

Resistance Testing

Isolates of P. graminea were obtained from UK barley seed harvested in 1990 and 1991 and from infected plants collected from certification control plots at the Scottish Agricultural Science Agency during the summer of 1993. Seeds were plated on agar as described previously and a small agar plug was taken from the actively growing margins of P. graminea colonies and transferred to fresh potato dextrose agar (PDA) plates. This procedure was repeated until the isolate was in pure culture. Diseased leaves were cut into 2cm sections and were surface sterilised by immersion in a solution of sodium hypochlorite (1-2% available chlorine) for 3 minutes. Leaf sections were drained and placed on PDA plates. Isolates were sub-cultured until a pure culture was produced. Cultures were maintained on PDA at 5°C in the dark. Isolates of P. graminea from Canada and Italy were included as controls.

TABLE 1. Number of isolates of P. graminea screened from 1990 to 1993 and concentrations of active ingredients.

Year	Number of isolates	Range of concentrations per fungicide screened (mg/l)	
		Organo-mercury	Imazalil
1990-91	195	2, 5, 10, 20, 40	None
1992-93	253	None	0.5, 2.5, 5.0

The sensitivity of the P. graminea isolates to organo-mercury and imazalil, was determined by transferring 1cm diameter discs of agar from the margins of actively growing colonies to Petri dishes containing PDA supplemented with concentrations of the fungicides. Concentrations of fungicides and numbers of isolates screened are given in Table 1.

RESULTS

Seed testing

Table 2 shows that P. graminea infection was highest in seed harvested in 1989 and 1990; in these years there was little difference in infection between seed intended for certification and farm-saved seed. Infection levels were considerably reduced in 1991 and 1992 after the introduction of the Code of Practice. No certified seed lot tested in 1992 failed to meet the 4% standard in the Code of Practice.

Table 2. Incidence of P. graminea in Scottish spring barley seed.

Year harvested	Number of samples tested		Percentage of samples infected		Percentage of samples with more than 4% infection	
	Certified	Farm-saved	Certified	Farm-saved	Certified	Farm-saved
1988	32	30	12	20	3	3
1989	53	45	74	64	35	37
1990	1052	953	80	85	26	44
1991	438	519	36	51	2	11
1992	98	234	4	16	0	1

Resistance Testing

All of the 195 P. graminea isolates collected from Scottish spring barley seed harvested in 1990 and 1991 were resistant to organo-mercury at a concentration of 10mg/l in agar. The mean colony growth at this concentration was 52% of the control. In contrast, isolates from Canada and Italy were sensitive to mercury; no growth occurred on agar containing 5mg/l MEMA (2-methoxyethylmercury acetate). Isolates of P. graminea from a single sample of Gaulois winter barley, grown in Lincolnshire in 1991, gave no growth at 10mg/l MEMA and their growth on agar containing 5mg/l MEMA was 20% of the control whereas for mercury resistant isolates 50-90% growth was recorded at this concentration.

None of the 253 isolates tested in 1993 was resistant to imazalil. No isolate grew on agar containing 5mg/l imazalil and in each case growth was inhibited by 90% or more in agar containing 2.5mg/l. The target dose for imazalil in commercial barley seed is approximately 40mg/l. There have been no records of leaf stripe in Scottish spring barley crops grown from imazalil treated seed.

DISCUSSION

Although P. graminea resistance to organo-mercury was reported during the 1980s, seed treatments based on organo-mercury continued to be used and in 1990 a high percentage of Scottish spring barley seed lots was infected. The incidence of infection was similar for certified and farm-saved seed harvested in 1989 and 1990. P. graminea infection increased from 1988 to 1990 and resistance testing confirmed that this seed infection was attributed to organo-mercury resistant strains of P. graminea.

The introduction of a voluntary Code of Practice, agreed between the seed trade and Scottish growers, required barley seed that was to be certified in 1991, and in subsequent years, to be tested for P. graminea infection. Where infection was greater than 4% seed stocks were required to be treated with a fungicide effective against P. graminea or discarded. The resultant seed crop in 1991 showed an increase in the percentage of certified and farm-saved samples that met the 4% standard. The continued requirement for testing under the Code and the increased use of effective treatments, especially those containing imazalil, resulted in a significant reduction in P. graminea infection from 1990 to 1992 when no stock of certified seed failed to meet the voluntary standard.

A number of commercially available cereal seed treatments include imazalil to control specifically P. graminea on barley. The results from resistance testing during 1993 showed that P. graminea isolates from Scottish spring barley were very sensitive to imazalil.

Seed-borne diseases such as leaf stripe, loose smut and bunt have the potential to multiply rapidly within a few seasons. Monitoring disease incidence and the sensitivity of pathogens to active ingredients in commercial fungicides will help to avoid the unacceptable levels of disease seen in the Scottish spring barley crop in 1990.

ACKNOWLEDGEMENTS

The Home Grown Cereals Authority provided financial support for the fungicide resistance work.

REFERENCES

Jones, D. R.; Slade, M. D.; Briks, K. A. (1989) Resistance to organo-mercury in Pyrenophora graminea. Plant Pathology, **38**, 509-513.

Rennie, W. J.; Tomlin, M. M. (1984) Barley leaf stripe working sheet No 6 (2 Ed) ISTA Handbook on Seed Health Testing, Zurich.

Richardson, M. J. (1986) An assessment of the need for routine use of organo-mercurial cereal seed treatment fungicides. Field Crops Research, **13**, 3-24.

Session 6
Resistance Risk Evaluation

Chairman D W HOLLOMON

Session Organiser D W HOLLOMON and K BRENT

PESTICIDE RESISTANCE AND THE EC PLANT PROTECTION PRODUCTS DIRECTIVES

C. FURK, D. D. SLAWSON

MAFF, Pesticides Safety Directorate, Rothamsted, Harpenden, Herts AL5 2SS

ABSTRACT

Data relating to the development of pesticide resistance required by both the UK Regulatory Authority and the EC's Authorization Directive (91/414/EEC as specified in Annex III) are discussed. The rationale for these requirements and their role in the regulatory process are outlined in this paper.

INTRODUCTION

The European Commission (EC) has harmonised the requirements for authorization/approval of plant protection products across the Community by the introduction of a series of Directives. Included in these Directives are requirements for resistance risk assessment and management strategies to prevent the selection of pesticide-resistant organisms. The two relevant Directives that refer to resistance are Council Directive 91/414/EEC - the Authorization Directive (Anon, 1991a) and Commission Directive 93/71/EEC - the Efficacy Section of Annex III (Anon, 1993). The latter specifies requirements under the heading of "Information on the occurrence or possible occurrence of the development of resistance". These requirements do not differ very much from the UK's own Data Requirements Handbook which has been in use since 1986 (Anon, 1986).

THE E.C. DIRECTIVES

The E.C. Directives state:

"Laboratory data and where it exists, field information relating to the occurrence and development of resistance or cross resistance in populations of harmful organisms to the active substance(s), or to related active substances, must be provided. Where such information is not directly relevant to the uses for which authorization is sought or to be renewed (different species of harmful organism or different crops), it must, if available, nevertheless be provided, as it may provide an indication of the likelihood of resistance developing in the target population.

Where there is evidence or information to suggest that, in commercial use, the development of resistance is likely, evidence must be generated and submitted as to the sensitivity of the population of the harmful organism concerned to the plant protection product. In such cases a management strategy designed to minimise the likelihood of resistance or cross-resistance developing in target species must be provided."

DEFINITION OF RESISTANCE

The UK Government's Regulatory Authority considers a working definition of pesticide resistance to be :

"Pesticide resistance by an organism is an inheritable change of a population in its sensitivity to an active ingredient which will be reflected in reductions in levels of field control during commercial use according to the product label. Such a change in sensitivity can be confirmed by bioassay using a validated or recognised technique".

To explain further, it is accepted that while resistance of an organism to an active ingredient can often be demonstrated in the laboratory it need not necessarily mean that a grower will notice any lowering in the level of control in the field. It is the record of the occurrence of both factors which is necessary before it can be accepted that resistance has become significant to the use of the active ingredient. Dekker (1985) stated that this situation must be achieved by explaining clearly by what criteria resistance is being claimed.

INTERPRETATION OF DATA REQUIREMENTS

The issue of pesticide resistance must be addressed in all applications for approval/authorization of a new plant protection active ingredient. The registration authorities require information in order to decide if approval should be given and to guide them to a conclusion as to the need for further data and the need for development of strategies to minimise the possible occurrence of resistance. It is expected that applicants take a comprehensive world-wide view of the situation. The requirement for "information not necessarily relevant" is to discover the potential for the chemical group to select for resistance or for the selection of resistance in the target organism. It is not certain that cross-resistance will occur to a new active ingredient simply because there has already been selection in a target organism to a non-related chemical group, but the possibility cannot be dismissed. The amount of information required will depend on the perceived risk of resistance based on the evidence available at the time approval/authorization is sought.

The perceived risk of resistance will be 'low' if there is no evidence of resistance both to the pesticide/pesticide group and in the target pest to other active ingredients. If, however, the target pest has shown the propensity to develop resistance to other pesticides or if resistance has already been recorded to the pesticide/pesticide group either in the target pest or in other pests, then the perceived resistance risk may be 'high'. While accepting that evidence of resistance in the laboratory does not necessarily mean that control problems will be encountered in the field, such evidence will nonetheless be taken into account when determining the risk of resistance.

In both instances, the applicant must provide the information on which the resistance risk can be determined. Test protocols and the 'baseline' sensitivity of the target organism should also be made available when providing results of monitoring studies. Furthermore, in cases where the perceived risk of resistance is 'high', the applicant will be required to propose a management strategy designed to minimise the development of resistance in commercial situations.

Protocols for testing pest sensitivity

There are various protocols already published describing methods for the testing of susceptibilities of organisms to pesticides. Most of them allow for the production of baseline data that can be used for monitoring resistance. The Food and Agricultural Organisation (FAO) has published a series of recommended methods for the detection and measurement of fungicide and insecticide resistance (e.g. Anon 1980, 1982a, b) and the European and Mediterranean Plant Protection Organisation (EPPO) has also published methods for monitoring fungicide resistance (Anon, 1991b, 1992). However, these bioassays are not appropriate for all situations and new tests may need to be devised. These tests must be validated to show that they are capable of detecting changes in the susceptibility of the target organism.

Baseline response data

In cases where prior evidence is available that resistance has occurred to the pesticide/pesticide group or in the target organism, the data requirements state that the applicant should produce baseline data as to the sensitivity of the target organism to the proposed active ingredient. The production of such data are invaluable in resistance studies as once resistance is suspected it can never be unequivocally established that data produced subsequently are from truly susceptible organisms. Details of laboratory bioassays (or field tests if more appropriate) and results submitted to the registration authorities should be made publicly available. Anybody wishing to monitor for the development of resistance at a later date will therefore have access to the methods and the baseline data against which to judge their results.

Management strategies

The applicant has to provide a management strategy to minimise the likelihood of the development of resistance. An applicant should be able to construct a suitable strategy using knowledge of the mode of action of the new active ingredient to the target organism and the likely resistance mechanisms of the organism. The strategy should be publicised and brought to the attention of farmers and advisers either via instructions on the product labels or by technical literature. Management strategies to minimise the development of resistance could include measures such as marketing of the active ingredient only in formulated mixtures with other non-related active ingredients, or by restricting the number or timing of applications in order to reduce selection pressure. It is advisable that the development of strategies for resistance management are borne in mind when monitoring for resistance.

REVIEW OF EXISTING PLANT PROTECTION PRODUCTS

The EC is to undertake a major review of all approved pesticides with extant approvals within the Community. Agrochemical companies see the exercise as a formidable re-registration task which will include the need to produce data to answer the Directive's requirements on resistance. Although the data requirements as laid down by the Directives suggest that product re-registration should include a consideration of resistance, it is the opinion of the UK Regulatory Authorities that action will not be required by the agrochemical companies for the initial review of the active ingredient for Annex I listing but may be required when the authorities highlight a problem during the authorisation process of products which follows in individual Member States.

CONCLUSION

We are not able to specify exactly what evidence is required to satisfy the data requirements relating to resistance mainly because of the varied and complex nature of resistance. The issue is very complicated in that interpretation of data is often difficult and new pesticide/target organism interactions may be unique. Therefore, testing techniques, sample sizes etc. could be very different from existing ones. Furthermore, strategies designed to prevent or delay the development of resistance are largely untested, with scientific opinion split on their validity.

We would therefore expect a dossier provided in support of an application for the approval/authorization of a new pesticide to contain the following:

i) Details of a protocol for testing pest sensitivity;
ii) Baseline response data (sensitivity/susceptibility); and
iii) in the event of evidence of a problem, a management strategy designed to minimise the likelihood of resistance.

REFERENCES

Anon., (1980) Recommended methods for the measurement of pest resistance to pesticides *FAO Plant Production and Protection Paper 21*

Anon., (1982 a) Recommended methods for the detection and measurement of resistance of agricultural pests to pesticides *FAO Plant Production and Protection,* **30**, 36-71.

Anon., (1982 b) Recommended methods for the detection and measurement of resistance of plant pathogens to fungicides *FAO Plant Production and Protection,* **30**, 141-143.

Anon., (1986) Data requirements for approval under the Control of Pesticides Regulations 1986. Ministry of Agriculture, Fisheries and Food, United Kingdom.

Anon., (1991 a) Council Directive (91/414/EEC) concerning the placing of plant protection products on the market. *Official Journal of the European Community,* **L230**, 19 August 1991, 1-32.

Anon., (1991 b) FRAC methods for monitoring fungicide resistance. *EPPO Bulletin,* **21**, 291-354.

Anon., (1992) FRAC methods for monitoring the sensitivity of fungal pathogens to phenylamide fungicides. *EPPO Bulletin,* **22**, 297-322.

Anon., (1993) Commission Directive (93/71/EEC) amending Council Directive 91/414/EEC concerning the placing of plant protection products on the market. *Official Journal of the European Community,* **L221**, 31 August 1993, 27-36.

Dekker, J., (1985) The fungicide resistance problem: will it grow worse?. *EPPO Bulletin,* **15**, 337-344.

EARLY EVALUATION OF FUNGICIDE RESISTANCE RISK

S.G. GEORGOPOULOS

Agricultural Department, Lapapharm Inc., 73 Menandrou Str.
Athens 104 37, Greece

ABSTRACT

Early estimates of the initial variation for sensitivity, and attempts to induce resistance mutations in the laboratory, are essential for predicting the probability of evolution of resistance to a new fungicide. If the appropriate variability is present, the seriousness of the problem that should be anticipated is a function of the type and rate of population response to exposure to the new chemical. Whether the response will be qualitative or quantitative can be determined by studying the genetics of resistance in the laboratory before field testing. Measurements of the effects of resistance alleles on fungicide sensitivity and on fitness in the laboratory and the greenhouse may provide indications of the rate of response.

INTRODUCTION

One of the aspects needing investigation during fungicide development is the likelihood of loss of effectiveness because of the evolution of resistance to new products by target fungi. Early assessment of this risk is desirable, so that decisions can be made, if not to stop further development of compounds perceived to have too high a risk, to at least proceed cautiously. Work on appropriate marketing and use strategies to ensure the longest possible useful life of each new product could then be carried out within a meaningful context. Practical resistance problems arise when field populations of target organisms can respond to exposure to a chemical by a substantial change in their sensitivity. This contribution will try to indicate what laboratory and greenhouse tests can be made early, before field testing of a new fungicide, in order to answer two important questions :

a. Whether the use of the compound might cause a population response, and of what type. This depends on the availability of appropriate variation and the genetic control of such variation.
b. What will be the rate of evolution of resistance to the new compound. Main determinants of this rate are selection pressure and fitness.

Recent treatments of the subject of resistance risk assessment include that of Keiding (1986) for pesticides in general and those of Gisi and Staehle-Csech (1988a, b) and Brent *et al*. (1990) specifically for fungicides. It is to be understood, of course, that the results of resistance risk assessments may help to set priorities in the development work, but can not be taken as conclusive forecasts on the resistance behaviour of new chemicals in all cases.

BACKGROUND KNOWLEDGE ON THE EVOLUTION OF FUNGICIDE RESISTANCE

It cannot be excluded that the genetic variability required for resistance to a particular type of toxicant may not be available to the intended target organism(s). Unavailability of appropriate genes may be the reason for which some protectant fungicides give apparently the same level of protection today as they did when first introduced (Georgopoulos & Skylakakis, 1986). Resistance to copper for example, has evolved in plant pathogenic bacteria which can no longer be controlled with the amount of copper available from fixed copper fungicides (Cooksey, 1990), but apparently not in the many fungal pathogens against which copper is used.

Genetic variation can pre-exist in a population, or arise *de novo* by mutation (or recombination) after environmental change. The change will then generate a selection pressure and the population response will be either qualitative or quantitative. A qualitative response to a fungicide is expected with major-gene control of resistance, causing discontinuous variation with at least two distinct, non overlapping subpopulations. Alternatively, if selection acts on continuous (polygenic) variation, the population response will be quantitative : there will be a decrease in mean sensitivity, but distinct subpopulations will be impossible to recognise even after long exposures. With major-gene control, the organism can achieve the highest possible resistance in one step, by mutation of one gene. In the polygenic system, the effects of individual genes, even if recognisable, are generally small. Highly resistant strains cannot be obtained in a single step, but only through recombination or sequential selection.

Of the fungicides already in use, the benzimidazoles and the phenylamides cause typical qualitative population responses because high resistance to these compounds is obtained by major-gene mutations (Shabi *et al.*, 1983 ; Crute & Harrison, 1988). Examples of a quantitative response are those to dodine, (Mckay & MacNeill, 1979) to ethirimol (Brent, 1982), and to the C-14 demethylation inhibitors (Heaney, 1988). Resistance to members of the latter group is generally polygenic, but major-gene resistance to triadimenol has been recognised in *Nectria haematococca* (Kalamarakis *et al.*, 1989) and discontinuous distribution for sensitivity to the same compound has been observed in populations of *Rhynchosporium secalis* (Kendall *et al.*, 1993).

Prediction of a qualitative or quantitative response is of considerable practical importance. Generally speaking, resistance controlled by major genes may be expected to lead to disease control failures faster than polygenic resistance. Thus far only fungicides causing qualitative population responses have been classified by practical experience as high-risk fungicides (Georgopoulos, 1987). With major gene resistance, the increase in the frequency of the mutant forms is exponential and, unless impractically large numbers of samples are tested, monitoring is unlikely to detect field resistance until it is too late. Monitoring, however, may provide early warning in cases of polygenic control of resistance where the decrease in mean sensitivity is linear with time. This has adequately been explained by Brent *et al*. (1990).

The distinction is also important in order to decide on the initial use rate of a new fungicide. With major gene resistance, lowering of the selection pressure by reducing the dosage may slow down the change to a predominantly resistant population. Individuals which will not be killed will be wild types and will contribute sensitive progeny. In contrast, reduction of selection pressure in cases of polygenic control may encourage resistance evolution. In polygenic resistance, each gene contributes in a minor way and survival of individuals with a few such genes at a lower dose will give them the opportunity to acquire additional genes by mutation and/or recombination. If a higher rate is then applied, this will intensify the response to selection. Using a high dosage initially, will eliminate individuals with low resistance and the frequency of recombinations of polygenes will greatly be reduced.

OBTAINING DATA ON THE INITIAL VARIATION IN SENSITIVITY

Though only a few isolates of each target organism are used in the initial screening work, a larger number must be tested for sensitivity to those compounds which are considered for development. The sample cannot be large enough to detect rare mutants. Novel, DNA-based detection methods which might permit testing large numbers of isolates, require knowledge of the molecular basis of resistance. Estimates of the initial variation in sensitivity currently rely on bioassays which are time consuming, but it is advisable to include, e.g. a few dozens of isolates in these assays. The methods may differ for each fungus-fungicide combination, but it is important that some general principles be observed (Georgopoulos, 1982).

In the case of non-obligate parasites, sensitivity can be measured on artificial media, and it is usually not difficult to determine the response of each isolate to a number of concentrations, e.g. by replica plating. The range of concentrations is decided on the basis of preliminary tests, but it is important not to ignore the solubility limitations of the compound studied (Georgopoulos, 1982). Volatility may make it necessary to separate treatments. In tests with artificial media it is best to use technically pure material, so that interference of other components of the formulation can be avoided. However, when sensitivity needs to be tested on treated plant material, some formulation is usually needed for such treatments.

If isolates from different areas and from fields with different fungicide use histories are included, indications on the type of problems to be anticipated may be obtained during this study of the initial variation in sensitivity to each fungicide candidate. If, for example, in spite of a small sample size, discontinuous variation is found, evidence not only of major-gene control of sensitivity differences, but also of cross-resistance to some of the previously used fungicides has been obtained. Testing laboratory or field strains with specific resistance to some of the known fungicides for sensitivity to the new compound can clarify the latter point. In contrast, when continuous, even if quite considerable, variation is found, this does not indicate cross resistance to previously used chemicals. Experience has shown that such variation may exist even without any previous selection. For example, populations of *V. inaequalis* never exposed to dodine were found to contain individuals with up to 100-fold differences in the concentration required to inhibit spore germination (Mckay & MacNeill, 1979). It is, therefore, important in baseline sensitivity testing to study the type as well as the extent of preexisting variation.

ISOLATION AND CHARACTERISATION OF RESISTANT MUTANTS

If little initial variation for sensitivity to a new compound is recognised, isolation and characterisation of resistant mutants in the laboratory will be needed for the additional tests of risk assessment. Where the potential fungicide target involves pathogens which are difficult to manipulate genetically, suitable species may be substituted. *Neurospora crassa* and *Aspergillus nidulans* are often preferred, but pathogenic fungi, such as *N. haematococca* and *Ustilago maydis*, have the advantage of allowing the effects of fungicide resistance genes on pathogenicity to be examined. To have a better chance of recognising substantial changes in sensitivity it is best to use an organism highly sensitive to the compound studied.

Mutations occur spontaneously at very low frequency which can substantially be increased by the use of physical or chemical mutagens. Since treatments increasing the mutation rate also result in the death of many of the treated cells, one must avoid the extremes: having too few survivors at a high dose or having rather few mutants among many survivors at a low dose of the mutagen. A treatment giving 90-95% lethality is usually appropriate. It is advisable to aim at the isolation of a number of resistant strains, so that a thorough study of the variability available to the organism can be attempted when desirable. This will require exposure of several millions spores, or other propagules, to the mutagen and plating of the survivors on medium containing the fungicide.

It is suggested that two fungicide concentrations are used, the lower one being at least twice the minimal concentration which completely prevents growth from spores of the original strain when plated at high density. Failure to obtain strains with stable resistance to the new compound in this way is strong evidence against at least a high risk. If resistant strains are obtained but are only slightly less sensitive than the original strain, they can be used in attempts to achieve higher resistance by sequential selection (De Waard, 1988). In case of an organism which does not lend itself for genetic analysis, success in such a stepwise increase may be taken as an indication of polygenic control of resistance. On the other hand, high resistance obtained in one step points towards major-gene control.

The type of population response to the use of a new fungicide can safely be predicted, if the genes responsible for changes in sensitivity are identified and their interactions studied. The pathogenic vegetative phase of most fungi is haploid, hence dominance or recessiveness of a mutant gene is irrelevant and examination of the phenotypes of the F_1 from a resistant (R) x sensitive (S) cross is sufficient. In diploids, such as *Phytophthora* sp., recognition of a Mendelian ratio requires selfing of the F_1 and examination of the phenotypes of the F_2 generation. In such fungi, it is very important to know the type of intra-allelic interaction. A recessive resistance gene will not affect the phenotype of the heterozygote and, therefore, resistance should be expected to evolve more slowly. If resistance is semi-dominant, one should be careful in recommending reduced application rates with the aim to lower the intensity of selection. A rate high enough to eliminate the heterozygotes, if possible, is highly advisable.

Analyses of RxS crosses usually show that each strain obtained by one-step selection carries one mutant gene for resistance: in case of a haploid organism a 1R : 1S ratio is found in the F_1. It cannot be excluded that one RxS cross may yield a higher ratio, up to 3:1. In such a case the resistant parent in the cross must carry two mutant genes. This finding indicates that resistance to the new compound is probably polygenic because the probability of inducing two resistance mutations in the same nucleus is very low, unless the genes involved are numerous. If no evidence of polygenic control is obtained in this way, analyses of RxR crosses will be required.

Even if resistant mutants differ considerably in the degree of resistance, they do not necessarily carry resistance genes at different loci. In *V. inaequalis*, for example, very high, high and moderate resistance to benzimidazoles is controlled by different alleles of the same major gene (Shabi *et al.*, 1983). Involvement of different loci can be accepted only if some sensitive recombinants are obtained from some RxR crosses. Recognition of several non-allelic genes for resistance to the same fungicide, however, does not show that resistance is polygenic. An early study of 100 mutants of *N. haematococca* resistant to aromatic hydrocarbon fungicides identified five chromosomal loci involved in resistance (Georgopoulos & Panopoulos, 1966), but recombinants carrying mutant genes at two or more of these loci were not less sensitive than single-gene mutants because of epistasis. Consequntly, variation was discontinuous. On the other hand, because it is not practically feasible to analyse a large number of RxR crosses, the fact that only a few loci involved in resistance to a new fungicide have been recognised does not exclude polygenic control.The characteristic of quantitative inheritance, and the cause of continuous variation is the positive interaction between non-allelic genes in the haploid nucleus and the additivity of their effects. An increase in the degree of resistance correlated with the number of mutant genes present has been conclusively shown, for example, in the case of resistance of *N. haematococca* to dodine (Kappas & Georgopoulos, 1970) and to fenarimol (Kalamarakis *et al.*, 1991) and of *A. nidulans* to imazalil (van Tuyl, 1977).

EVALUATION OF THE RATE OF RESISTANCE EVOLUTION

Having shown that the potential target organism(s) possess genes which can mutate to give resistance to a fungicide considered for development, and that the population response will be qualitative or quantitative, additional information is needed to predict the speed of response. The rate of resistance evolution may be affected by factors related to the environment (greenhouse or open field, conditions favorable for high disease pressure etc.) and disease management practices (sanitation, host resistance, etc.). Other factors being equal, the determinants of the rate at which the proportion of strains carrying given resistance alleles will increase are :

a. *intensity of selection* which is a function of the decrease in fungicide sensitivity caused by each gene and
b. *relative fitness*, i.e. the survival and subsequent reproductive success of the resistant strains in the absence of a discriminating fungicide concentration, as compared to the wild type.

Although selection pressures are usually referred to, they are difficult to measure. The degree of resistance of the various types of mutants is important, but in laboratory measurements, the degree of resistance is often overestimated because of the low water solubility of most fungicides (Georgopoulos, 1982). Even if an accurate numerical value is obtained, its significance may differ, depending on the properties of the chemical studied. Often, increasing the amount that can be made available for biological activity beyond a certain level is not possible in the field, so that even strains with low resistance are very difficult to control.

Relative fitness estimates are also difficult. In the laboratory, strains carrying particular alleles for resistance to a candidate fungicide may be compared to the wild type with respect to sporulation, time required for spore germination, germ tube elongation, linear growth and ability to survive under extreme conditions of temperature, humidity, osmotic pressure etc. Pathogenicity is, of course, a very important fitness determining characteristic. Not only the ability to infect a susceptible host, but also the rate of tissue colonisation, latent period, sporulation capacity and intensity must be considered. Information has been obtained by various workers by the method of mixed inoculations in laboratory or greenhouse tests. Reduced fitness of a laboratory selected strain may or may not be due to the resistance mutation itself. The problem can be resolved by examining the behavior of progeny from R x S crosses. Even if it is shown that recognised resistance genes affect important fitness determining characteristics, it can not be excluded that in nature rarer alleles, not affecting fitness, may be selected for by exposure to the new fungicide. Of interest in this regard, are recent observations of Faretra & Pollastro (1993) regarding high resistance to dicarboximides in *Botryotinia fuckeliana*. The high resistance allele is responsible for increased sensitivity to media of high osmolarity and this is considered the cause of low pathogenicity and fitness. In heterokaryons, however, in which nuclei carrying the allele for high resistance coexist with wild type nuclei, the high dicarboximide resistance phenotype is dominant, while that of osmotic hypersensitivity is partially recessive.

CONCLUSION

Tests needed for resistance risk assessment at the time a new compound is considered for development as an agricultural fungicide are rather simple. No particularly great effort and expenditure is required compared to the benefit of acquiring knowledge so important for decision making. Depending on the information that becomes available from the risk assessment tests, some more or less safe predictions may be possible. If major-gene resistance is obtained in the laboratory and the experiments do not show significantly lower fitness of the mutant strains, there can be little doubt that we have a high-risk fungicide. A low resistance risk is quite likely if repeated attempts fail to isolate strains with considerable resistance to the new compound. Intermediate situations of moderate risk may be anticipated if polygenes are shown to be involved, or major-gene mutations increase resistance at a more or less high cost in terms of fitness. The seriousness of problems likely to arise in such moderate-risk situations may vary, depending, amongst other factors, on the magnitude of this cost.

REFERENCES

Brent, K.J. (1982) Case study 4: Powdery mildews of barley and cucumber. In *Fungicide Resistance in Crop Protection*, J. Dekker and S.G. Georgopoulos (Eds). Pudoc, Wageningen, Netherlands, pp 219-230.

Brent, K.J. ; Hollomon, D.W. ; Shaw, M.W. (1990) Predicting the evolution of fungicide resistance. In *Managing Resistance to Agrochemicals*, M.B. Green, H.M. LeBaron and W.K. Moberg (Eds). ACS Symposium Series 421, pp. 303-319.

Cooksey, D.A. (1990) Genetics of bactericide resistance in plant pathogenic bacteria. *Annual Review of Phytopathology 28*, 201-219.

Crute, I.R. ; Harrison, J.M. (1988) Studies on the inheritance of resistance to metalaxyl in *Bremia lactucae* and on the stability and fitness of field isolates. *Plant Pathology 37*, pp. 231-250.

De Waard, M.A. (1988) Interactions of fungicide combinations. In *Fungicide Resistance in North America*, C.J. Delp (Ed.) APS Press, St. Paul, Minnesota, pp. 98-100.

Faretra, F. ; Pollastro, S. (1993) Genetics of sexual compatibility and resistance to benzimidazole and dicarboximide

fungicides in isolates of *Botryotinia fuckeliana (Botrytis cinerea)* from nine countries. *Plant Pathology 42*, 48-57.
Georgopoulos, S.G. (1982) Detection and measurement of fungicide resistance. In *Fungicide Resistance in Crop Protection,* J. Dekker and S.G. Georgopoulos (Eds). Pudoc, Wageningen, Netherlands pp. 24-31.
Georgopoulos, S.G. (1987) The development of fungicide resistance. *In Populations of Plant Pathogens - Their Dynamics and Genetics,* M.S. Wolfe and C.E. Caten (Eds). Blacwell, Oxford pp. 239-251.
Georgopoulos, S.G. ; Panopoulos, N.J. (1966) The relative mutability of the *cnb* loci in *Hypomyces. Can. J. Genet. Cytol. 8*, 347-349.
Georgopoulos, S.G. ; Skylakakis, G. (1986) Genetic variability in the fungi and the problem of fungicide resistance. *Crop Protection 5*, 299-305.
Gisi, U. ; Staehle-Csech, U. (1988a) Resistance risk evaluation of new candidates for disease control. In *Fungicide Resistance in North America,* C.J. Delp (Ed.) APS Press, St. Paul, Minnesota, pp. 101-106.
Gisi, U. ; Staehle-Csech, U. (1988b) Resistance risk evaluation of phenylamide and EBI fungicides. *Brighton Crop Protection Conference,Pests and Diseases* 1988, 1, 359-366.
Heaney, S.P. (1988) Population dynamics of DMI fungicide sensitivity changes in barley powdery mildew. In *Fungicide Resistance in North America.* C.J. Delp (Ed.) APS Press, St. Paul, Minnesota, pp. 89-92.
Kalamarakis, A.E. ; Demopoulos, V.P. ; Ziogas, B.N. ; Georgopoulos S.G. (1989) A highly mutable major gene for triadimenol resistance in *Nectria haematococca* var. *cucurbitae. Neth. J. Plant Path.* 95 (Suppl. 1), 109-120.
Kalamarakis, A.E. ; De Waard, M.A. ; Ziogas, B.N. ; Georgopoulos, S.G. (1991) Resistance to fenarimol in *Nectria haematococca* var. *cucurbitae. Pestic.Biochem. Physiol. 40*, 212-220.
Kappas, A. ; Georgopoulos, S.G. (1970) Genetic analysis of dodine resistance in *Nectria haematococca. Genetics 66*, 617-622.
Keiding, J. (1986) Prediction or resistance risk assessment. In *Pesticide Resistance - Strategies and Tactics for Management.* NRC Board on Agriculture, National Academy Press, Washington, DC. pp. 279-297.
Kendall, S.J. ; Hollomon, D.W. ; Cooke, L.R. ; Jones, D.R. (1993) Changes in sensitivity to DMI fungicides in *Rhynchosporium secalis. Crop Protection 12,* 357-362.
McKay, M.C.R. ; MacNeill, B.H. (1979) Spectrum of sensitivity to dodine in field populations of *Venturia inaequalis. Canadian Journal of Plant Pathology 1*, 76-78.
Shabi, E. ; Katan, T. ; Marton, K. (1983) Inheritance of resistance to benomyl in isolates of *Venturia inaequalis* from Israel. *Plant Pathology 32,* 207-211.
Van Tuyl, J.M. (1977) Genetics of fungal resistance to systemic fungicides. Meded. Landb. Hogesch. Wageningen 77-2, 1-137.

RESISTANCE RISK EVALUATION OF FLUDIOXONIL, A NEW PHENYLPYRROLE FUNGICIDE

U.W. HILBER, H. SCHÜEPP

Swiss Federal Research Station, Department of Plant Pathology, 8820 Wädenswil, Switzerland

F.J. SCHWINN

Institute of Botany, University of Basle, Hebelstrasse 1, 4056 Basle, Switzerland

ABSTRACT

Several authors have dealt with the question whether the resistance risk of new compounds can be assessed in laboratory and greenhouse tests. Risk assessment schemes were developed but only applied as retrospective analyses after build-up of resistance in the field. In this paper we focus on the evaluation of the inherent risk of the novel phenylpyrrole fungicide fludioxonil. We could demonstrate two biologically dissimilar resistance types: laboratory and field resistance. Fludioxonil laboratory resistant strains could be selected easily without the use of mutagens or UV irradiation. These spontaneously occurring laboratory resistant strains were consistently also resistant to dicarboximides. Classical genetic analysis showed that there is no cross resistance to fludioxonil in dicarboximide resistant field isolates. Reliability and pitfalls of the modified risk evaluation scheme of Gisi and Staehle-Csech (1988) are discussed.

INTRODUCTION

The idea of predicting the resistance risk of new candidate fungicides is not new. Dekker (1982) discussed the question "Can we estimate the fungicide resistance hazard in the field from laboratory and greenhouse tests?", Gisi and Staehle-Csech (1988) published a detailed protocol for the evaluation of the resistance risk of new fungicides and Brent *et al.* (1990) described the effort to combine genetic, monitoring, multifactorial and modelling approaches to develop the best judgement of risk before and during the early application of a fungicide. Although several authors have dealt with the theoretical aspects of the prediction of fungicide resistance the schemes which they proposed were only applied as retrospective analyses on existing resistance problems in the field. Currently several novel groups of fungicides such as phenylpyrroles, anilinopyrimidines and methoxyacrylates are in an advanced state of development and probably will enter the market within the next few years. More than twenty years of experience and a profound theoretical background in risk analysis can be used in the risk assessment concerning the new generation of fungicides. Active research has to be dedicated to the aspects: dynamics of fungal populations with respect to resistance, resistance risk assessment and resistance management. There is an ideal opportunity to apply our knowledge to predict the potential risk of resistance build-up and to develop and implement antiresistance strategies from the very beginning of the product's use in the field. We report on joint efforts and close cooperation in the risk assessment between industry, academia and registration authorities. In the present paper we give an overview of what has been done to evaluate the resistance risk of fludioxonil, a novel phenylpyrrole fungicide.

RESISTANCE RISK ASSESSMENT

As a whole the resistance risk is a combination of inherent resistance risk and management resistance risk (Staub and Sozzi, 1984). Management risk and inherent risk are equally important. However, in contrast to the inherent risk, the management risk can be influenced by various means. "High risk" fungicides, when used with proper strategies (= low management risk), may not cause more overall risk than "low risk" fungicides used improperly (= high management risk) (Gisi and Staehle-Csech 1988). In this paper we focus on the evaluation of the fungicide - pathogen related inherent resistance risk.

Fungicide

Fludioxonil is a novel, non-systemic phenylpyrrole fungicide. It is a derivative of the antibiotic pyrrolnitrin and is highly active against a broad spectrum of fungi among Ascomycetes and Basidiomycetes. Fludioxonil is being developed for foliar use with *Botryotinia fuckeliana* as its major target pathogen (Gehmann 1990). Jespers (1994) demonstrated that the primary mode of action of this class of fungicides is new: it is based on the inhibition of transport associated phosphorylation of glucose.

Pathogen

B. fuckeliana (syn. *Botrytis cinerea*) causal agent of grey mould is an economically important pathogen on a wide range of host plants and it causes considerable damage during storage and transportation. *B. fuckeliana* grows fast on artificial and complex media and the sexual stage can be induced under laboratory conditions (Faretra *et al.* 1988) which makes this pathogen accessible to classical genetic analysis. It has also been successfully transformed making it available for molecular genetic studies (Hilber *et al.* 1994b). Beyond the genetic variability caused by mutation and sexual reproduction, *B. fuckeliana* additionally shows a genetic flexibility that may be caused by the selection of different alleles within the heterokaryon. High genetic variability and flexibility, high reproduction rate, wide host range and the possibility of saprophytic growth on virtually any plant debris favour a high inherent risk for resistance. In the past resistant *B. fuckeliana* populations were selected only few years after the introduction of the benzimidazoles and the dicarboximides.

DESIGN OF RESISTANCE RISK ASSESSMENT

Gisi and Staehle-Csech (1988) proposed a step by step procedure for estimating the resistance risk of new fungicides. Risk analysis, however, highly depends on the combination fungicide - pathogen and protocols can only be guidelines that have to be modified in each new case. In our evaluation of the inherent resistance risk of the combination *B. fuckeliana* - fludioxonil we adopted the protocol from Gisi and Staehle-Csech and modified it as shown in Figure 1.

Test Methods - Baseline sensitivities

To be able to test a new compound in standard agar plate assays at least two requirements have to be met: a) the fungicide must be soluble in a solvent that can be mixed with agar (and does not influence the pathogen) and b) the fungicide must be active in the *in vitro* test. Fludioxonil meets both requirements. The active ingredient (technical grade) was dissolved in ethanol and then added to the previously cooled agar. Mycelial growth tests and germination tests have revealed that fludioxonil is a highly active inhibitor of conidia germination and of mycelial growth. Since germination is inhibited 19 times less than mycelial growth, and evaluation of the germination tests is not always easy, we favoured the mycelial growth test as the *in vitro test*. EC_{50} values range from 0.08 to 0.2 mg/l for conidia germination and from 0.003 to 0.016 mg/l for mycelial growth. The variation in the sensitivities of isolates never exposed to fludioxonil (baseline sensitivities) was low and in the range of the variation known for dicarboximide fungicides.

Monitoring

Fungicide resistance monitoring, testing the sensitivity of target organisms from field populations, is the basis not only of the management of fungicide resistance but also of the resistance risk evaluation itself. As a part of the risk assessment monitoring has to start during product development, and before the start of sales. Companies should, however, prolong monitoring throughout the product's life span to be able to judge the success of their antiresistance strategies, and to review strategies as soon as necessary.

Results obtained with dicarboximides showed that resistant fungal strains occurred at a frequency of 1×10^{-7} before the first application of the novel ingredient (Martinetti 1986). In the first stage of monitoring it is important to monitor a large population. Changes in sensitivity to fludioxonil were intensively monitored by Ciba Geigy and by ourselves. Up to now more than 1000 isolates were tested. None of them showed stably reduced sensitivity to fludioxonil.

Figure 1: Modified Gisi and Staehle-Csech risk evaluation scheme

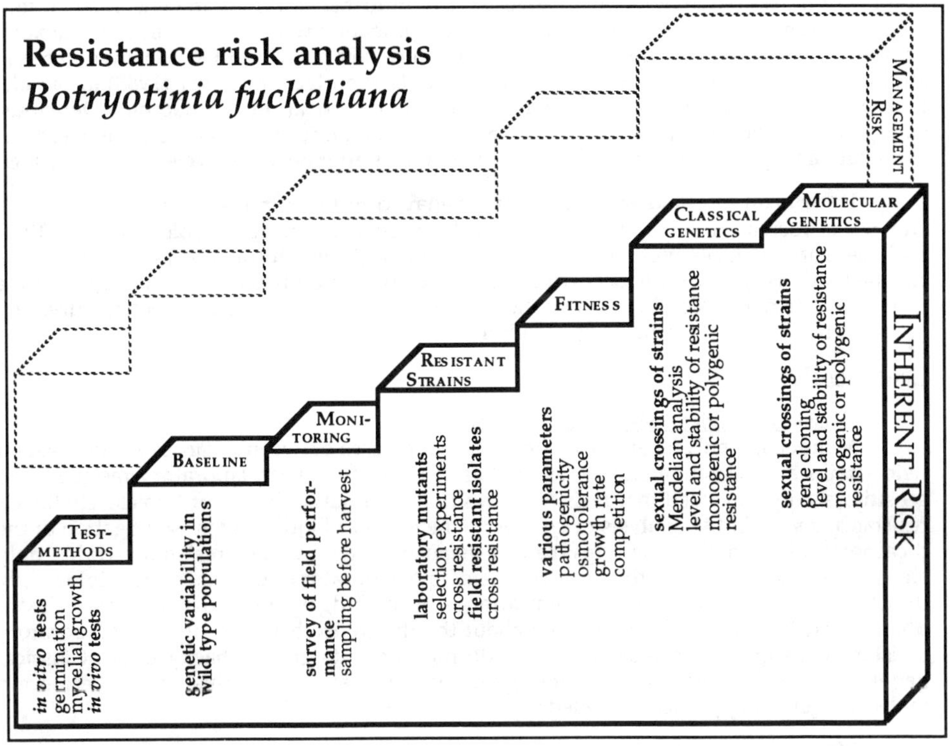

Resistant Strains

Early laboratory experiments with dicarboximides such as vinclozolin have shown that laboratory resistant strains could easily be selected by exposing conidia or mycelium to fungicide amended plates (Martinetti 1986, Schüepp *et al.* 1982), without mutagens or UV irradiation. Experiments in our laboratory have revealed that the same is true for fludioxonil (Hilber 1992, Hilber *et al.* 1993). Laboratory mutants that were selected in this way were cross resistant to both vinclozolin and fludioxonil. It did not matter whether these mutants were selected on vinclozolin or fludioxonil amended medium. "Training" experiments, involving repeated exposure to increasing, but sublethal concentrations of the fungicide did not yield resistant strains. We did not apply mutagens or UV radiation as the frequency of spontaneously occurring laboratory mutants was very high. Such experiments would have needed the selection of mutants on fungicide amended media after the mutagenic treatment which would not have allowed the distinction between mutagen induced mutants and spontaneously occurring ones.

Dicarboximide resistant strains that were isolated from rotten grapes in the field showed a different resistance pattern. These field resistant strains revealed only a moderate level of resistance to vinclozolin while they were sensitive to fludioxonil. Laboratory resistant strains, in contrast, were resistant to fludioxonil and to vinclozolin, but this type of resistance was not encountered in the field.

Fitness

Field application of an active ingredient creates a specific selection pressure for the fungal population in addition to the various naturally occurring selection pressures due to adverse physical, chemical or biological factors. Under laboratory conditions only a limited number of fitness parameters can be investigated (Hilber *et al.* 1994a). In general laboratory resistant strains that were selected on either agar amended with vinclozolin or fludioxonil showed a highly decreased osmotic stress tolerance compared to their sensitive parental strains. These results were in accordance with data reported by Beever (1983).In parallel laboratory resistant strains lost their pathogenicity on apple cv. Golden Delicious. Competition experiments revealed similar results: laboratory resistant strains were not competitive they were suppressed by sensitive isolates after only few cycles of coculturing.

Dicarboximide field resistant strains behaved in the opposite way: although they showed a decreased sensitivity to vinclozolin their osmotic stress tolerance and their pathogenicity on apple was equal to that of sensitive strains. It must be assumed that the resistant strains sporadically occurring before the population had been in contact with the active ingredient can slowly improve their fitness under the long lasting selection pressure accompanying regular fungicide applications.

Genetic analysis

Classical genetics - Mendelian analysis

Mendelian analysis of sexual progenies of crosses between dicarboximide resistant field strains and sensitive strains, as well as of crosses between laboratory resistant and sensitive strains, revealed that resistance was due to mutation in one or two closely linked resistance genes. In the analysis of crosses we never found independent segregation of the dicarboximide and the phenylpyrrole resistance although field resistant strains only showed resistance to dicarboximides. Field and laboratory resistance are biologically dissimilar. This is of considerable importance in the judgement of the resistance risk of phenylpyrroles. Although knowledge about the difference between field and laboratory resistance has greatly increased we still do not know its genetic background. Classical genetic analysis is not sensitive enough to give the answer in this case. Therefore a molecular genetic approach is needed.

Molecular genetics

Molecular genetic techniques have improved dramatically over the last decade. PCR, a major breakthrough in this field, and its various applications are commonly used techniques in many laboratories. Compared to well documented model organisms such as *Neurospora* or *Saccharomyces*, little is known about the genetics of *B. fuckeliana*. Molecular genetic exploration of this pathogen was anticipated to be cumbersome but, reports on the segregation of DNA polymorphisms by RAPD analysis (Van der Vlugt-Bergmanns *et al.* 1993) and successful transformation of *B. fuckeliana* (Hilber *et al.* 1994b), have opened new perspectives. A molecular genetic approach could answer remaining questions which were left open by the less sensitive classical (Mendelian) approach. Molecular genetic analysis is a further tool to be used for resistance risk assessment.

CRITICAL EVALUATION

The novel botryticide fludioxonil shows similarities with the chemically unrelated dicarboximide fungicides. Leroux and coworkers (1991, 1992) hypothesised that dicarboximides and phenylpyrroles might have the same mode of action. First experimental data suggested that fludioxonil has a "high inherent risk". Experience with dicarboximides and benzimidazoles has taught us that resistant *B. fuckeliana* subpopulations can be selected very rapidly under heavy selection pressure. Laboratory data revealed that fludioxonil resistant strains can be selected easily without the use of mutagens or UV irradiation. These spontaneously occurring laboratory mutants are cross resistant to dicarboximides and fludioxonil.

Laboratory results were, however, completely contradictory to the performance in the field. Efficacy of fludioxonil was excellent in all our field trials. We could clearly demonstrate two biologically dissimilar resistance types: laboratory and field resistance. In plots where dicarboximide fungicide were applied only a few times in the past, we repeatedly found an increase in the resistance frequency from low levels to 100% after one to two application of dicarboximides (Hilber *et al.* 1994a). Fludioxonil, however, did not alter the dicarboximide resistance frequencies which was evidence for lack of cross resistance in the field.

Classical genetic analysis confirmed that there is no cross resistance to fludioxonil in dicarboximide field resistant strains. As this situation is neither matched by the term cross resistance nor by the term multiple resistance a new term has to be found. A final answer to Dekker's question "Can we estimate the fungicide resistance hazard in the field from laboratory and greenhouse tests?" cannot be given. The model of Gisi and Staehle-Csech is a good guideline for the experimental design but as demonstrated in our analysis it contains pitfalls. An exact calculation of a risk as it is suggested by the Gisi and Staehle-Csech model, is not feasible. The laboratory data presented in this paper have to be interpreted with care. No parallels to the phenomenon of the laboratory resistance was observed in the field. We assume that we have to pay attention to two significantly different processes: the selection for resistance and the selection for fitness. Under natural conditions in the field, even slight differences in fitness with regard to various, not yet elucidated aspects can essentially improve the proliferation potential or the survival mechanisms of a specific strain. As demonstrated in laboratory conditions, selection for resistance is fast and constitutes, no doubt, a potential risk. Selection for fitness in resistant strains, however, seems to be very slow. This may explain why the build-up of a resistant fungal population having acquired normal fitness and thus being competitive with sensitive strains usually takes years. With the current state of knowledge we assume that the inherent risk of fludioxonil is medium. However, the management resistance risk is high as there are no fungicides available yet that are suitable partners. The limitation of applications to one or two sprays per season is difficult to be enforced. In the case of dicarboximides this strategy was not effective as demonstrated in field experiments (Hilber *et al.* 1994a). Industry and regulatory authorities have the obligation to enforce "true" resistance strategies which could be a mixture of phenylpyrroles and anilinopyrimidines. Compounds of both classes are in the final steps of registration. Management risk could be kept low resulting in a good chance of success in *Botrytis* management.

ACKNOWLEDGEMENTS

We thank B. Buchmann, K. Langenegger and M. Hilber- Bodmer for technical assistance and we are grateful to Ciba-Geigy Ltd., Basle, for financial support.

REFERENCES

Brent, K.J.; Hollomon, D.W.; Shaw, M.W. (1990) Predicting the evolution of fungicide resistance. In: *Managing resistance to agrochemicals - from fundamental research to practical strategies*, M.B. Green, H.M. LeBaron and W.K. Moberg (Eds.), Washington D.C., pp. 303-319.

Beever, R.E. (1983) Osmotic sensitivity of fungal variants resistant to dicarboximide fungicides. *Transactions of the British Mycological Society*, **80**, 327-331.

Dekker, J. (1982) Can we estimate the fungicide resistance hazard in the field from laboratory and greenhouse tests?. In: *Fungicide resistance in crop protection*, J. Dekker and S.G. Georgopoulos (Eds), Wageningen, pp. 128-138.

Faretra, F.; Antonacci, E.; Pollastro, S. (1988) Improvement of the technique used for obtaining apothecia of *Botryotinia fuckeliana* (*Botrytis cinerea*) under controlled conditions. *Annals of Microbiology*, **38**, 29-40.

Gehmann, K.; Nyffeler, R.; Leadbeater, A.J.; Nevill, D.; Sozzi, D. (1990) CGA 173506: A new phenylpyrrole fungicide for broad spectrum disease control. *Brighton Crop Protection Conference - Pests and Diseases 1990*, **2**, 399-406.

Gisi, U.; Staehle-Csech, U. (1988) Resistance risk evaluation of new candidates for disease control. In: *Fungicide resistance in North America*, C.J. Delp (Ed.), Minnesota, pp. 101-106.

Hilber, U.W. (1992) Comparative studies on genetic variability and fungicide resistance in *Botryotinia fuckeliana* (de Bary) Whetzel against vinclozolin and the phenylpyrrole CGA 173506. PhD Thesis, Basle. University of Basle.

Hilber, U.W.; Schüepp, H.; Schwinn, F.J. (1993) Resistance of *Botryotinia fuckeliana* (de Bary) to phenylpyrrole fungicides as compared to dicarboximides. In: *Proceedings of the 10th International Symposium on Systemic Fungicides and Antifungal Compounds*, H. Lyr and C. Polter (Eds.), Stuttgart, pp. 105- 111.

Hilber, U.W.; Schüepp, H.; Schwinn, F.J. (1994a) Development of resistance to dicarboximides in integrated control of *Botryotinia fuckeliana* (de Bary) Whetzel. *Journal of Plant Disease and Protection*, **101**, 66-73.

Hilber, U.W.; Bodmer, M.; Smith, F.D.; Köller, W. (1994b) Biolistic transformation of conidia of *Botryotinia fuckeliana. Current Genetics*, **25**, 124- 127.

Jespers, A.B.K. (1994) Mode of action of the phenylpyrrole fungicide fenpiclonil in *Fusarium sulphureum*. PhD Thesis, Wageningen. Agricultural University.

Schüepp, H.; Küng, M.; Siegfried, W. (1982) Dévelopment des souches de *Botrytis cinerea* dans les vignes de la Suisse alémanique. *Bulletin EPPO*, **12**, 157-161.

Staub, T.; Sozzi, D. (1984) Fungicide resistance: A continuing challenge. *Plant Disease.*, **68**, 1026-1031.

Leroux, P (1991) Mise en evidence d´une similitude d´action fongicide entre le fenpiclonil, l´iprodione et le tolclofos-methyl. *Agronomie*, **11**, 115-117.

Leroux, P.; Lanan, C.; Fritz, R. (1992) Similarities in the antifungal activities of fenpiclonil, iprodione and tolclofos-methyl against *Botrytis cinerea* and *Fusarium nivale. Pesticide Science*, **36**, 255-261.

Martinetti. G. (1986) Charakterisierung der genetischen Variabilität von *Botrytis cinerea* aufgrund von Fungizidresistenz und Enzymaktivität. PhD Thesis, Zürich. Swiss Federal Institute for Technology.

Van der Vlugt-Bergmanns, C.J.B.; Brandwagt, B.F.; Van't Klooster, J.W.; Wagemakers C.A.M.; Van Kan, J.A.L. (1993) Genetic variation and segregation of DNA polymorphisms in *Botrytis cinerea. Mycological Research*, **97**, 1193-1200.

Session 7
Concluding Remarks

THE WAY AHEAD

K.J. Brent

St Raphael, Norton Lane, Chew Magna, Bristol, BS18 8RX, UK.

In this brief concluding contribution I was asked to assume the difficult role of pathfinder, and to discuss future priorities for research and practice. I suppose I have reasonable background for this task, having addressed fungicide resistance problems for over 25 years from the interacting but differently biassed standpoints of industrial R and D, public-sector research and independent consultancy. But it remains a daunting task, and I have been glad to be able to draw freely on all the information and opinions given over the past three days.

What have we learnt from our quarter-century of dealing with the phenomenon of resistance to fungicides? One hears totally different views on this amongst informed workers, ranging from the gloomy 'virtually nothing, we still cannot predict it reliably or stop it' to the optimistic 'we understand the problem, we know what to do, active implementation is the keynote'.

Actually I think we have achieved a great deal. We are all much more alert to the threat of resistance, so that manufacturers do now make and act upon early appraisals of risk, involving base-line monitoring, cross-resistance studies and sometimes mutation experiments. Companies are devising and implementing use strategies which must aim at a difficult compromise: using materials to the best short to medium term commercial advantage, and conserving their effectiveness for the longer term. This of course calls for inter-company co-operation regarding related compounds. Over the past twelve years FRAC has worked in this role. The speed of action of FRAC and the adequacy of some of its recommendations have been questioned from time to time, but it has to be appreciated that the tasks of securing interchange of confidential information, reconciling conflicting technical, commercial and logistic factors, and achieving consensus views, are often very hard. I believe that FRAC has done its job remarkably well, and long may it continue. Public-sector research and advisory services have also been influential, and have interacted with FRAC to a limited extent. The announcement here by its Chairman that FRAC will consult more openly with the public sector, and will support some external research initiatives, is a good omen.

The publication of data, supported by full experimental details, and statistical analysis where appropriate, continues to be of paramount importance. It was good to see the new data presented here, particularly from the industrial side. More in the future, please - especially on phenylpyrroles, methoxyacrylates, anilinopyrimidines and other new groups. In connection with publication, we still need to be very careful about defining our terms, and particularly what we mean in each particular context by 'resistance'. Some participants proposed that we adopt a narrow definition of 'resistance' to denote the loss of effectiveness in practice. In my view, any attempt to impose new semantics is unnecessary, will fail, and will cause confusion. 'Resistance' has been used over many years as an umbrella term, covering laboratory mutants, field isolates and large populations. This started in bacteriology with sulphamide and antibiotic resistance, and then in entomology with DDT resistance. But when we need to be more precise, as we often do, then we must refer specifically to 'laboratory resistance', 'field resistance' (which may be very rare or slight), or 'practical resistance' (where field resistance is frequent and severe enough to interfere with effective disease control). Fortunately, precision in reporting is gradually improving. At this meeting, 'percentage resistance' in graphs and tables tended to be replaced by 'percentage of samples unaffected by 10 ug/ml in leaf-disc test', 'percentage of sites where resistant forms were detected', or whatever was actually measured. However, more care is still needed, and must be checked by editors.

Strategies have been implemented, and most of us believe that they have helped. Most crop diseases are still under good control, but we heard of some situations where the defences are now inadequate (e.g. *Botrytis cinerea* on grapes in the Champagne region) or stretched very thinly (e.g. *Mycosphaerella fijiensis* in bananas in C. and S. America). New fungicides are urgently needed. It is heartening to know that we have reached the end of the long lull in invention, that has existed since the explosive appearance of many novel fungicide groups in the late 1960s and 1970s. It was good to hear that the UK registration authority is now welcoming increased diversity of available chemicals. This is crucially important in the battle against resistance, and is also environmentally favourable. Chemical control will be needed for many years, and it is a worrying thought that falling agrochemical profits may hinder further invention.

Some participants were rather disappointed to find us still debating some of the questions that we were discussing 25 years ago, such as mixtures versus rotations, and low rates versus high rates. These strategy problems are very difficult to research under field conditions. It generally takes several years for clear differences in resistance build-up to occur, and also to determine whether particular use strategies are of lasting advantage. Even with large plots, invasion by external populations of the target pathogen, and loss of the original population, can easily occur. When conclusive results have been obtained, these have sometimes been conflicting. We heard theoretical arguments that lower rates may be safer with regard to major-gene resistance, and higher rates safer with regard to polygenic resistance. However the doses that actually reach the target organism vary enormously in space and time, giving complex exposure sequences. More field experimentation must be encouraged, so that soundly based judgement on the best strategies for particular circumstances can be achieved.

We still lack sufficient understanding of the underlying mechanisms of resistance. In particular more research is needed on the genetic basis, and on the field behaviour of resistance genes under different selection conditions. Use of the latest molecular techniques, including DNA probes and other diagnostic agents, will greatly help us to make a much closer examination of field phenomena. For example, what is happening in potato fields where metalaxyl resistance is readily detected in bulk spore samples, but control is sustained? Are resistant and sensitive forms intimately mixed, or are there heterokaryotic populations of resistant and sensitive nuclei within the coenocyte? Will selection differ against zoospore and direct sporangial infections?

On the other hand, how much monitoring do we need to do? Year by year monitoring is expensive, and as our knowledge grows is probably giving diminishing returns. Once good base-line data are obtained, then the priority should be on continual scrutinising of field performance, if possible in a more systematic, precise and well-reported way than at present, and involving regional field trials when appropriate. Performance monitoring is often anecdotal, and forms the weakest part of most resistance management programmes. If problems of control do arise, then sensitivity monitoring should be done as an explanatory measure. The importance of harmonising monitoring methods was stressed by some speakers. If reached by worker consensus this is fine, but the external imposition of standard methods must be avoided, first because different methods suit different circumstances, secondly because the best methods are hard to select and need to evolve by experience, and thirdly because results indicated by several different methods are actually more reliable.

Resistance management needs to be seen more clearly as a component of the broader concept of integrated crop management. When ICI first introduced ethirimol to control barley powdery mildew, I remember that we claimed that it would permit the plant breeder to stop worrying about disease resistance, and to focus better on yield and quality parameters. Even today many farmers will select a relatively disease-prone variety, relying totally on fungicide treatments to maintain crop health. Far more lasting will be the combined use of chemical and genetic defences in partnership, as exemplified in several of the papers. If only the plant breeders could find the genetic equivalent of mancozeb, at least in its stability and breadth of effect! Other husbandry components, such as disease-suppressive rotations and avoidance of excessive nitrogen fertilisation, will also form an increasingly important part of the integrated approach.

Overall, the resistance phenomenon has not been the disaster forecast by some observers, but it is an ever present threat with new cases arising and some old problems still continuing. There is much more to learn, of great practical and scientific interest, and I hope that we will all meet again in two or three years' time to discuss a lot more new and interesting data.

Syndicate Workshop Discussion Reports

Syndicate Workshop Session Discussion Reports

Five syndicates operated simultaneously, each with a chairman and assistant to capture discussion points. Each syndicate was set the following four questions under the general banner 'Resistance risk assessment and management'.

1. Resistance surveys and monitoring
   - can statistics help?

2. Resistance surveys and monitoring
   - can new technology help?

3. Predicting resistance development
   - can genetics help?

4. Anti-resistance strategies
   - mixtures or alternations

The chairman had the discretion to focus on one or more questions as appropriate.

Syndicate 1

Chairman: Dr. J.A. Lucas
Assistant: Dr. M.J. Hocart

Question 1. Resistance surveys and monitoring - can statistics help?

- What is the relevance of EC_{50} data? Values for fungicide sensitivity are calculated in different ways, sometimes based on spore populations from field samples, or genetically ill-defined mass isolates, while other estimates derive from single spore isolates, which may not be homokaryotic. In comparing data sets are we comparing like with like? What information are we trying to obtain?

- Sampling. How does one establish the baseline value, i.e. the fungicide sensitivity of the original non-selected population, against which changes in sensitivity of the population can be recognised. How many samples need to be tested?

Question 2. Resistance surveys and monitoring - can new technology help?

- DNA-based techniques may be powerful tools for surveying and monitoring fungicide resistance, provided the nature of the variation is known in detail. For example DNA probes can identify particular mutations conferring resistance. However development of such systems requires detailed analysis and hence they may not become available until after fungicide resistance has become established in the pathogen population.

- Theoretically the technology allows rapid screening of large samples. If PCR-based techniques can be made quantitative, their potential for screening would be greatly enhanced.

Question 3. Predicting resistance development - can genetics help?

First a question was asked: "Is there any such thing as polygenic resistance?"

Fungicides with multiple sites of action represent, in general, a low risk of resistance. With site-specific fungicides the question is not "Will resistance occur?" but "When will resistance occur?" The rate of resistance development depends upon the selection pressure applied, in terms of fungicide use, but also on the biology and epidemiology of the pathogen. We know comparatively little about natural variation in sensitivity to fungicides. Hence:

- Genetic information needs to be linked to the epidemiology and biology of the pathogen. Otherwise the predictive value of any genetic information will be poor.

Question 4. Anti-resistance strategies - mixtures or alternations?

Growers perceive fungicide resistance to be "not their problem" but rather a problem for the agrochemical industry.

The use of fungicide alternations was considered difficult to implement as growers are reluctant to apply chemicals that may be less effective than the best available. Experience in vineyards in France shows, however, that this strategy can be successfully adopted provided growers are sufficiently aware of the problem of resistance; grower education is therefore important.

For agrochemical companies fungicide mixtures represent the most pragmatic approach, since the failure of the vulnerable component of the mixture, through resistance, will not result in a serious loss of disease control. But is recommendation of fungicide mixtures an anti-resistance strategy, or sometimes a ploy for continued marketing of an obsolete product?

Anti-resistance strategies must be rooted in practical disease control. For <u>any</u> strategy to stand a chance of success it must be implemented by growers. Consequently effective strategies will not be too complex.

Syndicate 2

Chairman: Dr. J. Gilmour
Assistant: Rosemary Collier

By popular demand Syndicate 2 addressed the questions in reverse order.

Question 4. Anti-resistance strategies - mixtures or alternations?

1. We should not be considering mixtures or alternations but mixtures **and** alternations, preferably alternations of mixtures. In Greece no problems had occurred where such an approach had been used for 10 years in contrast to the rapid appearance of resistance when new molecules were used alone.

2. There was a general feeling in favour of mixtures over alternations but programmes on fruit were cited as examples against this. It was suggested that this approach had been adopted because of the early occurrence of resistance problems, the hi-tech approach to the crop, the high value of the crop and the need for a high level of disease control.

3. There was serious doubt as to whether a farmer would really have resistance at the forefront of his mind when deciding which fungicide to use.

4. In practice most mixtures are made for reasons other than combating resistance, particularly to broaden the spectrum of control and in some cases for synergy.

5. Agrochemical companies will make mixtures of fungicides if it will be of benefit to them, i.e. if there will be a benefit they can sell to the farmer.

6. EU legislation - the group was generally content that the new Directives would appear to allow enough scope in label recommendations for the adoption of effective anti-resistance programmes. It was accepted that for new molecules this would have to be based on theory, not on evidence.

7. Finally, the danger of generalisation was stressed. It was accepted there was very little experimental evidence about the respective merits of mixtures and alternating programmes.

Question 3. Predicting resistance development - can genetics help?

1. NO. Genetic analysis could be very useful after the event but was not thought to be useful in prediction.

2. The group suggested that more effort should be directed to looking at variability within the pathogen when the new molecule is first introduced, but this raises the question of what would we be looking for and at what frequency.

3. It was stressed that mode of action studies would not be an indicator of potential mechanisms of resistance. Therefore genetic studies of mode of action are unlikely to help.

4. Artificial mutant studies were said to be able to show "everything and nothing". Such laboratory studies have not been good indicators of problems and lack of problems in the field.

Question 2. Resistance surveys and monitoring - can new technology help?

1. Molecular markers would be useful but can be used only after a resistance problem has been identified.

2. There was concern about the cost of some of the new techniques. However it was stressed that quick results from field enquiries would sometimes be very valuable and repay the higher costs incurred.

3. It was suggested that new diagnostics for the presence of pathogens may be more beneficial. In support of this the example was cited from Chile where diagnostics for Botrytis are used in combination with a forecasting technique that has allowed vine growers to reduce their fungicide

programme from typically nine sprays to only two sprays per season.

Question 1. Resistance surveys and monitoring - can statistics help?

1. Some concern was expressed about the wide range of different techniques being used. It was thus not always clear whether the results obtained by different workers were the same or different.

2. It was pointed out that there are some recognised protocols (especially EC_{50}), though the need for different approaches was appreciated.

3. It was apparent that there was some need for more education and training in the use of the techniques that already exist.

Syndicate 3

Chairman: Dr. M.W. Shaw
Assistant: Dr. R. Beresford

Question 4. Anti-resistance strategies: mixtures or alternations?

There was broad agreement that both strategies were usable, but that mixtures were far more attractive, because they offered many advantages as well as a possible reduction in the rate at which resistance evolves. These included the fact that the strategy could be sure of implementation, because it depended on manufacturers rather than growers; that it provided coverage against a wider range of pathogens; that it could give "insurance" against failure of one of the components; and that it would often allow reduction in the rates of the individual components. For the most part, participants favoured mixtures of an "at risk" fungicide with an historically safe chemical such as chlorothalonil or dithiocarbamate. Appropriate application rates became a central theme of the discussion: what rates should be used in a mixture? If full rates were used, the main attraction of alternation was that it involved substantially less active ingredient overall. A majority opinion emerged, although there were dissenting voices. This opinion was that if the resistance to which a fungicide was at risk was monogenic and large, as with metalaxyl or MBC, the rate used should be as low as possible; if the resistance were continuously distributed and polygenically controlled, the rate should be as high as possible. (Our attention was drawn to the paper of Stevas in these proceedings, which supported the latter idea). However, it was also pointed out that pathogen biology needs to be considered separately for each individual case, and that "rate" by itself has little meaning: what matters is the percentage survival of the pathogen, and the spray coverage.

Question 3. Predicting resistance development - can genetics help?

Laboratory studies of the genetics of resistance were felt to be useful, but not very useful. Their main use was as a negative guide: if it was easy to get resistance in the laboratory, there must be deemed to be a considerable danger in the field. However, this could be misleading - the example of MBC resistance in *Rhynchosporium* was cited - and the failure to get laboratory resistance was no guarantee of safety in the field.

Statisticians know a lot about sample surveys, and one would be foolish to undertake one without taking statistical advice, mainly about sampling strategy. The sample size required depends on the question, and quite small samples can be useful for some questions. You are unlikely to get early warning of monogenic resistance arising in the field, but you might for polygenic, because of the different dynamics in the two cases.

There was very cautious support for the promise of new technology. Some of the new techniques are so specific that they might miss a novel resistance mechanism. More importantly, development of such techniques takes a long time, and the circumstances in which they will then be useful are restricted, as it will often be too late. DNA techniques seemed to offer little for polygenic resistance, simply because too many genes are supposed to be involved.

Syndicate 4

Chairman: Dr. J.K.M. Brown
Assistant: Dr. B.J. Nielsen

This workshop concentrated on questions about the population biology of fungicide resistance which are so far unresolved. Participants discussed current areas of controversy and the research that would be needed to resolve them. In discussing the question, "Predicting resistance development - can genetics help?" several participants pointed out that the prediction that strong control will necessarily select resistance is not always valid. For example, resistance to tridemorph has not developed in barley powdery mildew or black leaf streak of banana, despite the extensive use of this fungicide. Four points which might mitigate the development of resistance were raised. One is the pathogen's reproductive biology. It was argued, on one hand, that sexual reproduction could recombine resistance genes and so promote resistance, and on the other hand, that sex could break up combinations of resistance genes, thus slowing the evolution of resistance. This question clearly requires further study. Secondly, migration might overwhelm selection. The relative importance of these two factors is poorly understood in plant pathology in general. Thirdly, the mode of action of the fungicide may be such that resistant mutants occur at a very low frequency. This might be a factor in the continuing effectiveness of tridemorph. A final issue is the fitness of resistant isolates. There is little good data on whether or not resistance adversely affects fitness, and little understanding of the significance of fitness in the population genetics of many pathogens.

Much of the discussion of a second question, "Resistance surveys and monitoring - can new technology help?" was motivated by an outline of a test for benzimidazole resistance in scald of barley, using an oligonucleotide probe, presented by Dr Derek Hollomon (Wheeler et al., 1994). This method has greatly increased the throughput of tests for resistance, but several questions about the general applicability of molecular tests were raised. One problem is that, in order to develop such a test, the mechanism of resistance must be known. These methods may not, therefore, be capable of predicting whether or not resistance will occur in future, but may assist in monitoring the development of resistance once it has been detected, and in evaluating strategies for its control. A further limitation is that, in some cases, the phenotype - resistance or susceptibility - of an isolate would not be fully predicted by genetic variation in a single DNA sequence. Finally, some participants doubted that a molecular test would always save time or money. Despite this, the test for benzimidazole resistance in scald indicates the potential of these techniques.

By contrast, a discussion of the question, "Resistance surveys and monitoring - can statistics help?", quickly reached a consensus. Participants agreed that it should be possible to detect resistance which is serious enough to be a practical problem without elaborate analysis. To quote one participant, "if you need a mathematician to tell you if you have resistance, then you don't really have resistance". However, statistical analysis should be used in planning experiments, for instance in choosing dose rates and sampling schemes, and in summarising results, by presenting a few figures which describe the major conclusions, by testing the significance of the conclusions and by allowing comparisons with control isolates used in different experiments.

REFERENCE

Wheeler, I.; Kendall, S.; Butters, J.; Hollomon, D. (1994) Rapid detection of benzimidazole resistance in *Rhynchosporum secalis* using allele-specific oligonucleotide probes. *These proceedings.*

Syndicate 5

Chairman: Dr. C.E. Caten
Assistant: Dr. S. Heaney

Although Syndicate 5 was a relatively small group of around 12, we had a lively and varied discussion with the result that only two of the four questions posed by the Symposium organisers were addressed.

We started by considering whether statistics can help interpret data from resistance surveys. Such surveys usually aim to answer two questions:

1. Is there a shift in sensitivity between the test population and some baseline, control population?

2. If there is a shift, is it of practical significance?

There was broad agreement that while suitable statistical procedures to tackle these questions exist, they are not sufficiently employed. One reason for this may be the lack of familiarity of many biologists with all but the most basic statistical procedures. Resistance survey data frequently pose special statistical problems, e.g. the parameters (LC_{50}, MIC, etc.) are themselves complex and may not be normally distributed, sampling methods and sample sizes may be critical, bioassays contain many potential sources of error. For all these reasons it is important to first define the objectives of the study and then adopt methods appropriate to these objectives. Statistics can not compensate for a poorly designed survey and therefore, where results are likely to be complex, statisticians should be consulted from the outset of the study. Our conclusions were summed up by one member of the group as "statistics can help but only if used prophylactically".

The group then turned its attention to whether genetics can help predict resistance development. There was unanimous agreement that *in vitro* mutagenesis experiments can reveal the potential for resistance. Furthermore, it was agreed that the nature of the genetic control of resistance, whether major gene or polygenic, is a useful indicator of the likely speed of development and magnitude of resistance problems, and could suggest the use of different anti-resistance strategies. However, it was noted that the type of genetic control is a property of the particular strains being crossed; not the character or organism in question. Thus resistance to fungicide X in species Y may be determined by a major gene in one cross, but under polygenic control in another. While user-friendly model systems will continue to be important in studies of the genetics of fungicide resistance, there is a need to check the results from such studies on actual target pathogens. Up to this point the group had been of one mind in its view of the value of genetics. (Perhaps this was to be expected from a group dominated by fungal geneticists!) However, this cosy equanimity was destroyed when the question of the effect of regular sexual recombination on the development of polygenic resistance was raised. Will it facilitate the acquisition of quantitative resistance by combining individual resistance polygenes, as some members thought, or slow it down by breaking up those synergistic combinations that do arise, as others felt? Given the lateness of the hour, we were happy to agree on the need for more studies on this point, before retiring gracefully to the bar.